Regulatory Networks in Prokaryotes

JMMB Symposium Series Volume 6

Edited by:

Peter Dürre
Mikrobiologie und Biotechnologie
Universität Ulm
D-89069 Ulm
Germany

and

Bärbel Friedrich
Humboldt-University Berlin
Institute of Biology/Microbiology
Chausseestr. 117
D-10115 Berlin
Germany

horizon scientific press

Horizon Press
32 Hewitts Lane
Wymondham
Norfolk NR18 0JA
U.K.

www.horizonpress.com

British Library Cataloguing-in-Publication Data

A catalogue record for this book is available from the British Library

ISBN: 1-898486-49-2

Printed and bound in Great Britain by Biddles Limited

Dedication

This volume is dedicated to the memory of Werner Klipp, a member of this DFG priority program, who died in May 2002

Other Titles of interest

- **Bioremediation:** A Critical Review (2003)
  Editors: Ian M Head, Ian Singleton, and Mike Milner
- **Peptide Nucleic Acids:** Protocols and Applications (2nd Edition) (2003)
  Editor: Peter E. Nielsen
- **Ebola and Marburg Viruses:** Molecular and Cellular Biology (2003)
  Editors: Hans-Dieter Klenk and Heinz Feldmann
- **Genomics of GC-Rich Gram-Positive Bacteria** (2002)
  Editor: Antoine Danchin
- **Genomic Technologies:** Present and Future (2002)
  Editors: David Galas and Stephen McCormack
- **Probiotics and Prebiotics:** Where are We Going? (2002)
  Editor: Gerald W. Tannock
- **Frontiers in Computational Genomics** (2003)
  Editors: Michael Y. Galperin and Eugene V. Koonin
- **Transgenic Plants:** Current Innovations and Future Trends (2003)
  Editor: C.N. Stewart
- **Bioinformatics and Genomes:** Current Perspectives (2003)
  Editor: Miguel A. Andrade
- **Vaccine Delivery Strategies** (2003)
  Editors: W. Goebel and Guido Dietrich
- **Computational Genomics** (2003)
  Editor: Richard P. Grant
- **Chromosome Mapping and Sequencing** (2003)
  Editor: Ian Dunham
- **Multiple Drug Resistant Bacteria:** Emerging Strategies (2003)
  Editor: Carlos Amabile-Cuevas

Contents

Contributors

S. Achebach
Institut für Mikrobiologie und Weinforschung
Johannes Gutenberg-Universität Mainz
D-55099 Mainz
Germany

Karlheinz Altendorf
Universität Osnabrück
Fachbereich Biologie/Chemie
Abteilung Mikrobiologie
D-49069 Osnabrück
Germany

Birgit Baumgarth
Lehrstuhl für Genetik
Fakultät für Biologie
Universität Bielefeld
Postfach 100131
D-33501 Bielefeld
Germany

Anke Becker
Lehrstuhl für Genetik
Fakultät für Biologie
Universität Bielefeld
Postfach 100131
D-33501 Bielefeld
Germany

Dagmar Beier
Lehrstuhl für Mikrobiologie
Theodor-Boveri-Institut für Biowissenschaften
Universität Würzburg
D-97074 Würzburg
Germany

Michael Bernhard
Humboldt-University Berlin
Institute of Biology/Microbiology
Chausseestr. 117
D-10115 Berlin
Germany

Peter Alexander Bertram-Drogatz
Lehrstuhl für Genetik
Fakultät für Biologie
Universität Bielefeld
Postfach 100131
D-33501 Bielefeld
Germany

Hans-Matti Blencke
Institut für Mikrobiologie Biochemie und Genetik
Friedrich-Alexander-Universität Erlangen-Nürnberg
Staudtstr. 5
D-91058 Erlangen
Germany

Andreas Bock
Lehrstuhl für Mikrobiologie
Theodor-Boveri-Institut für Biowissenschaften
Universität Würzburg
D-97074 Würzburg
Germany

August Böck
Institute of Genetics and Microbiology
University of Munich
Maria-Ward-Strasse 1a
D-80638 Munich
Germany

Alex Böhm
Department of Biology
University of Konstanz
D-78457 Konstanz
Germany

Michael Böhringer
Cell and Molecular Biology
Morton 4-685
Northwestern University Medical School
303 E. Chicago Avenue
Chicago, IL 60611
USA

Winfried Boos
Department of Biology
University of Konstanz
D-78457 Konstanz
Germany

Reinhold Brückner
Mikrobiologie
University of Kaiserslautern
Paul Ehrlich Str. 23
D-67663 Kaiserslautern
Germany

Ina Budde
Max-Planck-Institut für terrestrische Mikrobiologie
D-35043 Marburg
Germany

Thorsten Buhrke
Humboldt-University Berlin
Institute of Biology/Microbiology
Chausseestr. 117
D-10115 Berlin
Germany

Michael K. Dahl
Department of Biology
M605
University of Konstanz
D-78457 Konstanz
Germany

Katja Decker
Department of Biology
University of Konstanz
D-78457 Konstanz
Germany

Christian Detsch
Institut für Mikrobiologie Biochemie und Genetik
Friedrich-Alexander-Universität Erlangen-Nürnberg
Staudtstr. 5
D-91058 Erlangen
Germany

Ulrich Dobrindt
Institut für Molekulare Infektionsbiologie
Universität Würzburg
Röntgenring 11
D-97070 Würzburg
Germany

Thomas Drepper
Ruhr-Universität Bochum
Fakultät für Biologie
Lehrstuhl für Biologie der Mikroorganismen
D-44780 Bochum
Germany

Thomas Dünnebier
Institut für Industrielle Genetik
Universität Stuttgart
D-70569 Stuttgart
Germany

Peter Dürre
Mikrobiologie und Biotechnologie
Universität Ulm
D-89069 Ulm
Germany

Gunter Fischer
Max-Planck-Gesellschaft
Weinbergweg 22
D-06120 Halle/Saale
Germany

Annette Flotho
Department of Biological Chemistry
Harvard Medical School
Boston
MA 02115
USA

Bärbel Friedrich
Humboldt-University Berlin
Institute of Biology/Microbiology
Chausseestr. 117
D-10115 Berlin
Germany

Thilo M. Fuchs
Lehrstuhl für Mikrobiologie
Theodor-Boveri-Institut für Biowissenschaften
Universität Würzburg
D-97074 Würzburg
Germany

Roman Grabbe
Institut für Mikrobiologie und Genetik
Grisebachstr. 8
D-37077 Göttingen
Germany

Jutta Gregor
Institut für Mikrobiologie und Molekularbiologie
Universität Giessen
Heinrich-Buff-Ring 26-32
D-35392 Giessen
Germany

Dagmar Gregor
Institut für Mikrobiologie und Genetik
TU Darmstadt
D-64287 Darmstadt
Germany

Roy Gross
Lehrstuhl für Mikrobiologie
Theodor-Boveri-Institut für Biowissenschaften
Universität Würzburg
D-97074 Würzburg
Germany

Silke Gross
Ruhr-Universität Bochum
Fakultät für Biologie
Lehrstuhl für Biologie der Mikroorganismen
D-44780 Bochum
Germany

Jörg Hacker
Institut für Molekulare Infektionsbiologie
Universität Würzburg
Röntgenring 11
D-97070 Würzburg
Germany

Regine Hakenbeck
University of Kaiserslautern
Department of Microbiology
Paul Ehrlich Strasse 23
D-67663 Kaiserslautern
Germany

Klaus Hantke
Mikrobiologie/Membranphysiologie
Universität Tübingen
Auf der Morgenstelle 28
D-72076 Tübingen
Germany.

Michael Hecker
Institut für Mikrobiologie und Molekularbiologie
F.-L. Jahn-Strasse 15
D-17487 Greifswald
Germany

Regine Hengge-Aronis
Institut für Biologie
Mikrobiologie
Freie Universität Berlin
D-14195 Berlin
Germany

Annette Hofacker
Institut für Mikrobiologie und Genetik
TU Darmstadt
D-64287 Darmstadt
Germany

Dieter Jahn
Technische Universitaet Braunschweig
Institut für Mikrobiologie
Spielmannstrasse 7
D-38106 Braunschweig
Germany

Ivana Jankovic
Nestlé Research
Vers-chez-les-Blanc
CH-1000 Lausanne 26
Switzerland

Kirsten Jung
Universität Osnabrück
Fachbereich Biologie/Chemie
Abteilung Mikrobiologie
D-49069 Osnabrück
Germany

Kristina Kaminski
Max-von Pettenkofer Institut für Hygiene und Medizinische Mikrobiologie
D-80336 München
Germany

Wolfgang Klein
Philipps Universität Marburg
Biochemie - FB Chemie
Hans-Meerwein-Strasse
D-35032 Marburg
Germany

Werner Klipp
Ruhr-Universität Bochum
Fakultät für Biologie
Lehrstuhl für Biologie der Mikroorganismen
D-44780 Bochum
Germany

Kai Klopprogge
Institut für Mikrobiologie und Genetik
Grisebachstr. 8
D-37077 Göttingen
Germany

Gabriele Klug
Institut für Mikrobiologie und Molekularbiologie
Universität Giessen
Heinrich-Buff-Ring 26-32
D-35392 Giessen
Germany

Jochen König
Lehrstuhl für Mikrobiologie
Theodor-Boveri-Institut für Biowissenschaften
Universität Würzburg
D-97074 Würzburg
Germany

Ingrid Korsa
Institute of Genetics and Microbiology
University of Munich
Maria-Ward-Strasse 1a
D-80638 Munich
Germany

Günter Kramer
Institut für Biochemie und Molekularbiologie
Albert-Ludwigs-Universitat Freiburg
D-79104 Freiburg
Germany

Sung-Jae Lee
Department of Biology
University of Konstanz
D-78457 Konstanz
Germany

Oliver Lenz
Humboldt-University Berlin
Institute of Biology/Microbiology
Chausseestr. 117
D-10115 Berlin
Germany

Susanne Leonhartsberger
Institute of Genetics and Microbiology
University of Munich
Maria-Ward-Strasse 1a
D-80638 Munich
Germany

Mark van der Linden
DKFZ
D-61920 Heidelberg
Germany

Holger Ludwig
Institut für Mikrobiologie Biochemie und Genetik
Friedrich-Alexander-Universität Erlangen-Nürnberg
Staudtstr. 5
D-91058 Erlangen
Germany

Bernd Masepohl
Ruhr-Universität Bochum
Fakultät für Biologie
Lehrstuhl für Biologie der Mikroorganismen
D-44780 Bochum
Germany

Mohamed A. Marahiel
Philipps Universität Marburg
Biochemie - FB Chemie
Hans-Meerwein-Strasse
D-35032 Marburg
Germany

Thorsten Mascher
University of Kaiserslautern
Department of Microbiology
Paul Ehrlich Strasse 23
D-67663 Kaiserslautern
Germany

Michelle Merai
University of Kaiserslautern
Department of Microbiology
Paul Ehrlich Strasse 23
D-67663 Kaiserslautern
Germany

Matthias Merzbacher
Institut für Mikrobiologie Biochemie und Genetik
Friedrich-Alexander-Universität Erlangen-Nürnberg
Staudtstr. 5
D-91058 Erlangen
Germany

Jess Meyer
Mikrobiologie
University of Kaiserslautern
Paul Ehrlich Str. 23
D-67663 Kaiserslautern
Germany

Stephan Nakotte
LipoNova GmbH
Feodor Lynen Str. 23
D 30625 Hannover
Germany

Annette Paschen
Ruhr-Universität Bochum
Fakultät für Biologie
Lehrstuhl für Biologie der Mikroorganismen
D-44780 Bochum
Germany

Alice Pawlowski
Ruhr-Universität Bochum
Fakultät für Biologie
Lehrstuhl für Biologie der Mikroorganismen
D-44780 Bochum
Germany

Ralf Peist
Department of Biology
University of Konstanz
D-78457 Konstanz
Germany

Anne-Laure Perraud
Lehrstuhl für Mikrobiologie
Theodor-Boveri-Institut für Biowissenschaften
Universität Würzburg
D-97074 Würzburg
Germany

Felicitas Pfeifer
Institut für Mikrobiologie und Genetik
TU Darmstadt
D-64287 Darmstadt
Germany

Katharine Piechaczek
Institut für Molekulare Infektionsbiologie
Universität Würzburg
Röntgenring 11
D-97070 Würzburg
Germany

Petra Plösser
Institut für Mikrobiologie und Genetik
TU Darmstadt
D-64287 Darmstadt
Germany

Alfred Pühler
Lehrstuhl für Genetik
Fakultät für Biologie
Universität Bielefeld
Postfach 100131
D-33501 Bielefeld
Germany

Ingmar Quester
Lehrstuhl für Genetik
Fakultät für Biologie
Universität Bielefeld
Postfach 100131
D-33501 Bielefeld
Germany

Karsten Raabe
Ruhr-Universität Bochum
Fakultät für Biologie
Lehrstuhl für Biologie der Mikroorganismen
D-44780 Bochum
Germany

Kai-Uwe Riedel
Ruhr-Universität Bochum
Fakultät für Biologie
Lehrstuhl für Biologie der Mikroorganismen
D-44780 Bochum
Germany

Bettina Rohde
Max-Planck-Institut für Terrestrische Mikrobiologie
D-35043 Marburg
Germany

Silvia Rüberg
Lehrstuhl für Genetik
Fakultät für Biologie
Universität Bielefeld
Postfach 100131
D-33501 Bielefeld
Germany

Steffen Schaffer
Forschungszentrum Jülich
IBT 1
Postfach 1913
D-52425 Jülich
Germany

Birgit Scharf
Lehrstuhl für Genetik
Universität Regensburg
D-93040 Regensburg
Germany

Angelika Schierhorn
Max-Planck-Gesellschaft
Weinbergweg 22
D-06120 Halle/Saale
Germany

Anja Schlegel
Department of Biology
University of Konstanz
D-78457 Konstanz
Germany

Matthias Schmalisch
Institut für Mikrobiologie Biochemie und Genetik
Friedrich-Alexander-Universität Erlangen-Nürnberg
Staudtstr. 5
D-91058 Erlangen
Germany

Rüdiger Schmitt
Lehrstuhl für Genetik
Universität Regensburg
D-93040 Regensburg
Germany

Ruth A. Schmitz
Institut für Mikrobiologie und Genetik
Grisebachstr. 8
D-37077 Göttingen
Germany

Max Schobert
Technische Universitaet Braunschweig
Institut für Mikrobiologie
Spielmannstrasse 7
D-38106 Braunschweig
Germany

Edward Schwartz
Humboldt-University Berlin
Institute of Biology/Microbiology
Chausseestr. 117
D-10115 Berlin
Germany

Angela V. Smirnova
Max-Planck-Institut für terrestrische Mikrobiologie
35043 Marburg
Germany

Jessica Stips
Institut für Mikrobiologie und Genetik
Grisebachstr. 8
D-37077 Göttingen
Germany

Jörg Stülke
Lehrstuhl für Mikrobiologie
Institut für Mikrobiologie Biochemie und Genetik
Friedrich-Alexander-Universität Erlangen-Nürnberg
Staudtstr. 5
D-91058 Erlangen
Germany

Kai Thormann
Dept. Civil and Environmental Engineering
Terman Engineering Center
Stanford University
Stanford
CA 94305-4020
USA

Matthias S. Ullrich
Max-Planck-Institut für terrestrische Mikrobiologie
D-35043 Marburg
Germany

G. Unden
Institut für Mikrobiologie und Weinforschung
Johannes Gutenberg-Universität Mainz
D-55099 Mainz
Germany

Rolf Wagner
Institut für Physikalische Biologie
Heinrich-Heine-Universität Düsseldorf
Universitätsstr. 1
D-40225 Düsseldorf
Germany

Ling Wang
Max-Planck-Institut für terrestrische Mikrobiologie
D-35043 Marburg
Germany

Helge Weingart
Max-Planck-Institut für terrestrische Mikrobiologie
D-35043 Marburg
Germany

Verena Weiss
Department of Biology
University of Konstanz
D-78457 Konstanz
Germany

Dorothea Zähner
University of Kaiserslautern
Department of Microbiology
Paul Ehrlich Strasse 23
D-67663 Kaiserslautern
Germany

Y. Zeuner
Institut für Mikrobiologie und Weinforschung
Johannes Gutenberg-Universität Mainz
D-55099 Mainz
Germany

Brigitte Zickner
Mikrobiologie und Biotechnologie
Universität Ulm
D-89069 Ulm
Germany

Peter Zimmermann
Institut für Mikrobiologie und Genetik
TU Darmstadt
D-64287 Darmstadt
Germany

Walter G. Zumft
Lehrstuhl für Mikrobiologie
Universität Karlsruhe
D-76128 Karlsruhe
Germany

Chapter 1

Regulatory Networks in Prokaryotes: Variations on a Theme

Bärbel Friedrich

Most unicellular organisms thrive in natural environments, where they are exposed to nutrient deprivation and stress. To survive and to propagate successfully, even under unfriendly conditions, a high degree of metabolic flexibility combined with sophisticated regulation is necessary. It was the goal of a joint research project to explore regulatory networks in prokaryotes including some organisms that had scarcely been investigated on the molecular level. We set out to discover new regulatory strategies, to dissect variations of common themes and to get further insights into well-studied mechanisms.

Halobacterium represents a rare archaeal genus, which is amenable to genetic manipulation. Species of *Halobacterium* produce gas vesicles as a means of motility, allowing cells to float up and down in response to light, oxygen and salt concentration. A transcriptional activator is described in detail, which controls the expression of two gene clusters for gas vesicle synthesis (Pfeifer). Chemotaxis and motility in the symbiotic nitrogen-fixing bacterium *Sinorhizobium meliloti* differ remarkably from the well-studied tactic behavior of enterics. The flagella rotate only clockwise. Directional changes are achieved by modulating the speed of rotation. It is shown that the tactic response is mediated by two regulators (CheY1, CheY2) of which CheY1 is considered as a sink for phosphoryl groups and CheY2 as the protein which interacts with the motor (Schmitt). The same species forms two sets of exopolysaccharides, succinoglycan (EPS I) and galactoglucan (EPS II) which are involved in the invasion of root nodules. EPS II synthesis is part of the PhoR regulatory network and stimulated during phosphate limitation. Two newly identified regulators, MucR and ExpG, are specifically involved in the regulatory cascade (Pühler/Becker).

Host-related interactions of plant and human pathogens were also studied. *Pseudomonas syringae* produces the phytotoxin coronatine as a major virulence factor. Expression of the Cor-encoding genes is regulated by two response regulators. CorR together with its cognate histidine protein kinase CorS is involved in thermoregulation. The second response regulator CorP is proposed to modulate the activity of CorRS (Ullrich). Expression of virulence in *Bordetella pertussis*, the etiological agent of whooping cough, is rather unstable and reversibly affected by changing environmental conditions such as temperature. It is demonstrated that BvgAS, an unorthodox two-component system, transmits the signal by a multistep phospho-relay reaction providing a fine-tuned coordinate expression of the virulence regulon (Gross). The role of the minor $tRNA_5^{Leu}$, encoded by *leuX*, on the expression of the virulence factors in an uropathogenic strain of *Escherichia coli,* was explored by proteome analysis. The data suggest that this tRNA has a global impact on the posttranscriptional regulation of the virulence network (Hacker). New mechanisms of β-lactam resistance in *Streptococcus pneumoniae* are obviously related to the function of regulatory components implicating the protein kinase CiaH as a major player in monitoring the integrity of the cell wall. Mutations in *ciaH* are pleiotropic and lead to the occurrence of antibiotic resistance and the failure to express genetic competence. Thus the *cia* system governs a complex regulatory network (Hakenbeck).

Most of the bacterial species take an active role in acquiring nutrients. Iron is an essential component of several metabolic reactions, although abundant, it is not readily available to the cell due to its extraordinary insolubility. The production and secretion of iron-chelating siderophores is one strategy for increasing the availability of the metal. Once taken up, siderophores release iron into the cytoplasm. An unusual Fe-S cluster-containing protein (FhuF) is thought to participate in iron mobilization. *fhuF* gene transcription sensitively responds to the level of Fur and Fe^{2+} which repress iron uptake genes. A Fur-related protein (Zur) is shown to repress the high affinity zink uptake system (Hantke). Cells of *E. coli* react to changes in the osmolality of their environment by synthesizing various proteins to counteract deleterious extremes of ionic strength. Under potassium-limiting conditions and in response to osmotic upshift a high affinity K^+-uptake system is produced. Expression of this *kdpFABC* operon is under control of the response regulator KdpE and the transmembrane histidine kinase KdpD, which was initially considered as a turgor sensor. Autophosphorylation data favor intracellular parameters as triggering agents in KdpD signal perception (Altendorf/Jung).

The availability of nitrogen is the stimulus of the multigene Ntr network. Elegant studies with enterics uncovered the now classical two-component regulatory system, NtrB/NtrC. Dependent on the supply of nitrogen, the protein kinase NtrB either phosphorylates its cognate response regulator NtrC or promotes rapid dephosphorylation of NtrC-P. Both phosphotransfer and phosphatase activity reside in the H-domain of the NtrB transmitter module. Further inspection revealed that a small peptide of 25 amino acids of the H-domain stimulates dephosporylation of NtrC-P (Weiss).

The balance between autokinase and phosphatase activity of NtrB is controlled by an additional regulator PII, which is modified by uridylylation in response to nitrogen. A PII paralogue (GlnK) transmits the nitrogen status in N_2-fixing cells of *Klebsiella pneumoniae* by interaction with the regulatory couple NifA/NifL. NifA activates transcription of the *nif* operons and NifL inhibits its activity in response to nitrogen and oxygen. There is evidence that the FAD-containing NifL undergoes conformational changes, which affect NifA activity in response to the redox-state of the cofactor (Schmitz). Regulation of N_2 fixation in the phototroph *Rhodobacter capsulatus* is even more complex since the bacterium produces an alternative nitrogenase in addition to the molybdenum-containing enzyme. Three levels of regulation were uncovered which provide subtle mechanisms to adapt the synthesis and activity of the two nitrogenases to the external supply of ammonia, molybdenum, light and oxygen (Klipp).

Many microbes have the capacity to adjust quickly to changing conditions in their environment by using alternative modes of energy conservation. If species of the purple bacterium *Rhodobacter* encounter high oxygen tension, they obtain their energy by aerobic respiration. At low O_2 tension, however, the formation of pigments, photosynthetic complexes and intracytoplasmic membranes is highly elevated and the cells shift to energy conservation by photosynthesis. A central role in this overriding coordinate activation of photosynthetic gene expression is assigned to the RegB/RegA two-component regulatory system (Klug). Molecular hydrogen is an attractive fuel for organisms that can shift from chemoorganic to chemolithotrophic metabolism. To benefit from H_2, occurring only in traces in oxic environments, aerobic chemolithotrophes have evolved a highly specific H_2-sensing system. This signal transduction chain consists of a regulatory [NiFe]-hydrogenase which recognizes H_2 and transmits a redox signal to an uncommon two-component system, in which the non-phosphorylated form of the response regulator activates transcription (Friedrich).

Fluctuation in O_2 availability is a significant switch in the mode of energy transduction. One of the major transcriptional regulators controlling this respiratory change is Fnr. Oxygen destroys the activity of Fnr by a reaction which leads to the conversion of a Fe-S cluster and in turn the formation of apoFnr. It was demonstrated that apoFnr can be reconstituted to active Fnr by using glutathione as a reducing agent (Unden). Fnr is one of a set of regulatory components, which include the master regulator FhlA, responsible for integrating the synthesis and degradation of formate, a key compound of anaerobic metabolism in enterics, into a complex regulatory network. This network provides an excellent example of how the cells adjust the carbon flux to their metabolic needs (Böck). Transcription factors closely related to the family of Fnr/Crp proteins have been discovered also in other groups of proteobacteria, including *Pseudomonas, Paracoccus* and *Rhodobacter.* The Dnr/Nnr regulators respond to extremely low concentrations of nitric oxide and elicit a key signal for expressing nitrite and nitric oxide reductases. Both are constituents of a multigene denitrification apparatus (Zumft). Hemes are integral components of energy-conserving electron transport chains and cofactors of many redox proteins. It is conceivable that a switch in energy metabolism coincides with the cellular heme content. In non-photosynthetic proteobacteria a wide range of regulatory strategies are employed by the cells to adapt heme biosynthesis to the metabolic requirements of a given species (Jahn).

Operon structures and transcription factors in obligate anaerobes like *Clostridium acetobutylicum* are becoming gradually apparent. This gram-positive bacterium produces solvents, which compensate for the harmful effect of acids that accumulate during the first stage of fermentation. The metabolic switch from acid to solvent production correlates with the onset of sporulation. A regulatory role for the principal transcription factor SpoOA linking solvent gene expression and sporulation is discussed (Dürre).

Several decades of intensive effort have focused on questions of how a cell senses the appropriateness of each catabolic pathway under a given circumstance and how the cells make the decision what enzymes to make and at what quantities. The molecular program underlying carbon catabolite repression (CRP) especially in non-enterics is still a matter of debate. *E. coli* has evolved a sophisticated regulatory network for the utilization of maltose. The maltose regulon is controlled by MalT, a transcriptional activator, which acts together with the inducer maltotriose and ATP. The expression of *malT* is controlled by the cAMP/CAP complex and the repressor Mlc, whose activity is regulated by the transport status of the glucose-specific phosphotransferase system (PTS). The second level of regulation affects the activity of MalT. MalT is activated by maltotriose and inactivated by a widely unknown interaction with three different enzymes including MalK, a component of the maltodextrin transport system (Boos). The gram-positive *Staphylococcus xylosus* has developed a completely different strategy to assure the priority status of glucose as a carbon and energy source. The utilization of CRP-sensitive substrates such as maltose, lactose and xylose is repressed by the control protein CcpA, which is activated by seryl-phosphorylated HPr, the phosphocarrier protein of the PTS system. Mutant analysis has shown that glucose-specific CRP in *S. xylosus* is also affected by a second route of glucose entry, the uptake protein GlcU and the glucose kinase GlkA (Brückner).

In addition to acting as a negative regulator, CcpA appears to be instrumental also in transcription activation. CcpA mutants of *B. subtilis* show a severe inhibition of growth on minimal medium. The data suggest that under these circumstances CcpA is required for the expression of cellular functions such as glycolysis, excretion of metabolites and ammonium assimilation (Stülke).

CRP in gram-positive bacteria seems to be more complex than originally thought. With the *B. subtilis* trehalose operon it was shown that a CcpA-independent CRP mechanism exists presumably based on inducer-exclusion (Dahl).

Compensatory adjustments to changes in growth rate are of a truly global character. Microbes use small basic proteins to condense and modify the architecture of their DNA. Two classes of DNA-interacting proteins have been distinguished: histone-like proteins, which indirectly affect gene expression by altering the reaction rate of RNA polymerase and proteins that specifically bind to defined elements in the promoter region and primarily assist integrating various global stimuli into networks that control growth rates. Representatives of the two classes of proteins, abundant in *B. subtilis,* are discussed in detail: the histone-like HBsu and the transition-state regulator AbrB (Marahiel). Cells of *E. coli* that are exposed to a rapid substrate downshift or amino acid deprivation stop synthesizing protein and stable RNA. This central process called stringent response correlates with the accumulation of guanosine nucleotides. The regulatory nucleotides (p)ppGpp modulate RNA polymerase activity to transcribe from specific promoters, including rRNA promoters, characterized by a GC-rich discriminator element and further structural requirements. Based on newly collected data a model is presented of how (p)ppGpp affects RNA polymerase during various steps of initiation and elongation (Wagner). Associated with the changing growth rate is the expression of another global regulator, the alternative sigma factor σ^{38} of RNA polymerase encoded by *rpoS*. It appears that RpoS is a master regulator in the most complex network known so far. It regulates the expression of a large number of genes, whose products are involved in the survival under extreme stress situations like starvation, high osmolality, low pH and high temperature. Detailed investigations have shed new light on the control of the cellular level of RpoS by translational regulation and proteolysis, on the function of RpoS as an initiation factor and on the basis of the promoter selectivity of RpoS (Hengge-Aronis).

Chapter 2

Regulation of Gas Vesicle Formation in Halophilic Archaea

Felicitas Pfeifer, Dagmar Gregor, Annette Hofacker, Petra Plößer and Peter Zimmermann

Abstract

The halophilic archaea *Halobacterium salinarum* and *Haloferax mediterranei* produce gas vesicles depending on the growth phase and on environmental factors such as light, salt, or oxygen. Fourteen different *gvp* genes (*gvpACNO* and *gvpDEFGHIJKLM*) are involved in their formation, and the regulation of *gvp* gene expression occurs at the transcriptional and translational level. *Haloferax volcanii* offers a clean genetic background for the functional analysis of gas vesicle genes by transformation experiments. Such experiments show that the promoter of the *gvpA* gene encoding the major gas vesicle structural protein is activated by the endogenous basic leucine-zipper protein GvpE. On the other hand, the GvpD protein, which contains a p-loop motif, is involved either directly or indirectly in the repression of the *gvpA* promoter activity. Eight of the fourteen p-*gvp* genes (p-*gvpAO* and p-*gvpFGJKLM*) enable gas vesicle formation in *Hf. volcanii* transformants and thus constitute the minimal p-vac region.

Gas vesicles are proteinaceous flotation devices produced by some bacteria and archaea to allow these microorganisms to float to the surface of their aqueous environment where oxygen and light are present and can be used for energy metabolism (Walsby, 1994). Among the halophilic archaea, *Halobacterium salinarum*, *Haloferax mediterranei*, and the haloalkaliphilic *Natronobacterium vacuolatum* produce gas vesicles. These are spindle- or cylinder-shaped structures ranging in size from 0.2 to 1.5 μm in length and 0.2 μm in diameter. The major structural protein GvpA forms the 4.6 nm-wide ribs that are arranged perpendicular to the long axis. These ribs are formed by a helix of low pitch rather than a stack of hoops (Offner *et al.*, 1998). The gas vesicles are filled by passive diffusion with gases dissolved in the environment (Walsby, 1994). Apart from the hydrophobic 7-8 kDa GvpA protein there is a second, but minor constituent (GvpC) located at the outer surface of the ribs and strengthening the entire gas vesicle structure (Englert and Pfeifer, 1993; Halladay *et al.*, 1993).

Hb. salinarum PHH1 contains spindle-shaped gas vesicles during all stages of growth, whereas *Hb. salinarum* PHH4 (a derivative of *Hb. salinarum* PHH1 containing a smaller plasmid) and *Hf. mediterranei* produce cylinder-shaped gas vesicles only during the stationary growth phase (Table 1). Environmental factors such as anaerobic conditions or darkness enhance the amount of gas vesicles in *Hb. salinarum* PHH4 (unpublished observations). Also the salt content of the medium affects gas vesicle formation: the moderately halophilic *Hf. mediterranei* produces gas vesicles when grown in media with a salt content exceeding 17% (Englert *et al.*, 1990). Since all these factors influence gas vesicle formation, the genes involved offer an interesting system to study gene regulation and also signal transduction.

Fourteen different *gvp* (*g*as *v*esicle *p*rotein) genes were identified that cluster in a 9-kb DNA region termed vac region (Englert *et al.*, 1992a) or *gvp* gene cluster (Jones *et al.*, 1991; DasSarma *et al.*, 1994). These genes are arranged as *gvpACNO* and, upstream and oppositely oriented, *gvpDEFGHIJKLM* (Figure 1). We use the designations p-vac, c-vac and mc-vac (and also c-*gvp*, p-*gvp* and mc-*gvp* for the respective genes) to distinguish the different vac regions and *gvp* genes according to their halobacterial origin. The largest vac region, mc-vac (*mediterranei c*hromosomal), is found in the chromosome of *Hf. mediterranei* (Figure 1; Englert *et al.*, 1992a). Two different vac regions are present in *Hb. salinarum* PHH1, namely the *c*hromosomal c-vac region and the *p*lasmid-born p-vac region located on the 150 kb plasmid pHH1 (Englert *et al.*, 1992a). The total genome sequence determined for the related *Halobacterium* species NRC-1 indicates similar *gvp* gene clusters, with two *gvp1* gene clusters almost identical to p-vac located on the large plasmids pNRC100 (191 kb) and pNRC200 (365 kb) (DasSarma *et al.*, 1994; Ng *et al.*, 1998; Ng *et al.*, 2000). The *gvp1* gene cluster is part of an 145 kb region shared by both plasmids. The *gvp2* gene cluster found in addition on pNRC200 is almost identical to c-vac but lacks the c-*gvpM* gene and should thus not be functional (Ng *et al.*, 2000). The various vac regions are summarized in Table 1.

Relationship Between Different *Halobacterium* Species

Although the purpose of this review is to summarize our data on the regulation of gas vesicle gene expression, the relationship between the different *Halobacterium* species used to investigate gas vesicle formation should be discussed in some detail, since there are

Table 1. The *gvp* gene clusters and presence of gas vesicles during growth.

	gvp gene cluster(s)	Presence of gas vesicles	
		Exponential	Stationary
Hb. salinarum PHH1	p-vac and c-vac	++	++
Hb. salinarum PHH4	c-vac	—	+
Hb. salinarum NRC-1	2x *gvp1*; *gvp2**	++	++
Hf. mediterranei, 17–30% salt	mc-vac	—	++
Hf. volcanii	—	—	—
Nb. vacuolatum, 15–30% salt	nv-vac	+	++

**gvp1* = p-vac; *gvp2* = c-vac, but lacking the c-*gvpM* gene.

confusions due to the frequent renaming of the halobacterial strains and species. The two *Halobacterium* strains PHH1 and NRC-1 both originate from a "*Hb. halobium*" strain kept in the strain collection of Walther Stoeckenius at UC San Francisco (USA). The American Type Culture Collection, the DSMZ in Germany (origin of "*Hb. halobium*" PHH4 = DSM 670), Ford Doolittle (origin of "*Hb. halobium*" NRC-1), Dieter Oesterhelt (origin of "*Hb. halobium*" NRC817 = "*Hb. halobium*" PHH1 strain) and also Felicitas Pfeifer obtained samples from this collection at different points in time. A later comparison of these strains based on hybridization patterns of various insertion elements and the presence of pHH1-specific plasmid sequences suggests that the further cultivation of these strains (usually accompanied by purification *via* single colonies) resulted in unexpected genome variations which turned out to be a characteristic feature of "*Hb. halobium*" (Pfeifer *et al.*, 1981a; Pfeifer *et al.*, 1981b; Sapienza and Doolittle, 1981; Pfeifer, 1988; Pfeifer and Blaseio, 1989; Pfeifer and Ghahraman, 1993). These alterations are mainly due to the action of multiple insertion elements (ISH elements) found predominantly in the plasmid population, and in islands of (AT)-rich DNA in the chromosome of "*Hb. halobium*" (Pfeifer *et al.*, 1983; Pfeifer and Betlach, 1985; Pfeifer, 1988; Charlebois and Doolittle, 1989). The plasmid population of "*Hb. halobium*" PHH1 is highly dynamic: pHH1 usually incurred multiple insertions or deletions in phenotypic mutants such as gas vesicle minus (Vac^-) mutants (Pfeifer *et al.*, 1981b; Pfeifer *et al.*, 1988; Pfeifer and Blaseio, 1989).

Also the two large plasmids of the *Halobacterium* species NRC-1 (pNRC100 and pNRC200) are related to pHH1. All three plasmids share an almost identical replicon region, the p-vac region (= *gvp1* gene cluster)

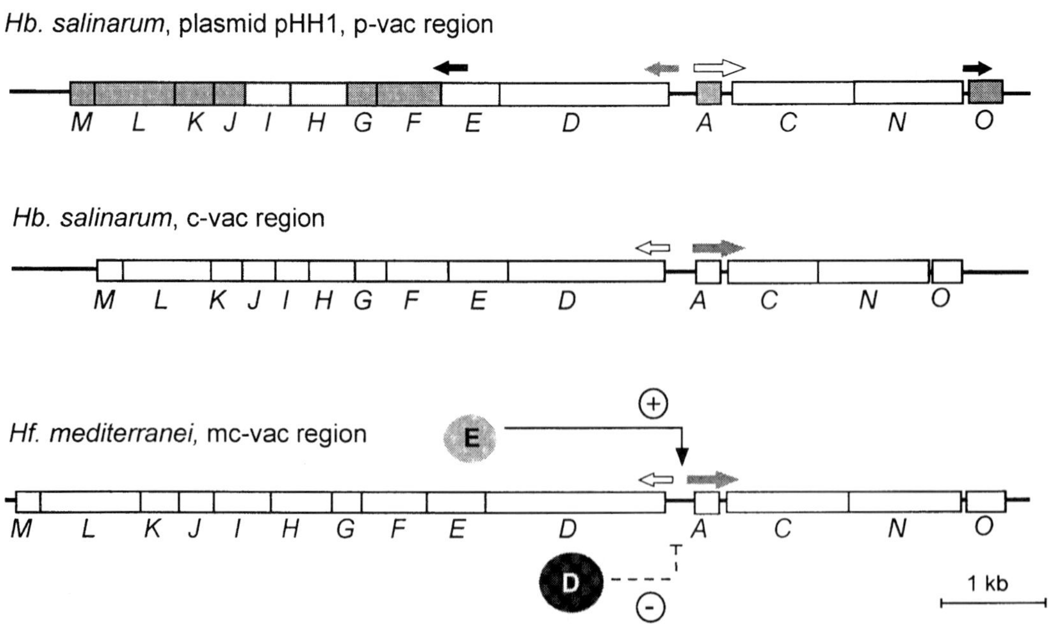

Figure 1. Genetic maps of the p-vac and the c-vac region of *Hb. salinarum* PHH1, and the mc-vac region of *Hf. mediterranei*. The fourteen *gvp* genes are depicted by boxes labeled *A* and *C* through *O*. The eight *gvp* genes constituting the minimal p-vac region (Offner *et al.*, 2000) are shaded in grey. Arrows above the maps indicate start sites and direction of transcription. Black arrows depict transcription during the exponential growth phase, open arrows transcription during all phases of growth, and arrows in grey indicate transcription that occurs only during the stationary growth phase. The two regulator proteins GvpE (activator) and GvpD (repressor) are shown with the mc-vac region; GvpE acts directly at the mcA and mcD promoter, whereas the action of GvpD could also be indirect.

and adjacent sequence including various ISH elements (with ISH3 = ISH27, ISH8 = ISH26, and ISH9 = ISH28) (Ng *et al.*, 1991; Pfeifer and Ghahraman, 1993). DasSarma's group introduced new designations for ISH elements already published for "*Hb. halobium*" PHH1 (Ebert *et al.*, 1987; Pfeifer and Blaseio, 1990). The plasmid pNRC100 also contains an inverted repeat of 33 kb not found in pHH1, and both plasmids might in addition harbour unrelated sequences (Ng *et al.*, 1991; Pfeifer and Ghahraman, 1993). All these "*Hb. halobium*" strains and also the "*Hb. salinarium*" and "*Hb. cutirubrum*" species contain sequences homologous to pHH1 (Pfeifer *et al.*, 1981a; Pfeifer and Ghahraman, 1993). All these strains and species (including "*Hb. halobium*" NRC-1) were renamed "*Hb. salinarium*" (Tindall, 1992). However, this designation was changed again to *Hb. salinarum* due to the grammatically incorrect form "*salinarium*" (Ventosa and Oren, 1996). Since 1996 we use the designation *Hb. salinarum* for all *Halobacterium* strains that share a common ancestor (including "*Hb. halobium*" NRC817 and "*Hb. halobium*" NRC-1), and distinguish them on the basis of their major plasmids (pHH1, pHH4 and pNRC).

Eight of the Fourteen *gvp* Genes Constitute the Minimal p-vac Region

The minimal number of genes required for gas vesicle formation has been determined by transformation experiments using the gas vesicle negative species *Haloferax volcanii* as recipient (Offner and Pfeifer, 1995; Offner *et al.*, 1996; Offner *et al.*, 2000). *Hf. volcanii* is easy to transform, and this species offers a clean genetic background with respect to the *gvp* genes in contrast to *Hb. salinarum.* Single p-*gvp* genes were deleted and the remaining p-*gvp* genes analyzed for their ability to drive gas vesicle formation in the transformants. A deletion of either one of the six genes p-*gvpC*, p-*gvpD*, p-*gvpE*, p-*gvpH*, p-*gvpI*, or p-*gvpN* still leads to the formation of gas vesicles in such *Hf. volcanii* transformants (see Figure 1). Minor amounts of gas vesicles are also detectable in transformants containing a construct with the eight remaining p-*gvpFGJKLMAO* genes, indicating that these genes constitute the minimal p-vac region required for gas vesicle formation (Offner *et al.*, 2000).

Another method to mutate specific *gvp* genes has been employed by inserting foreign DNA into the various genes of the *gvp1* gene cluster and studying the effect of each mutation in pNRC100-negative *Hb. salinarum* NRC-1 transformants (DasSarma *et al.*, 1994). However, the presence of the pNRC200 plasmid containing the second *gvp1* and also the *gvp2* gene cluster are not mentioned in this publication. Thus, the results are very difficult to interpret. Compared to our deletion analyses the insertional mutations revealed different results for 6 out of 14 *gvp* genes (Pfeifer *et al.*, 1997). Difficulties with the insertional mutations also arise from possible polar mutations, and the authors never proved the absence of the mutated Gvp proteins, especially in those cases where the insert occurs close to the 3′ or 5′ end of a gene (discussed in detail by Pfeifer *et al.*, 1997). However, a deletion of a *gvp* gene clearly destroys its function, and if the control transformant harbouring the *gvp* gene under investigation still produces gas vesicles, the phenotype of the Δ*gvp* transformant provides a good indication of the effect of the mutation.

Among the six *gvp* genes not essential for gas vesicle formation are the two genes *gvpD* and *gvpE* encoding gas vesicle regulatory proteins (Offner and Pfeifer, 1995). The ΔDE transformants produce reduced numbers of spindle-shaped gas vesicles compared to the wild-type. Also ΔN transformants contain only minor amounts of gas vesicles. ΔC transformants harbouring a p-vac region with a deletion of p-*gvpC* form large amounts of irregularly shaped gas vesicles (Offner *et al.*, 1996). Complementation of the ΔC transformants with the p-*gvpC* gene again results in spindle-shaped gas vesicles, indicating that GvpC not only stabilizes the gas vesicle structure, but is also important for the shape determination. ΔI transformants contain very long (up to 2.7 μm) cylindrical gas vesicles, suggesting that GvpI is directly or indirectly involved in the determination of the gas vesicle length. Gas vesicles synthesized without GvpH are altered in strength, since they disaggregate into ribs when prepared for electron microscopy (Offner *et al.*, 2000).

The Mode of Transcription of the Various Vac Regions is Similar, but not Identical

Transcription in halophilic archaea depends on a single DNA-dependent RNA polymerase comprising 12 subunits that are homologous to the multiple subunits of the eukaryotic RNA polymerase II. The archaeal promoter consists of a TATA box centered around position −28 upstream of the transcription start site, and the initiation of transcription requires the TATA box binding protein TBP and the transcription factor TFB that is homologous to the eukaryotic transcription factor TFIIB (Hausner *et al.*, 1996; Qureshi *et al.*, 1995; Thomm, 1996). Many archaeal promoters mapped to date contain the TFB recognition element BRE adjacent to the TATA box (Bell *et al.*, 1999). Multiple divergent genes encoding the TFB and TBP proteins have been identified in *Haloferax volcanii* (Thompson *et al.*, 1999), and seven *tfb* and five *tbp* genes are also found in the genome sequence of *Hb. salinarum* NRC-1 (Ng *et al.*, 2000).

Despite the identical arrangement of the 14 *gvp* genes in the three vac regions, the transcription of these genes shows variations (see Figure 1). The p-vac region of plasmid pHH1 is expressed from four promoters located in front of p-*gvpA*, p-*gvpD*, p-*gvpF*, and p-*gvpO.* Three of these promoters are predominantly active during the exponential growth phase, whereas the p-*gvpDE* mRNA occurs only during the stationary growth phase (see Figure 1) (Offner and Pfeifer, 1995; Offner *et al.*, 1996). Large amounts of p-*gvpA* mRNA encoding the major gas vesicle structural protein are synthesized throughout growth, together with minor amounts of the p-*gvpACNO* cotranscript. Two promoters located in front of the chromosomal c-*gvpA* and c-*gvpD* genes drive the expression of the c-vac region (Englert *et al.*, 1992a). However, in *Hb. salinarum* PHH1 wild-type containing the p-vac and the c-vac region, only minor amounts of the relatively unstable c-*gvpDEFGHIJKLM* transcript are observed throughout growth, whereas the c-*gvpACNO* genes are not transcribed. This is the reason why the spindle-shaped gas vesicles of *Hb. salinarum* PHH1 are exclusively formed by pGvpA, and cGvpA is not involved. The cGvpD protein appears during exponential growth, whereas the cGvpE protein required for the transcription of the c-*gvpACNO* operon (see below) is not detectable, indicating that the first two reading frames of the c-*gvpDEFGHIJKLM* transcript are translated at different points in time (Krüger and Pfeifer, 1996).

The cylinder-shaped gas vesicles formed by cGvpA of the c-vac region are only observed in *Hb. salinarum* PHH4 lacking the p-vac region (Pfeifer and Blaseio, 1989). The c-*gvpDEFGHIJKLM* mRNA (and smaller products) are detectable throughout growth, whereas the c-*gvpA* promoter is completely inactive during the exponential growth phase (Krüger and Pfeifer, 1996). Also in this case, the c-*gvpD* and c-*gvpE* genes are differentially expressed: the cGvpD protein occurs during the early exponential growth phase, whereas the cGvpE protein is detectable during the stationary growth phase. At the same time, a large amount of c-*gvpA* mRNA and minor amounts of the c-*gvpACNO* transcript occur and gas vesicles are formed (see Figure 1). The expression of the related *gvp2* gene cluster on pNRC200 of *Hb. salinarum* NRC-1 (Ng *et al.*, 2000) has not been investigated so far. Since this cluster lacks the essential c-*gvpM* gene, *gvp1*-negative mutants should be gas vesicle negative. However, this *gvp2* cluster could still be expressed.

Also, the mc-vac region is transcribed from two promoters, leading to minor amounts of the mc-*gvpDEFGHIJKLM* and mc-*gvpA* mRNA during exponential growth (Röder and Pfeifer, 1996). Large amounts of mc-*gvpA* mRNA only occur in the stationary growth phase, together with minor amounts of the mc-*gvpACNO* transcript (Figure 1). In contrast to the c-vac region, the promoter of the mc-*gvpD* gene is activated during the stationary growth phase. The resulting strong transcription leads to shorter transcripts such as the mc-*gvpDEF* and mc-*gvpD* mRNAs, and large amounts of transcripts smaller than the mc-*gvpD* gene (Röder and Pfeifer, 1996). The early termination (or processing) of the *gvpDEFGHIJKLM* transcript ensures that the products encoded by the *gvp* genes located downstream of mc-*gvpF* are only translated for a short time during exponential and early stationary growth.

Transcription of *gvp* Genes in *Hf. volcanii* Transformants and Promoter Analyses

Hf. volcanii transformants containing the entire p-vac or mc-vac region produce very similar *gvp* transcription patterns during the growth phases as determined for the respective wild-type strains. Thus, *Hf. volcanii* appears to be a useful host for comparative analyses of the different *gvp* promoters. *Hf. volcanii* transformants containing the p-*gvpA* gene by itself express this gene constitutively, demonstrating the high basal activity of the the pA promoter. In contrast, transformants with the mc-*gvpA* gene produce very minor amounts of mc-*gvpA* mRNA (Röder and Pfeifer, 1996), whereas the c-*gvpA* gene is not expressed at all (Krüger and Pfeifer, 1996). The latter results demonstrate that the promoters of the c-*gvpA* and mc-*gvpA* genes need activation to yield the high amount of *gvpA* mRNA found in the transformants harbouring the entire vac region. The TATA box sequence of both promoters (mcA and cA) indicate less conservation compared to the archaeal consensus TATA box sequence, especially with respect to the central TA nucleotides, whereas the TATA box of the pA promoter is highly conserved (Figure 2). The factor required for the activation of the *gvpA* promoters is the product of the *gvpE* gene located in the *gvpDEFGHIJKLM* gene cluster (Röder and Pfeifer, 1996; Krüger *et al.*, 1998).

	BRE TATA archaeal consensus: RNWAAWNYTTAWG	promoter activity in vac region context	
		exp	stat
pA	TTACAGGAGACATAACGACTGGTGAAACCATACACATCCTTATGTGATGCCCGAGTATAGTTAGAGAT*	++	++
cA	TTACGAGCGCCGAAACGGGGGTTGAACTCACAACGGCGGTTTTCCGGACACTCCCTGTAGTTGCGGGT*	-	++
mcA	TTACGAGAGGTGAAACGGTTGCTGAACCAACACGAATGATTTTGTTACTTGCCAACACGTTTTCAGAT*	(-)	++
mcD	GTTCAGCAACCGTTTCACCTCTCGTAATGAGTCAAACATCTAAGTAGTTTGGCAGATGAATGACA*	(+)	++
pD	TTTCACCAGTCGTTATGTCTCCTGTAATGAGTCGTCATTCTAAGTACGGAGAGTGTAAAGCTTCTTAG*	-	+
cD	TCGTAATAGTTCGCTGATTTGTACAAAATCTTGAACAAGATAAAAATGCGTTCTTATTGTTTCAACACT*	+	+
pF	TTTGGGGCAGACCTGAGTCCGGGTACAGTATACCCGCATTTAAATGACCTTGCAGTCGAAGGTGTACT*	++	-
pO	AATAGAATCCGCGATCGACGACATGGAAGTCGCCCTTTCTTAAGATCCGGGGTCTCTACATAGAAGC*	++	(+)

Figure 2. Sequences of the various *gvp* promoters. The start site of the transcription (+1) is marked by a star. The TATA box centered around −28 is underlined, and the consensus sequence of the TATA box and of the putative BRE element are given above (R = A or G, N = any base, W = A or T, Y = C or T). The plus and minus on the right describe the relative activity of the promoters during the exponential (exp) and stationary (stat) growth phase as determined by the amount of *gvpA* mRNA seen in Northern analyses using RNA derived from the respective wild-type strains. Minus = no activity; (−) = minor activity; (+) = low activity; + = intermediate activity; ++ = high activity.

Transformants containing the c-*gvpA* or mc-*gvpA* gene together with the respective *gvpE* gene produce high amounts of *gvpA* mRNA and also GvpA protein, suggesting that the *gvpE* gene encodes a transcriptional activator (see below).

Quantitative Analyses of *gvpA* Promoter Activities Using a β-Galactosidase Gene as Reporter

The *bgaH* reading frame encoding the β-galactosidase of *Haloferax alicantei* (Holmes and Dyall-Smith, 2000) can be used as reporter gene to investigate the basal and induced activities of the various *gvpA* promoters at a more quantitative level. *Hf. volcanii* lacks detectable β-galactosidase activity and offers in this respect a clean genetic background. Transformants expressing the β-galactosidase turn blue when sprayed with Xgal on agar plates, and the activity of this enzyme can be quantified by a standard ONPG assay (Holmes and Dyall-Smith, 2000). The application of this system for *Hb. salinarum* has recently been reported (Patenge *et al.*, 2000).

Transformants containing fusions between the promoter regions of the various *gvpA* genes and the *bgaH* reading frame (A-*bgaH*) yield β-galactosidase activities that correlate well with the amount of the respective *gvpA* mRNA (Gregor and Pfeifer, 2001). The cA-*bgaH* fusion gene is not expressed due to the completely inactive cA promoter, the mcA-*bgaH* gene reveals low β-galactosidase activity, and the pA-*bgaH* fusion gene leads to a constitutive production of β-galactosidase in *Hf. volcanii* transformants. Transformants harbouring an A-*bgaH* construct together with the homologous *gvpE* reading frame expressed under the control of the ferredoxin (*fdx*) gene on the expression vector pJAS35 yield severalfold higher β-galactosidase activities, indicating that all three *gvpA* promoters are activated by GvpE (Gregor and Pfeifer, 2001). Each A-*bgaH* gene was also tested for the activation by the heterologous GvpE proteins. From these experiments it appears that cGvpE is the strongest and pGvpE the weakest transcriptional activator. Each GvpE protein is able to activate the pA or the mcA promoter, whereas the cA promoter in cA-*bgaH* is only activated by the homologous cGvpE protein (Gregor and Pfeifer, 2001). A chimeric pAcA promoter consisting mainly of the cA promoter sequence (including the TATA box) but containing 21 nucleotides 5′ to the TATA box substituted with the respective p-*gvpA* promoter sequences, still lacks the basal promoter activity. Nevertheless, this chimeric promoter aquired the ability for activation by all three GvpE proteins. These results imply that the 21 nt sequence upstream of the TATA-box is important for the promoter activation mediated by GvpE (Gregor and Pfeifer, 2001).

The Transcriptional Activator GvpE Resembles a Basic Leucine-Zipper Protein

A closer inspection of the GvpE protein sequence implies that GvpE resembles a basic leucine-zipper protein typically involved in the regulation of gene expression in eukaryotes. A molecular modeling study of the carboxy-terminal portion of cGvpE indicates an amphiphilic helix (AH6) suitable for the formation of the leucine-zipper structure within a cGvpE dimer. Close to the N-terminal end of the AH6 helix is a cluster of basic amino acids that could constitute the DNA binding site (DNAB, Krüger *et al.*, 1998). A model of the cGvpE dimer docked onto DNA suggests that the side-chains of the basic amino acids found in DNAB could perfectly interact with the negatively charged phosphate groups of the DNA backbone. Mutations in this DNAB motif generated cGvpE proteins that were unable to activate the c-*gvpA* promoter in *Hf. volcanii* transformants, indicating that these amino acids are required for cGvpE activity (Krüger *et al.*, 1998). Other mutations affecting conserved amino acids in the hydrophobic surface of AH6 should prevent the dimer formation of the putative leucine-zipper. Most of these cGvpE "zipper mutants" are unable to activate the c-*gvpA* promoter *in vivo*, underlining the importance of this structure for the transcriptional activator activity (Plößer and Pfeifer, 2002). Biochemical analyses of the isolated cGvpE proteins and cGvpE mutants are in progress and will provide further insights into cGvpE dimer formation, the DNA binding activity of cGvpE, and the importance of distinct amino acids for the cGvpE activator function.

GvpD is Involved in the Repression of Gas Vesicle Formation

The involvement of *gvpD* in the repression of gas vesicle formation was first observed in *Hf. volcanii* transformants containing an mc-vac region with an internal deletion in mc-*gvpD* (ΔD transformants, Englert *et al.*, 1992b). These ΔD transformants are gas vesicle overproducers (Vac^{++}), whereas ΔD/D transformants, containing ΔD and in addition the mc-*gvpD* reading frame under the control of the *fdx* promoter in pJAS35, do not form gas vesicles and are Vac^{-} (Figure 3) (Pfeifer *et al.*, 1994). The function of mcGvpD in repression can also be observed at the transcriptional level when the amount of mc-*gvpA* transcripts is investigated in these transformants. The Vac^{++} ΔD transformants contain high amounts of mc-*gvpA* mRNA predominantly in the stationary growth phase, whereas the amount of this transcript is significantly reduced in the Vac^{-} ΔD/D transformants, implying that mcGvpD is able to prevent the GvpE-mediated mc-*gvpA* promoter activation (Pfeifer *et al.*, 2001).

The deduced amino acid sequences of all three GvpD proteins indicate two interesting features that are important for this repressor function: a putative nucleotide-binding site (p-loop motif) near the N-terminus, and two basic regions (Pfeifer *et al.*, 2001). Alterations of conserved amino acids in the p-loop motif of mcGvpD (36LVN*GAP*G*T*G*KT*44, conserved amino acids are in italics) reveal mutated proteins (D_{mut}) that have

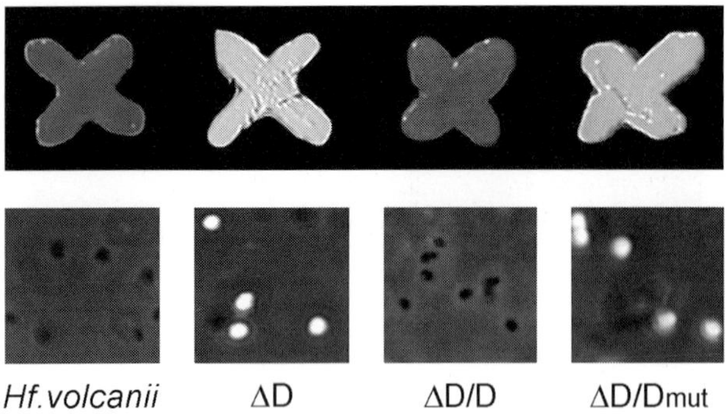

Figure 3. Phenotype of *Hf. volcanii* and various transformants. Top: colonies streaked in cross-shape on agar medium; bottom: cells (length 1 μm) observed by phase-contrast microscopy. Vac^{++} colonies and cells appear white due to the high level of light reflection of the gas vesicles. The Vac^{-} cells appear dark, and the Vac^{-} colonies show a dark red colour like the colonies of *Hf. volcanii* wild-type.

been tested *in vivo* in $\Delta D/D_{mut}$ transformants for their ability to reduce the amounts of gas vesicles in these transformants from Vac^{++} to Vac^{-}. All D_{mut} proteins with an alteration of an amino acid at a conserved position in the p-loop motif are unable to repress gas vesicle formation (see Figure 3). Thus, the nucleotide binding (and hydrolysis?) is important for the mcGvpD repressor function. Mutations in basic region 1 (position 201–221 in the mcGvpD sequence) also abolish the repressing activity of mcGvpD, whereas the alteration of basic region 2 (position 494–496) leads to ambiguous results: the alteration of RRR to AAA in this region results in a "super"-repressor protein, completely blocking the mc-*gvpA* expression in $\Delta D+D_{AAA}$ transformants (Pfeifer *et al.*, 2001). In contrast, a mcGvpD protein containing an alteration of RRR to ADA in this region is unable to repress. A positive charge in basic region 2 is obviously not important for the mcGvpD repressor function.

The question as to whether additional mc-*gvp* genes are required for the repression of the mc-*gvpA* promoter activity was addressed by constructing transformants harbouring the constructs ADE, AΔDE or AΔDE+D (Pfeifer *et al.*, 2001). Construct ADE contains the three genes mc-*gvpA* and mc-*gvpDE* in the native arrangement and expressed from the endogenous promoter; the AΔDE construct incurred an internal deletion within the mc-*gvpD* reading frame, and this construct is complemented with mc-*gvpD* expressed under *fdx* promoter control in the AΔDE+D transformant. The amount of mc-*gvpA* mRNA is high in AΔDE transformants, slightly lower in ADE transformants, and significantly reduced in the AΔDE+D transformants. The slightly lower amount of mc-*gvpA* mRNA in ADE compared to AΔDE transformants presumably reflects the relatively weak expression of mc-*gvpD* under the endogenous promoter control, whereas the strong reduction of the mc-*gvpA* mRNA in the AΔDE+D transformants is most likely due to an earlier and stronger synthesis of mc-*gvpD* mRNA (and mcGvpD) under *fdx* promoter control. Since the amount of mc-*gvpA* mRNA is not reduced to the minor amounts observed during exponential growth with transformants containing the entire mc-vac region, these results suggest that other mcGvp proteins are involved in mc-*gvpA* regulation. However, the lack of these other mcGvp proteins can be complemented by higher amounts of mcGvpD protein in case of the AΔDE+D transformants. It is not known so far whether mcGvpD acts directly or indirectly at the mc-*gvpA* promoter, since it is also possible that mcGvpD inactivates mcGvpE, or acts at the mc-*gvpE* mRNA level by reducing the amount of mcGvpE produced during exponential growth.

Conclusions

The genes involved in gas vesicle formation offer an interesting system to study gene regulation at the transcriptional (and also the translational level) in halophilic archaea. Despite the eukaryotic RNA polymerase and promoter structure, most archaeal gene regulator proteins characterized to date are of the bacterial type. Examples are a repressor protein involved in the regulation of nitrogen fixation in *Methanococcus maripaludis* (Cohen-Kupiec *et al.*, 1997), or the regulator of the arginine fermentation in *Hb. salinarum* (Ruepp and Soppa, 1996; Soppa *et al.*, 1998). So far, the only example of a regulatory protein with structural similarities to eukaryotic basic leucine-zipper regulators appears to be the GvpE protein described here. The activities of various *gvpA* (and *gvpD*) promoters depend on this protein, and the repression of these promoters involve the action of the GvpD protein. Whether and where the GvpE protein binds in the *gvpA* promoter region is currently under investigation; these experiments will give further insight into the importance of specific DNA sequences. How the environmental factors light, oxygen and salt are transduced to these regulators and influence *gvp* transcription is not known to date. Some of these factors have been investigated with respect to the amount of *gvp* gene expression. Higher amounts of mc-*gvpA* mRNA (and also of other mc-*gvp* mRNAs) are found in response to high salt concentrations with *Hf. mediterranei* (Englert *et al.*, 1990; Röder and Pfeifer, 1996), and aeration caused a 5-fold higher expression of (p)-*gvpA* in the case of *Hb. salinarum* NRC-1 (Yang and DasSarma, 1990). The investigation of the signal transduction pathways involved is in progress and should help to uncover the general regulatory networks affecting gene expression in halophilic archaea.

Acknowledgements

This work received financial support from the Deutsche Forschungsgemeinschaft as a project in the program on "Regulatory networks in bacteria" (PF 165/6-1 through 6-3). We thank Kathryn Nixdorff and Jobst Gmeiner for critical reading of this manuscript.

References

Bell, S., Jaxel, C., Nada, M., Kosa, P., and Jackson, S. 1998. Temperature, template topology, and factor requirements of archaeal transcription. Proc. Natl. Acad. Sci. USA 95: 15218–15222.

Charlebois R., and Doolittle, W.-F. 1989. Transposable elements and genome structure in halobacteria. In: Mobile DNA, D. Berg and M. Howe (eds.), ASM, Washington DC, p. 297–307.

Cohen-Kupiec, R., Blank, C., and Leigh, J.A. 1997. Transcriptional regulation in Archaea: *In vivo* demonstration of a repressor binding site in a methanogen. Proc. Natl. Acad. Sci. USA 94: 1316–1320.

DasSarma, S., Arora, P., Lin, F., Molinari, E., and Yin, L. 1994. Wild-type gas vesicle formation requires at least ten genes in the *gvp* gene cluster of *Halobacterium halobium* plasmid pNRC100. J. Bacteriol. 176: 7646–7652.

Ebert, K., Hanke, C., Delius, H., Goebel, W., and Pfeifer, F. 1987. A new insertion element, ISH26, from *Halobacterium halobium.* Mol. Gen. Genet. 206: 81–87.

Englert, C., and Pfeifer, F. 1993. Analysis of gas vesicle gene expression in *Haloferax mediterranei* reveals that GvpA and GvpC are both gas vesicle structural proteins. J. Biol. Chem. 268: 9329–9336.

Englert, C., Horne, M., and Pfeifer, F. 1990. Expression of the major gas vesicle protein in the halophilic archaebacterium *Haloferax mediterranei* is modulated by salt. Mol. Gen. Genet. 222: 225–232.

Englert, C., Krüger, K., Offner, S., and Pfeifer, F. 1992a. Three different but related gene clusters encoding gas vesicles in halophilic archaea. J. Mol. Biol. 227: 586–592.

Englert, C., Wanner, G., and Pfeifer, F. 1992b. Functional analysis of the gas-vesicle gene cluster of the halophilic archaeon *Haloferax mediterranei* defines the vac-region boundary and suggests a regulatory role for the *gvpD* gene or its product. Mol. Microbiol. 6: 3543–3550.

Gregor, D., and Pfeifer, F. 2001. Use of a halobacterial *bgaH* reporter gene to analyse the regulation of gene expression in halophilic archaea. Microbiol. 147: 1745–1754.

Halladay, J., Jones, J., Lin, F., MacDonald, F., and DasSarma, S. 1993. The rightward gas vesicle operon in *Halobacterium* plasmid pNRC100: Identification of the *gvpA* and *gvpC* gene products by use of antibody probes and genetic analysis of the region downstream of *gvpC.* J. Bacteriol. 175: 684–692.

Hausner, W., Wettach, J., Hethke, C., and Thomm, M. 1996. Two transcription factors related with the eucaryal transcription factors TATA-binding protein and transcription factor IIB direct promoter recognition by an archaeal RNA polymerase. J. Biol. Chem. 27: 30144–30148.

Holmes, M., and Dyall-Smith, M. 2000. Sequence and expression of a halobacterial β-galactosidase gene. Mol. Microbiol. 36: 114–122.

Jones, J., Young, D., and DasSarma, S. 1991. Structure and organization of the gas-vesicle gene cluster on the *Halobacterium halobium* plasmid pNRC100. Gene 102: 117–122.

Krüger, K., and Pfeifer, F. 1996. Transcript analysis of the c-vac region, and differential synthesis of the two regulatory gas-vesicle proteins GvpD and GvpE in *Halobacterium salinarium* PHH4. J. Bacteriol. 178: 4012–4019.

Krüger, K., Hermann, T., Armbruster, V., and Pfeifer, F. 1998. The transcriptional activator GvpE for the halobacterial gas vesicle genes resembles a basic region leucine-zipper regulatory protein. J. Mol. Biol. 279: 761–771.

Ng, W.L., Kothakota, S., and DasSarma S. 1991. Structure of the gas vesicle plasmid in *Halobacterium halobium:* inversion isomers, inverted repeats, and insertion sequences. J. Bacteriol. 173: 1958–1964.

Ng, W.L., Ciufo, S., Smith, T., Bumgarner, E., Baskin, D., Faust, J., Dall, B., Loretz, C., Seto, J., Slagel, J., Hood, L., and DasSarma, S. 1998. Snapshot of a large dynamic replicon in a halophilic archaeon: Megaplasmid or minichromosome? Genome Res. 8: 1131–1141.

Ng, W.L., Kennedy, S.P., Mahairas, G., Berquist, B., Pan, M., Shukla, H., Lasky, S., Baliga, N., Thorsson, V., Sbrogna, J., 32 others, and DasSarma, S. 2000. Genome sequence of *Halobacterium* species NRC-1. Proc. Natl. Acad. Sci. USA 97: 12176–12181.

Offner, S., and Pfeifer, F. 1995. Complementation studies with the gas vesicle-encoding p-vac region of *Halobacterium salinarium* PHH1 reveal a regulatory role for the p-*gvpDE* genes. Mol. Microbiol. 16: 9–19.

Offner, S., Wanner, G., and Pfeifer, F. 1996. Functional studies of the *gvpACNO* operon of *Halobacterium salinarium* reveal that the GvpC protein shapes gas vesicles. J. Bacteriol. 178: 2071–2078.

Offner, S., Ziese, U., Wanner, G., Typke, D. and Pfeifer, F. 1998. Structural characteristics of halobacterial gas vesicles. Microbiol. 144: 1331–1342

Offner, S., Hofacker, A., Wanner, G., and Pfeifer, F. 2000. Eight of fourteen *gvp* genes are sufficient for the formation of gas vesicles in halophilic archaea. J. Bacteriol. 182: 4328–4336.

Patenge, N., Haase, A., Bolhuis, H., and Oesterhelt, D. 2000. The gene for a halophilic β-galactosidase (*bgaH*) of *Haloferax alicantei* as a reporter gene for promoter analyses in *Halobacterium salinarum.* Mol. Microbiol. 36: 102–113.

Pfeifer, F. 1988. Genetics of halobacteria. In: Halophilic bacteria, F. Rodriguez-Valera (ed.), CRC Press, Inc. Boca Raton, USA. p. 105–133.

Pfeifer, F., and Betlach, M. 1985. Genome organization of *Halobacterium halobium:* A 70 kb island of more (AT) rich DNA in the chromosome. Mol. Gen. Genet. 198: 449–455.

Pfeifer, F., and Blaseio, U. 1989. Insertion elements and deletion formation in a halophilic archaebacterium. J. Bacteriol. 171: 5135–5140.

Pfeifer, F., and Blaseio, U. 1990. Transposition burst of the ISH27 insertion element family in *Halobacterium halobium.* Nucleic Acids Res. 18: 6921–6925.

Pfeifer, F., and Ghahraman, P. 1993. Plasmid pHH1 of *Halobacterium salinarium*: characterization of the replicon region, gas vesicle gene cluster and insertion elements. Mol. Gen. Genet. 238: 193–200.

Pfeifer, F., Weidinger, G., and Goebel, W. 1981a. Characterization of plasmids in halobacteria. J. Bacteriol. 145: 369–374.

Pfeifer, F., Weidinger, G., and Goebel, W. 1981b. Genetic variability in *Halobacterium halobium.* J. Bacteriol. 145: 375–381.

Pfeifer, F., Betlach, M., Martienssen, R., Friedman, J., and Boyer, H.W. 1983. Transposable elements of *Halobacterium halobium.* Mol. Gen. Genet. 191: 182–188.

Pfeifer, F., Blaseio, U., and Ghahraman, P. 1988. Dynamic plasmid populations in *Halobacterium halobium.* J. Bacteriol. 170: 3718–3724.

Pfeifer, F., Offner, S., Krüger, K., Ghahraman, P., and Englert, C. 1994. Transformation of halophilic archaea and investigation of gas vesicle synthesis. Syst. Appl. Microbiol. 16: 569–577.

Pfeifer, F., Krüger, K., Röder, R., Mayr, A., Ziesche, S., and Offner, S. 1997. Gas vesicle formation in halophilic archaea. Arch. Microbiol. 167: 259–268.

Pfeifer, F., Zotzel, J., Kurenbach, B., Röder, R., and Zimmermann, P. 2001. A p-loop motif and two basic regions in the regulatory protein GvpD are important for the repression of gas vesicle formation in the archaeon *Haloferax mediterranei.* Microbiol. 147: 63–73.

Plößer, P., and Pfeifer, F. (2002). A6ZIP protein from holophilic archaea: structural features and dimer formation of cGvpE from Halobacterium salinavum. Mol. Microbiol. 45: 511–520.

Qureshi, S., Baumann, P., Rowlands, T., Khoo, B., and Jackson, S. 1995. Cloning and functional analysis of the TATA binding protein from *Sulfolobus shibatae.* Nucleic Acids Res 23: 1775–1781.

Röder, R., and Pfeifer, F. 1996. Influence of salt on the transcription of the gas vesicle genes of *Haloferax mediterranei* and identification of the endogenous transcriptional activator gene. Microbiol. 142: 1715–1723.

Ruepp, A., and Soppa, J. 1996. Fermentative arginine degradation in *Halobacterium salinarum*: Genes, gene products and transcripts of the *arcRACB* gene cluster. J. Bacteriol. 178: 4942–4947.

Sapienza, C., and Doolittle, W.F. 1982. Unusual genome organization of the Halobacterium genome. Nature 295: 384–389.

Soppa, J., Link, T.A., Ruepp, A., Vatter, P., and zur Mühlen, A. 1998. Regulation of gene expression in *Halobacterium salinarum*: The *arcRACB* gene cluster and the TATA box-binding protein. In: Microbiology and Biogeochemistry of Hypersaline Enviroments. A. Oren (ed.), CRC press. p. 249–263.

Thomm, M. 1996. Archaeal transcription factors and their role in transcription initiation. FEMS Microbiol. Rev. 18: 159–171.

Thompson, D.K., Palmer, J.R., and Daniels, C.J. 1999. Expression and heat-responsive regulation of a TFIIB homologue from the archaeon *Haloferax volcanii.* Mol. Microbiol. 33: 1081–1092.

Tindall, B. 1992. The family *Halobacteriaceae.* In: The procaryotes. A. Balows, H.G. Trüper, M. Dworkin, W. Harder, and K.H. Schleifer (eds.), Springer Verlag, New York. p 768–808.

Ventosa, A., and Oren, A. 1996. *Halobacterium salinarum* nom. corrig., a name to replace *Halobacterium salinarium* (Elzari-Volcani) and to include *Halobacterium halobium* and *Halobacterium cutirubrum.* Int. J. System. Bacteriol. 46: 347.

Yang, C., and DasSarma, S. 1990. Transcriptional induction of purple membrane and gas vesicle synthesis in the archaebacterium *Halobacterium halobium* is blocked by a DNA gyrase inhibitor. J. Bacteriol. 172: 4118–4121.

Walsby, A.E. 1994. Gas vesicles. Microbiol. Rev. 58: 94–144.

Chapter 3

Sensory Transduction and Motion Control in *Sinorhizobium meliloti*

Birgit Scharf and Rüdiger Schmitt

Abstract

Molecular mechanisms that govern chemotaxis and motility in the nitrogen-fixing soil bacterium, *Sinorhizobium meliloti*, are distinguished from the well-studied taxis systems of enterobacteria by new features. (i) In addition to six transmembrane chemotaxis receptors, *S. meliloti* has two cytoplasmic receptor proteins, McpY (*m*ethyl-accepting *c*hemotaxis *p*rotein Y) and IcpA (*i*nternal *c*hemotaxis *p*rotein A). (ii) The tactic response is mediated by two response regulators, CheY1 and CheY2, but no phosphatase, CheZ. Phosphorylated CheY2 (CheY2-P) is the main regulator of motor function, whereas CheY1 assumes the role of a 'sink' for phosphate that is shuttled from CheY2-P back to CheA. This phospho-transfer from surplus CheY2-P to CheA to CheY1 replaces CheZ phosphatase. (iii) *S. meliloti* flagella have a complex structure with three helical ribbons that render the filaments rigid and unable to undergo polymorphic transitions from right- to left-handedness. Flagella rotate only clockwise and their motors can increase and decrease rotary speed. Hence, directional changes of a swimming cell occur during slow-down, when several flagella rotate at different speeds. Two novel motility proteins, the periplasmic MotC and the cytoplasmic MotD, are essential for motility and rotary speed variation. Our data suggest that MotC, by binding to MotB, stabilizes the energizing MotA-MotB proton channel, while MotD (increasing speed) and CheY2-P (decreasing speed) compete for binding to the cytoplasmic surface of the motor (FliM) in an antagonistic fashion, which leads to variations in flagellar rotary speed.

Overview

Motile bacteria are propelled by helical flagellar filaments that enable them to swim towards an optimum environment. Mechanisms of swimming, chemoreception, and signal transduction have been most intensely studied in *Escherichia coli* and *Salmonella typhimurium* (Macnab, 1996; Stock and Surette, 1996). An increasing interest in non-enteric bacteria in recent years led to new insights into structures and molecular mechanisms that differ from the enterobacterial paradigm (Armitage and Schmitt, 1997; Schmitt, 2002). Members of the α-subgroup of bacteria, such as *Agrobacterium*, *Caulobacter*, *Rhodobacter* and *Sinorhizobium* exhibit variations on the basic chemosensory pathway as a consequence of more than 500 million years separate evolution (Ochman and Wilson, 1987) and as an adaptation to their different environments. The rhizosphere bacterium, *S. meliloti*, induces nodule formation in the rootlets of *Medicago sativa* for nitrogen fixation. Motility in viscous soil and detection of attracting root exudates from host plants requires complex patterns of chemosensing, signalling, swimming and motor control (Figure 1). In *S. meliloti*, all the genes involved in these processes are combined in one large cluster referred to as the "flagellar regulon". By contrast, the corresponding *E. coli* genes are dispersed onto three different chromosomal regions. The flagellar regulon of *S. meliloti* contains all genes necessary for chemotaxis, biosynthesis, motor rotation, as well as regulatory genes. Gene expression is controlled as a three-level hierarchy and involves transcription factors different from the enterobacterial system (Sourjik *et al.*, 2000).

Chemosensing

Sensing a change in the external concentration of an attractant or repellent by an *E. coli* cell relies on a conformational change of dimeric membrane-spanning receptors, the MCPs (*m*ethyl-accepting *c*hemotaxis *p*roteins; Mowbray and Sandgren, 1998). *S. meliloti* is similarly capable of sensing and reacting to numerous attractants, including sugars, amino acids, flavenoids, and other stimuli from the host plant (Götz *et al.*, 1982; Malek, 1989). Seven distinct *mcp*-genes encoding McpT – McpZ serve as transducers for attractants (Figure 1). A deletion of any one of these results in a reduction of chemotaxis towards a set of attractants, whereas deletion of all receptors causes a complete loss of the tactic response. This stepwise reduction of the response may result from "cross-talk" between several receptors. The McpY receptor is distinguished by its cytoplasmic location and the possession of two tandem PAS domains. This is reminiscent of the *E. coli* Aer receptor, which mediates aerotaxis (taxis towards oxygen) via an FAD-binding PAS domain (Taylor and Zhulin, 1999). However, purified McpY showed no detectable affinity for FAD, and an McpY-knockout mutant was not impaired in aerokinesis (acceleration towards oxygen; Zhulin *et al.*, 1995). Therefore, a functional role for McpY remains to be established. There is a second soluble sensor, IcpA (*i*nternal *c*hemotaxis *p*rotein, formerly named Orf1; Greck *et al.*, 1995), present in *S. meliloti*. This protein is

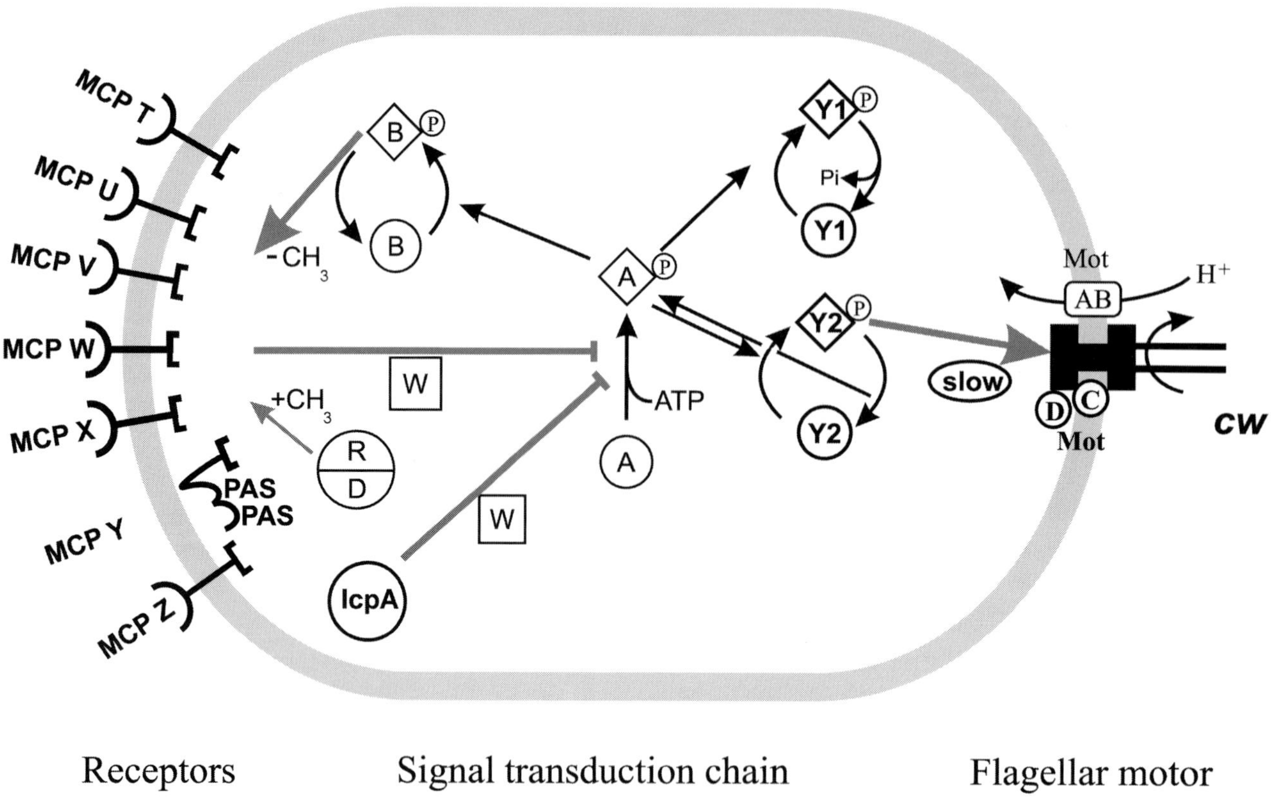

Figure 1. Schematic view of signal transduction from chemotaxis receptors (Mcp, Icp) to the flagellar motor of *S. meliloti*. Abbreviations: MCP, *m*ethyl-accepting *c*hemotaxis *p*rotein; IcpA, *i*nternal *c*hemotaxis *p*rotein; letters denote cytoplasmic chemotaxis proteins; A, histidine autokinase; B, demethylase; D and R, methyltransferase; W, positive regulator of CheA; Y1 and Y2, response regulators; P, phosphoryl-group; MotA and MotB, transmembrane proteins forming the energy-supplying proton channels; MotC, periplasmic motor protein; MotD, cytoplasmic motor protein; CW, clockwise sense of flagellar filament rotation. Differences to the enterobacterial paradigm are marked by bold symbols and lettering. For details see text.

encoded by the first gene in the chemotaxis (*che*) operon (Sourjik *et al.*, 1998). The encoded 57-kDa protein contains a conserved signaling domain, but neither a periplasmic ligand-binding domain nor hydrophobic transmembrane domains nor methylaccepting glutamyl residues, which are typical features of "classical" MCP's. Overexpressed IcpA interferes with *E. coli* chemotaxis, probably by binding essential components, such as CheW and CheA. IcpA is a good candidate for sensing internal stimuli that reflect the metabolic state of a cell.

Signaling Pathway

Unlike in *E. coli*, the signal transduction chain of *S. meliloti* contains two different response regulators, CheY1 and CheY2, but no CheZ phosphatase (Figure 1). This is the basis for a new CheY2-P dephosphorylation mechanism that appears to be ubiquitous among α-subgroup proteobacteria. Phosphorylation of CheY2 and CheY1 by the CheA autokinase is the crucial step in signal transduction. CheY2-P is the chief regulator of motor function, its action being modulated by CheY1 (Sourjik and Schmitt, 1996). The absence of attractant elicits ATP-dependent autophosphorylation of CheA, which in turn rapidly transfers phosphoryl groups to CheY1 and CheY2; CheY2-P directs the slow-down of motor rotary speed (see below). In the presence of an attractant, CheA is inactive; surplus CheY2-P then retro-transfers its phosphoryl group to CheA, which in turn phosphorylates CheY1 (Sourjik and Schmitt, 1998) and CheB (demethylase). CheY1 molecules in a cell are in 10fold excess over CheA; moreover, CheY1-P autodephosphorylates constantly ($\tau/2 \sim 10$ sec). CheY1 thus acts as a sink for phosphoryl groups from CheY2-P and emulates the role of a CheZ phosphatase. This mechanism was first detected in *S. meliloti* and now appears as a general feature of the signal transduction chain among members of the α-subgroup of proteobacteria. In this scheme (Figure 1), CheY2-P and CheY1-P differ in that the former is capable of back-shuttling the phosphoryl group to CheA, whereas the latter is not (Sourjik and Schmitt, 1998). An interpretation compatible with other phopsho-relay systems (Stock and Da Re, 2000) predicts that the high free energy ($\Delta G^{\circ\prime} \sim -14$ kcal mol^{-1}) of CheA histidyl phosphate is retained throughout by the aspartyl phosphate of CheY2-P (E1 state), thus facilitating the back-transfer of phosphate to CheA (Figure 1). By contrast, CheY1-P may spontaneously assume the conformation that predisposes the low-free energy state (E2), of the aspartyl-phosphate bond ($\Delta G^{\circ\prime} \sim +2$ kcal mol^{-1}). The molecular structures of CheY2 (inactive) and CheY2-P (active) have been determined by high-resolution NMR. Conformational changes observed upon phosphorylation of CheY2 involve side-chain 'flipping' at Phe59, a twist that thermodynamically favors phosphoryl group transfer to the adjacent Asp58 (active site). Moreover, both CheY2 and CheY2-P, interact with the P2 binding domain of CheA by engaging distinctive, partially overlapping sites. The binding of CheY2-P to CheA is new and suggests a shift of the reaction equilibrium towards retrophosphorylation of CheA.

Rotary Speed Variation of the Flagellar Motor

Flagellar filaments and their mode of rotation differ greatly between enteric bacteria (γ-proteobacteria) and *S. meliloti* (α-proteobacteria), due to their long phylogenetic distance (Ochman and Wilson, 1987) and the need of adaptation to different biotopes. The soil bacterium *S. meliloti* is driven by rigid, right-handed flagella allowing the cell to swim efficiently in viscous media (Götz *et al.*, 1982). Flagellar filaments are assembled from four closely related flagellin subunits (FlaA-FlaD) encoded by linked, but independently transcribed genes (Pleier and Schmitt, 1989; Sourjik *et al.*, 1998; Scharf *et al.*, 2001). The complex fine structure of *S. meliloti* filaments is dominated by three helical ribbons that lock the filament in the right-handed conformation thus disabling it from switching hand (Schmitt *et al.*, 1974; Trachtenberg *et al.*, 1987; Trachtenberg and Hamel, 1992). Concomitantly, the motor of *S. meliloti* does not reverse its sense of rotation. Rather, a swimming cell is propelled by the unidirectional clockwise (CW) rotation of its right-handed flagellar filaments. The flagellar motors of

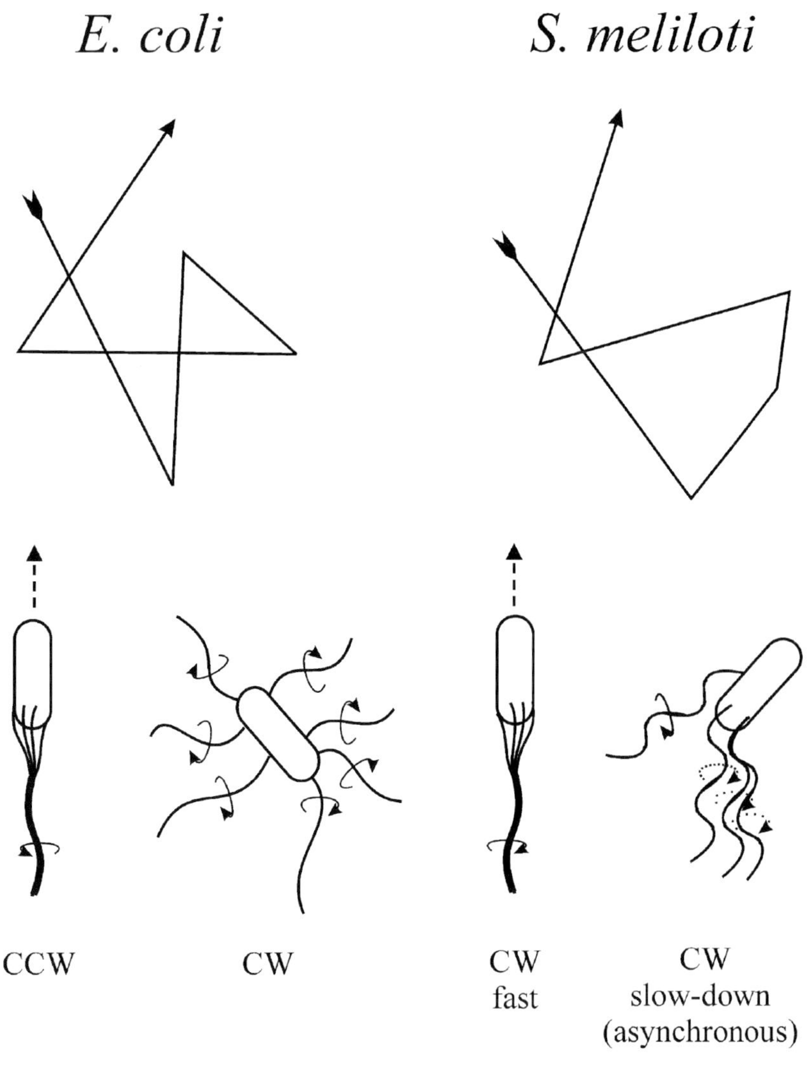

Figure 2. Comparison of the swimming patterns (top) of *E. coli* and *S. meliloti* and the corresponding modes of flagellar filament rotation (bottom). Note that the "biased random walk" in each case alternates between straight "runs" and directional changes. To this end, an *E. coli* cell reverses the rotation of its left-handed helical flagella from CCW (counterclockwise = run) to CW (clockwise = tumble), whereas the *S. meliloti* right-handed flagella persist in CW-rotation and increase rotational speed for straight runs (bundle formation) or decrease the rotational speed of individual flagella in an asynchronous fashion, which causes the propulsive bundle to dissociate and thus turns the swimming cell into a new direction ("caterpillar drive").

S. meliloti can increase and decrease their rotary speed (chemokinesis). The decrease of the rotary speed of the four to six pertrichously inserted flagella is asynchronous, which forces individual filaments of a flagellar bundle to fly apart, thus causing a change of swimming direction. The different modes, by which *E. coli* and *S. meliloti* cells accomplish directional changes are illustrated in Figure 2. The former briefly reverse the sense of flagellar rotation, which results in a tumbling motion and a change of swimming direction; the latter reduce the rotary speed of individual flagella without changing the CW sense of rotation, which results in the dissociation of the propulsive bundle and a new swimming direction (Scharf, 2002). This speed variation of the *S. meliloti* flagellar motor is a new feature, which has a molecular corollary in two new motility proteins, MotC and MotD, found in addition to MotA/MotB that form the energizing proton channels (Figure 1). MotC is a periplasmic protein interacting with the channel protein MotB, as shown by yeast two-hybrid analyses and affinity assays. We have shown before that binding of MotC to MotB is an essential prerequisite for the MotA-MotB channel function and proton flow, and that the dissociation of MotC from MotB (or mutations altering the 1:1 stoichiometry of MotB:MotC) interfere with or interrupt the proton flow (Platzer *et al.*, 1997). The implicated function of MotC as an indispensible "stabilizer" of MotB is an interesting feature of the flagella-energizing proton channel that needs further exploration. The cytoplasmic MotD was shown to increase rotary speed by interacting with FliM monomers that form the cytoplasmic surface of the rotor (C-ring), to which also the active response regulator, CheY2-P, binds to effect a decrease of rotary speed (Figure 1). MotD is thought to compete with CheY2-P in an antagonistic fashion for binding to the $\sim$ 34 copies of FliM. We propose that the degree of slowing or accelerating flagellar rotary speed may depend on the ratio of CheY2-P (slowing) to MotD (accelerating) molecules bound to FliM at any instant.

Acknowledgements

Work from our laboratory reported herein was supported by grants from the Deutsche Forschungsgemeinschaft (Scha 94/1-1 and Schm 68/34-1).

References

Armitage, J.P., and Schmitt, R. 1997. Bacterial chemotaxis: *Rhodobacter sphaeroides* and *Sinorhizobium meliloti* – variations on a theme? Microbiology 143: 3671–3682.

Götz, R., Limmer, N. Ober, K., and Schmitt, R. 1982. Motility and chemotaxis in two strains of *Rhizobium* with complex flagella. J. Gen. Microbiol. 128: 789–798.

Greck, M., Platzer, J., Sourjik, V., and Schmitt, R. 1995. Analysis of a chemotaxis operon in *Rhizobium meliloti*. Mol. Microbiol. 15: 989–1000.

Macnab, R.M. 1996. Flagella and motility, p.123–145. In F. Neidhardt, R. Curtiss III, J.L. Ingraham, E.C.C. Lin, K.B. Low, B. Magasanik, W.S. Reznikoff, M. Riley, M. Schaechter, and H.E. Umbarger (eds.), *Escherichia coli* and *Salmonella*: Cellular and Molecular Biology. ASM Press, Washington, D.C.

Malek, W. 1989. Chemotaxis in *Rhizobium meliloti* strain L5.30. Arch. Microbiol. 15: 611–612.

Mowbray, S.L., and Sandgren, M.O.J. 1998. Chemotaxis receptors: A progress report on structure and function. J. Struct. Biol. 124: 257–275.

Ochman, H., and Wilson, A.C. 1987. Evolution in bacteria: evidence for a universal substitution rate in cellular genomes. J. Mol. Evol. 26: 74–86.

Platzer, J., Sterr, W., Hausmann, M., and Schmitt, R. 1997. Three genes of a motility operon and their role in flagellar rotary speed variation in *Rhizobium meliloti*. J. Bacteriol. 179: 6391–6399.

Pleier, E., and R. Schmitt. 1989. Identification and sequence analysis of two related flagellin genes in *Rhizobium meliloti*. J. Bacteriol. 171: 1467–1475.

Sourjik, V., Muschler, P., Scharf, B., and Schmitt R. 2000. VisN and VisR are global regulators of chemotaxis, flagellar, and motility genes in *Sinorhizobium (Rhizobium) meliloti*. J. Bacteriol. 182: 782–788.

Scharf, B. 2002. Real-time imaging of fluorescent flagellar filaments of *Rhizobium lupini* H13-3: flagellar rotation and pH-induced polymorphic transitions. J. Bacteriol. 184: 5979–5986.

Scharf, B., Schuster-Wolf-Bühring, H., Rachel, R., and Schmitt, R. 2001. Mutational analysis of the *Rhizobium lupini* H13-3 and *Sinorhizobium meliloti* flagellin genes: importance of flagellin A for flagellar filament structure and transcriptional regulation. J. Bacteriol. 183: 5334–5342.

Schmitt, R. 2002. Sinorhizobial chemotaxis: a departure from the enterobacterial paradigm. Microbiol. 148: 627–631.

Schmitt, R., Bamberger, I., Acker, G., and Mayer, F. 1974. Feinstrukturanalyse der komplexen Geißeln von *Rhizobium lupini*. Arch. Microbiol. 100:145–162.

Sourjik, V., and Schmitt, R. 1996. Different roles of CheY1 and CheY2 in the chemotaxis of *Rhizobium meliloti*. Mol. Microbiol. 22: 427–436.

Sourjik, V., and Schmitt, R. 1998. Phosphotransfer between CheA, CheY1, and CheY2 in the chemotaxis signal transduction chain of *Rhizobium meliloti*. Biochemistry 37: 2327–2335.

Sourjik, V., Sterr, W., Platzer, J., Bos, I., Haslbeck, M., and Schmitt, R. 1998. Mapping of 41 chemotaxis, flagellar and motility genes to a single region of the *Sinorhizobium meliloti* chromosome. Gene 223: 283–290.

Stock, J., and Da Re, S. 2000. Signal transduction: response regulators on and off. Curr. Biol. 10: R420–424.

Stock, J.B., and Surette, M.G. 1996. Chemotaxis, p.1103–1129. In F. Neidhardt, R. Curtiss III, J.L. Ingraham, E.C.C. Lin, K.B. Low, B. Magasanik, W.S. Reznikoff, M. Riley, M. Schaechter, and H.E. Umbarger (eds.), *Escherichia coli* and *Salmonella*: Cellular and Molecular Biology. ASM Press, Washington, D.C.

Taylor, B.L., and Zhulin, I.B. 1999. PAS domains: internal sensors of oxygen, redox potential, and light. Microbiol. Mol. Biol. Rev. 63: 479–506.

Trachtenberg, S., DeRosier, D.J., and Macnab, R.M. 1987. Three-dimensional structure of the complex flagellar filament of *Rhizobium lupini* and its relation to the structure of the plain filament. J. Mol. Biol. 195: 603–620.

Trachtenberg, S., and Hamel, I. 1992. The rigidity of bacterial flagellar filaments and its relationship to filament polymorphisms. J. Struct. Biol. 109: 18–27.

Zhulin, I.B., Lois, A.F., and Taylor, B.L. 1995. Behavior of *Rhizobium meliloti* in oxygen gradients. FEBS L. 367: 180–182.

Chapter 4

Regulation of Succinoglycan and Galactoglucan Biosynthesis in *Sinorhizobium meliloti*

Anke Becker, Silvia Rüberg, Birgit Baumgarth, Peter Alexander Bertram-Drogatz, Ingmar Quester, and Alfred Pühler

Abstract

***Sinorhizobium meliloti* (*Rhizobium meliloti*) 2011 has the ability to produce the two acidic exopolysaccharides succinoglycan (EPS I) and galactoglucan (EPS II). EPS I is a branched heteropolysaccharide composed of octasaccharide repeating units, whereas EPS II is a linear heteropolysaccharide consisting of disaccharide subunits. The *exo-exs* and *exp* gene clusters are involved in the biosynthesis of EPS I and EPS II, respectively. EPS I and EPS II biosynthesis genes are differentially expressed resulting in a complex regulation of EPS production in *S. meliloti*. The phosphate concentration was identified as an important factor affecting the expression of *exp* genes.**

Introduction

Rhizobia are Gram-negative soil bacteria that are capable of inducing the formation of nitrogen fixing nodules on the roots of their leguminous host plants. This process involves specific recognition and progressive differentiation of both bacterial and host cells (Hirsch, 1992). *Sinorhizobium meliloti* (*Rhizobium meliloti*), the microsymbiont of *Medicago, Melilotus,* and *Trigonella* plants, is one of the best studied root nodule bacterium. *S. meliloti* produces the acidic exopolysaccharide succinoglycan. It is composed of octasaccharide repeating units containing one galactose and seven glucose residues joined by β-1$\rightarrow$3, β-1$\rightarrow$4 and β-1$\rightarrow$6 glycosidic linkages (Figure 1) (Reinhold *et al.*, 1994). The repeating unit can be decorated by acetyl, pyruvyl and succinyl groups. Succinoglycan often referred to as EPS I is produced in a high molecular weight (HMW) and low molecular weight (LMW) form (Wang *et al.*, 1999). Apart from the biosynthesis of EPS I, *S. meliloti* has the cryptic ability to produce a second EPS, designated galactoglucan or EPS II (Glazebrook and Walker, 1989). EPS II contains glucose and galactose residues in the ratio of one to one joined by α-1$\rightarrow$3 and β-1$\rightarrow$3 glycosidic bonds. It is decorated by acetyl- and pyruvyl groups.

Bacterial exopolysaccharides are essential for the establishment of a nitrogen-fixing symbiosis between *S. meliloti* and its host plant. At least LMW EPS I or LMW EPS II are required for nodule invasion of alfalfa plants by *S. meliloti* (Battisti *et al.*, 1992; *González et al.*, 1996; Wang *et al.*, 1999). The genome of *S. meliloti* consists of one chromosome and two megaplasmids. Two gene clusters directing the biosynthesis of EPS I and EPS II were identified on megaplasmid 2 (Long *et al.*, 1988; Müller *et al.*, 1988; Glazebrook and Walker, 1989).

The 27 kb *exo-exs* gene cluster that contains 19 *exo* and 2 *exs* genes directs the biosynthesis of EPS I (Buendia *et al.*, 1991; Müller *et al.*, 1993; Becker *et al.*, 1993a,b,c; 1995) (Figure 2A). In addition, five loci affecting EPS I biosynthesis were identified on the chromosome. These include *exoR*, *exoS* (Doherty *et al.*, 1988) and *mucR* (Zhan *et al.*, 1989; Keller *et al.*, 1995) which are implicated in regulation of EPS production, as well as *exoC* (Finan *et al.*, 1986) and *exoD* (Reed *et al.*, 1991). Biosynthesis of EPS II is directed by the 30 kb *exp* gene cluster that comprises 22 genes (Becker *et al.*, 1997) (Figure 2B), but the biosynthetic pathway of EPS II has not yet been elucidated in detail. The *exo* and *exs* genes encode proteins required for the synthesis of precursors and the repeating unit, decoration, polymerization, export and degradation of EPS I as well as for the regulation of EPS I production (Leigh and Walker, 1994; Becker and Pühler, 1998). The repeating unit of EPS I is assembled on a lipid carrier resulting in polyprenyl diphosphate octasaccharide intermediates that are subsequently polymerized (Reuber and Walker, 1993).

Regulation of Succinoglycan and Galactoglucan Production in *S. meliloti*

Under standard culture conditions in a complex medium, the *S. meliloti* 2011 wild type strain produces EPS I and synthesizes only traces of EPS II. Zhan *et al.* (1991) reported that phosphate-limiting conditions stimulate the production of EPS II by the wild type. In addition, extra copies of the *exp* gene cluster resulted in the biosynthesis of EPS II (Glazebrook and Walker, 1989). The production of EPS II was also observed in the presence of a mutation in either of the chromosomally located genes *expR* (Glazebrook and Walker, 1989; Pellock *et al.*, 2002) or *mucR* (Zhan *et al.*, 1989; Keller *et al.*, 1995). An additional regulatory gene, *expG* (*mucS*), was identified on megaplasmid 2 in the *exp* gene cluster (Astete and Leigh, 1996; Becker *et al.*, 1997; Rüberg *et al.*, 1999). The *expG* gene product shares weak homologies with transcriptional regulators of the MarR family that bind DNA through

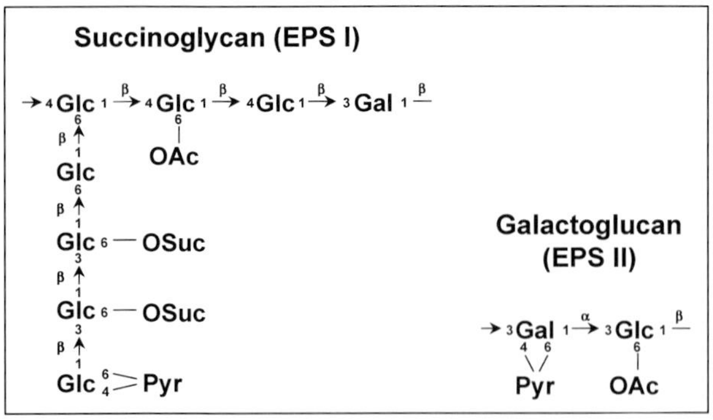

Figure 1. Repeating unit structure of succinoglycan and a galactoglucan produced by *S. meliloti.*

a helix-turn-helix motif (Becker *et al.*, 1997). *expG* is required for the stimulation of EPS II production by extra copies of the *exp* gene cluster and under phosphate limitation for the activation of the expression of the *expA* and *expE* operon (Astete and Leigh, 1996; Rüberg *et al.*, 1999). Using *exp-lacZ* transcriptional fusions an increase of *exp* gene transcription under phosphate limitation was detected for all *exp* complementation groups. This increase was reduced in strains characterized by a deletion of *expG* (Rüberg *et al.*, 1999). Extra copies of the *expG* gene also resulted in enhanced transcription of the *expA, expD* and *expE* operons, but did not influence the transcription of *expG* and *expC* (Rüberg *et al.*, 1999). Recently, we demonstrated that the ExpG protein binds to the *expA*, *expG*/*expD* and *expE* promoter regions. The high level of *exp* gene transcription in a *mucR* mutant was further elevated under phosphate-limiting conditions (Rüberg *et al.*, 1999). The *phoB* gene was found to be required for the activation of *exp* gene transcription by low phosphate concentrations (Rüberg *et al.*, 1999). In the *expA, expD, expG* and *expE* promoter regions sequences with similarities to PHO boxes known as the PhoB binding site in phosphate-regulated promoters in *Escherichia coli* were identified (Rüberg *et al.*, 1999). Two PHO box-like sequences were identified in each of these promoter regions. Both of these motifs were required for phosphate-dependent regulation of *exp* transcription.

Mutations in *mucR* resulted in the constitutive biosynthesis of EPS II and in a reduction of EPS I production (Zhan *et al.*, 1989; Keller *et al.*, 1995; Bertram-Drogatz *et al.*, 1998). The MucR protein (Keller *et al.*, 1995) is highly homologous to *Agrobacterium* Ros proteins which constitute repressors of the *vir*C and *vir*D operons as well as activators of EPS production (D'Souza-Ault *et al.*, 1993). Moreover, the *mucR* and *ros* genes are negatively autoregulated (D'Souza-Ault *et al.*, 1993; Keller *et al.*, 1995). A potential zinc finger motif of the C_2H_2 type was identified in the MucR and Ros amino acid sequences. D'Souza-Ault *et al.* (1993) demonstrated that the *A. tumefaciens* Ros protein can specifically bind to the *virC*/*virD* promoter region. The expression of *exp-lac*Z transcriptional fusions located in either of the four *exp* complementation groups was stimulated in the *S. meliloti mucR* mutant background (Keller *et al.*, 1995; Becker *et al.*, 1997; Rüberg *et al.*, 1999). In contrast, transcription of the *exo-lacZ* fusions was only slightly affected by the *mucR* mutation (Keller *et al.*, 1995; Bertram-Drogatz *et al.*, 1998). It was demonstrated that MucR can specifically bind to the *mucR*, *exoY* and *exoH* promoter regions (Bertram-Drogatz *et al.*, 1997, 1998). This implies that MucR itself excerts a negative autoregulation as well as a weak activation of *exoY* and a weak repression of *exoH* transcription (Bertram-Drogatz *et al.*, 1997, 1998).

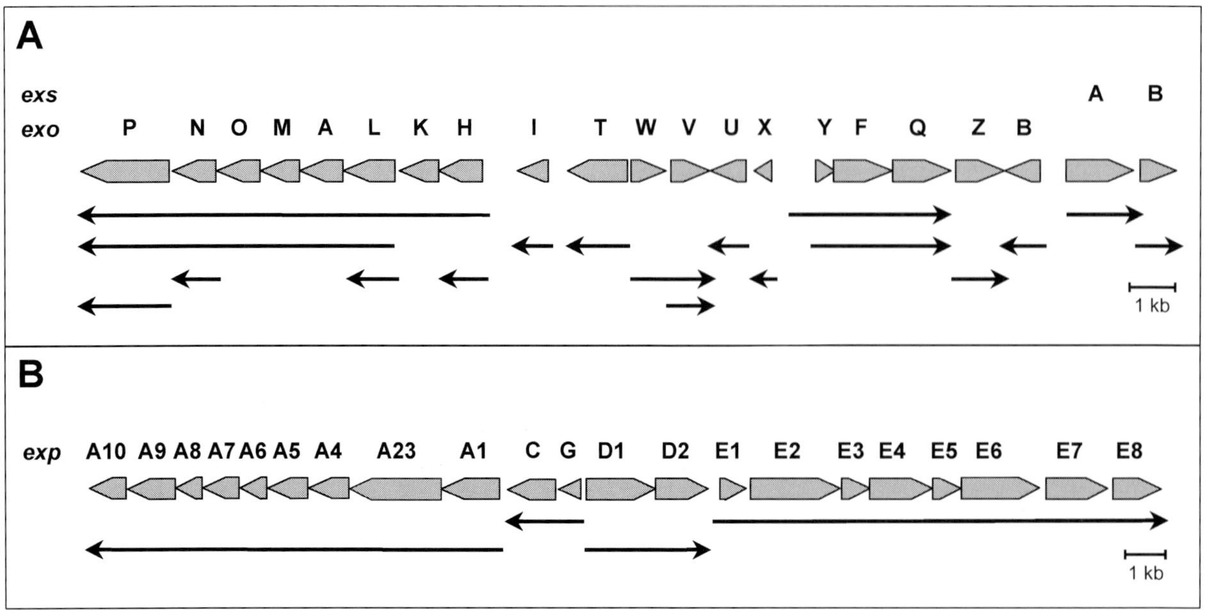

Figure 2. Organization and operon structure of the *S. meliloti exo-exs* and *exp* gene clusters. Operons are indicated by arrows.

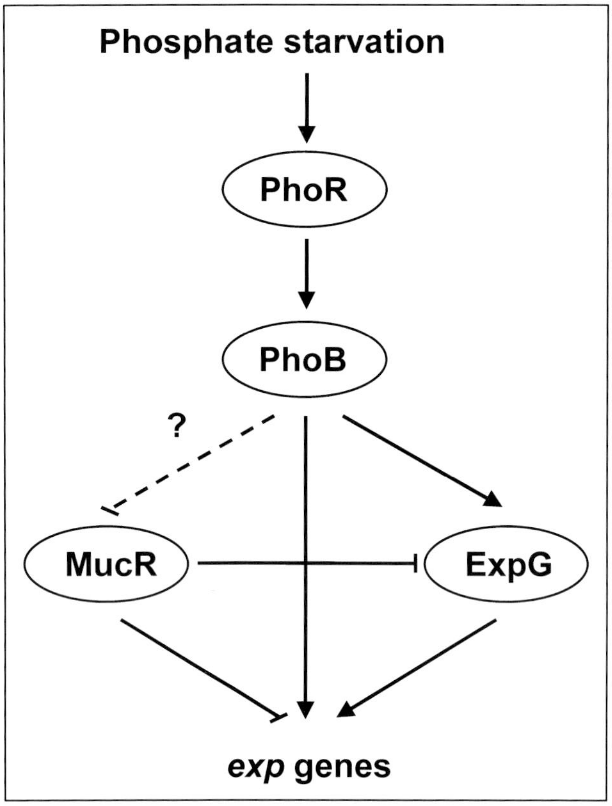

Figure 3. Model for the regulation of *exp* gene expression in *S. meliloti*. Lines with an arrowhead and cross hatch indicate positive and negative regulation, respectively.

Adaptation of EPS production to various environmental conditions requires a complex regulatory network involving cross talk between different regulatory components. Several regulators affecting the production of EPS I and EPS II were identified. Results from analyses of the role of MucR, ExpG and PhoB in the regulation of *exp* gene expression resulted in a model for this regulatory network (Figure 3) and can serve as a starting point to understand the molecular mechanisms of the differential regulation of EPS production in *S. meliloti*.

References

Astete, S.G., and Leigh, J.A. 1996. *mucS*, a gene involved in activation of galactoglucan (EPS II) synthesis gene expression in *Rhizobium meliloti*. Mol. Plant-Microbe Interact. 9: 395–400.

Battisti, L., Lara, J.C., and Leigh, J.A. 1992. Specific oligosaccharide form of the *Rhizobium meliloti* exopolysaccharide promotes nodule invasion in alfalfa. Proc. Natl. Acad. Sci. 89: 5625–5629.

Becker, A., Kleickmann, A., Arnold, W., and Pühler, A. 1993a. Analysis of the *Rhizobium meliloti exo*H/*exo*K/*exo*L fragment: ExoK shows homology to excreted endo-β-1,3-1,4-glucanases and ExoH resembles membrane proteins. Mol. Gen. Genet. 238: 145–154.

Becker, A., Kleickmann, A., Keller, M., Arnold, W., and Pühler, A. 1993b. Identification and analysis of the *Rhizobium meliloti exo*AMONP genes involved in exopolysaccharide biosynthesis and mapping of promoters located on the *exo*HKLAMONP fragment. Mol. Gen. Genet. 241: 367–379.

Becker, A., Kleickmann, A., Küster, H., Keller, M., Arnold, W., and Pühler, A. 1993c. Analysis of the *Rhizobium meliloti* genes *exo*U, *exo*V, *exo*W, *exo*T and *exo*I involved in exopolysaccharide biosynthesis and nodule invasion: *exo*U and *exo*W probably encode glucosyltransferases. Mol. Plant-Microbe Interact. 6: 735–744.

Becker, A., Küster, H., Niehaus, K., and Pühler, A. 1995. Extension of the *Rhizobium meliloti* succinoglycan biosynthesis gene cluster: identification of the *exsA* gene encoding an ABC transporter protein, and the *exsB* gene which probably codes for a regulator of succinoglycan biosynthesis. Mol. Gen. Genet. 249: 487–497.

Becker, A., Rüberg, S., Küster, H., Roxlau, A.A., Keller, M., Ivashina, T., Cheng, H.-P., Walker, G.C., and Pühler, A. 1997. The 32 kb *exp* gene cluster of *Rhizobium meliloti* directing the biosynthesis of galactoglucan: genetic organization and properties of the encoded gene products. J. Bacteriol. 179: 1375–1384.

Becker, A., and Pühler, A. 1998. Production of exopolysaccharides. In: The *Rhizobiaceae*. Molecular Biology of Model Plant-Associated Bacteria. H.P. Spaink, A. Kondorosi, and P.J.J. Hooykaas, (eds.), Kluwer Academic Publishers, Dordrecht, The Netherlands. p. 97–118.

Bertram-Drogatz, P.A., Rüberg, S., Becker, A., and Pühler, A. 1997. The regulatory protein MucR binds to a short DNA region located upstream of the *muc*R coding region in *Rhizobium meliloti*. Mol. Gen. Genet. 254: 529–538.

Bertram-Drogatz, P.A., Quester, I., Becker, A., and Pühler, A. 1998. The *Sinorhizobium meliloti* MucR protein, which is essential for the production of high-molecular-weight succinoglycan exopolysaccharide, binds to short DNA regions upstream of *exoH* and *exoY*. Mol. Gen. Genet. 257: 433–441.

Buendia, A.M., Enenkel, B., Köplin, R., Niehaus, K., Arnold, W., and Pühler, A. 1991. The *Rhizobium meliloti exo*Z/*exo*B fragment of megaplasmid 2: ExoB functions as a UDP-glucose-4′-epimerase and ExoZ shows homology to NodX of *Rhizobium leguminosarum* biovar *viciae* strain TOM. Mol. Microbiol. 5: 1519–1530.

Doherty, D., Leigh, J., Glazebrook, J., and Walker, G.C. 1988. *Rhizobium meliloti* mutants that overproduce the *R. meliloti* acidic calcofluor-binding exopolysaccharide. J. Bacteriol. 170: 4249–4256.

D'Souza-Ault, M.R., Cooley, M.B., and Kado, C.I. 1993. Analysis of the Ros repressor of *agrobacterium virC* and *virD* operons: molecular intercommunication between plasmid and chromosomal genes. J. Bacteriol. 175: 3486–3490.

Finan, T.M., Kunkel, B., de Vos, G.F., and Signer, E.R. 1986. Second symbiotic megaplasmid in *Rhizobium meliloti* carrying exopolysaccharide and thiamine synthesis genes. J. Bacteriol. 167: 66–72.

Glazebrook, J., and Walker, G.C. 1989. A novel exopolysaccharide can function in place of the calcofluor-binding exopolysaccharide in nodulation of alfalfa by *Rhizobium meliloti*. Cell 56: 661–672.

González, J.E., Reuhs, B.L., and Walker, G.C. 1996. Low molecular weight EPS II of *Rhizobium meliloti* allows nodule invasion in *Medicago sativa*. Proc. Natl. Acad. Sci. USA 93: 8636–8641.

Hirsch, A.M. 1992. Developmental biology of legume nodulation. New Phytol. 122: 211–237.

Keller, M., Roxlau, A., Weng, W.M., Schmidt, M., Quandt, J., Niehaus, K., Jording, D., Arnold, W., and Pühler, A. 1995. Molecular analysis of the *Rhizobium meliloti mucR* gene regulating the biosynthesis of the exopolysaccharide succinoglycan and galactoglucan. Mol. Plant-Microbe Interact. 8: 267–277.

Leigh, J.A., and Walker, G.C. 1994. Exopolysaccharides of *Rhizobium*: synthesis, regulation and symbiotic function. Trends Genet. 10: 63–67.

Long, S., Reed, J.W., Himawan, J., and Walker, G.C. 1988. Genetic analysis of a cluster of genes required for synthesis of the calcofluor-binding exopolysaccharide of *Rhizobium meliloti*. J. Bacteriol. 170: 4239–4248.

Müller, P., Hynes, M., Kapp, D., Niehaus, K., and Pühler, A. 1988. Two classes of *Rhizobium meliloti* infection mutants differ in exopolysaccharide production and in coinoculation properties with nodulation mutants. Mol. Gen. Genet. 211: 17–26.

Müller, P., Keller, M., Weng, W.M., Quandt, J., Arnold, W., and Pühler, A. 1993. Genetic analysis of the *Rhizobium meliloti exo*YFQ operon: ExoY is homologous to sugar transferases and ExoQ represents a transmembrane protein. Mol. Plant-Microbe Interact. 6: 55–65.

Pellock, B.J., Teplitski, M., Brinay, R.P., Duid Baver, W., Walker, G.C. 2002. A Lux R homolog controls production of symbiotically active extra cellular poly saccharide II by *Sinorhizobium meliloti*. J. Bacteriol. 184: 5067–5076.

Reed, J.W., Glazebrook, J., and Walker, G.C. 1991. The *exoR* gene of *Rhizobium meliloti* affects RNA levels of other *exo* genes but lacks homology to known transcriptional regulators. J. Bacteriol. 173: 3789–3794.

Reinhold, B.B., Chan, S.Y., Reuber, T.L., Marra, A., Walker, G.C., and Reinhold, V.N. 1994. Detailed structural characterization of succinoglycan, the major exopolysaccharide of *Rhizobium meliloti* Rm1021. J. Bacteriol. 176: 1997–2002.

Reuber, T.L., and Walker, G.C. 1993. Biosynthesis of succinoglycan, a symbiotically important exopolysaccharide of *Rhizobium meliloti*. Cell 74: 269–280.

Rüberg, S., Pühler, A., and Becker, A. 1999. Biosynthesis of the exopolysaccharide galactoglucan in *Sinorhizobium meliloti* is subject to a complex control by the phosphate-dependent regulator PhoB and the proteins ExpG and MucR. Microbiology 145: 603–611.

Wang, L.-X., Wang, Y., Pellock, B., and Walker. G.C. 1999. Structural characterization of the symbiotically important low-molecular-weight succinoglycan of *Sinorhizobium meliloti*. J. Bacteriol. 181: 6788–6796.

Zhan, H., Levery, S.B., Lee, C.C., and Leigh, J.A. 1989. A second exopolysaccharide of *Rhizobium meliloti* strain SU47 that can function in root nodule invasion. Proc. Natl. Acad. Sci. USA 86: 3055–3059.

Zhan, H., Lee, C.C., and Leigh, J.A. 1991. Induction of the second polysaccharide (EPSb) in *Rhizobium meliloti* strain SU47 by low phosphate concentrations. J. Bacteriol. 173: 7391–7394.

Chapter 5

Control of Temperature-Responsive Synthesis of the Phytotoxin Coronatine in *Pseudomonas syringae* by the Unconventional Two-Component System CorRPS

Angela V. Smirnova, Ling Wang, Bettina Rohde, Ina Budde, Helge Weingart, and Matthias S. Ullrich

Abstract

The phytopathogen *Pseudomonas syringae* produces the phytotoxin coronatine (COR) as a major virulence factor. COR and its precursor, coronafacic acid, function as molecular mimics of the plant signaling molecule jasmonate. A 32.8-kb plasmid-borne gene cluster mediates COR biosynthesis, which is optimal at 18°C and non-detectable at 28°C, the optimal growth temperature for *P. syringae*. The thermoregulation is mediated at the transcriptional level by an unconventional two-component regulatory system consisting of a histidine protein kinase, CorS, and two transcriptional activators, CorR and CorP. Dissection of this regulatory triad revealed that CorR binds to its target sequences in a thermoresponsive manner and that its DNA-binding activity is controlled by CorS. A Preliminary model for thermo-sensing by CorS is proposed based on its membrane topology and the analysis of translational fusions of CorS to reporter enzymes at different temperatures. CorP lacks a typical helix-turn-helix motif but possibly functions as a modulator of CorR or CorS activity. The thermoregulation of COR biosynthetic genes is widespread among various COR-producing *P. syringae* strains. Post-translational processes also contribute to the thermo-responsiveness of COR production. Additionally, COR synthesis in *P. syringae* is influenced by nutrient availability, *rpoN* encoding the alternative sigma factor σ^{54}, and HrpV, a negative regulator of *hrp* gene expression, suggesting a complex regulatory network governing phytotoxin synthesis.

Introduction

Besides an intensively studied type III secretory system, the Hrp (*h*ypersensitive *r*esponse and *p*athogenicity) system, which delivers so-called avirulence gene products to the host cell causing pathogenicity, virulence factors such as phytotoxins, exopolysaccharides, and extracellular enzymes significantly contribute to the disease development in pathogenic plant-microbe interactions (Alfano and Collmer, 1996). It remains to be determined how and under which environmental conditions most of these virulence factors impact the disease outcome. Opportunistic plant pathogens like our model organism, *Pseudomonas syringae*, preferably infect their host plants under conditions of high humidity and low temperature. Plants, as poikilothermic organisms do not maintain a given temperature. This complicates a direct comparison of thermoresponsive processes in plant pathogens with the well-studied thermoregulation of virulence factors in human and animal pathogens (Hurme and Rhen, 1998). While temperatures of 37–41°C signal to animal pathogens their arrival inside the warm-blooded host, numerous virulence factors of phytopathogens like *Erwinia amylovora, Agrobacterium tumefaciens, E. chrysanthemi*, and *P. syringae* are preferentially expressed, secreted, or assembled at lower temperatures, such as 18–22°C (Hugovieux-Cotte-Pattat *et al.*, 1992; Wei *et al.*, 1992; Rowley *et al.*, 2000; Banta *et al.*, 1998; Van Dijk *et al.*, 1999). In the current mini-review, we report on the dissection of one of these thermoresponsive systems, the regulation of biosynthesis of the phytotoxin coronatine in *P. syringae.*

The Phytotoxin Coronatine

Coronatine (COR) (Figure 1B) consists of a polyketide moiety, coronafacic acid (CFA), and a cyclized amino acid derivative, coronamic acid (CMA), which are fused to each other by an amide bond (Mitchell, 1982; Bender *et al.*, 1993). COR is synthesized by five different pathovars of *P. syringae* and functions in a non-host specific manner causing chlorosis, hypertrophy of storage tissue, compression of thyllakoids, thickening of plant cell walls, and accumulation of plant-borne protease inhibitors (Sakai *et al.*, 1979; Mitchell, 1982; Palmer and Bender, 1995). COR and CFA structurally and functionally resemble jasmonic acid, a plant compound involved in signal transduction during plant defense mechanisms and global cell differentiation processes, and its precursor, 12-oxophytodienoic acid. COR and CFA are believed to mimic these compounds at the molecular level (Feys *et al.*, 1994; Weiler *et al.*, 1994; Krumm *et al.*, 1995). COR production enhances the virulence of *P. syringae* strains on tomato plants, *Arabidopsis thaliana*, and Chinese cabbage plants

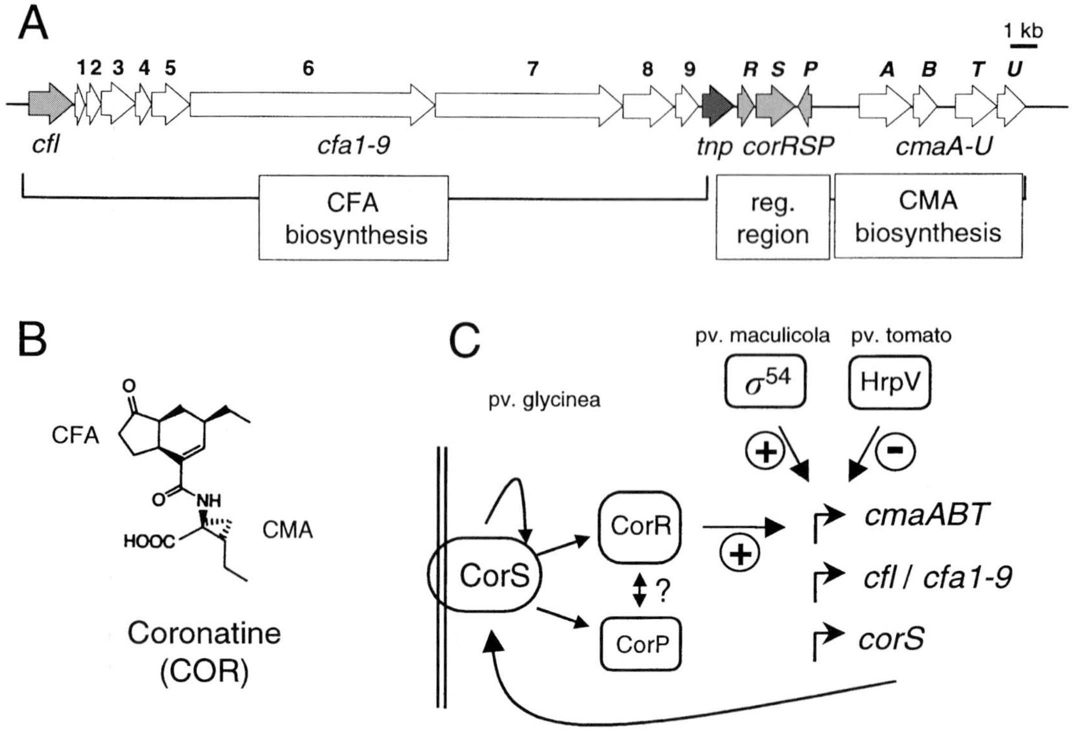

Figure 1. (**A**) Functional map of the COR biosynthetic gene cluster of PG4180 indicating the DNA regions required for CFA and CMA biosynthesis, respectively, and for transcriptional regulation. **(B)** Structure of COR which consists of CFA and CMA fused by amide bond. **(C)** Hypothetical scheme for transcriptional regulation of COR genes. A modified two-component regulatory system consisting of CorS, CorR, and CorP was characterized in *P. syringae* pv. glycinea. Additional regulatory factors found in *P. syringae* pvs. maculicola and tomato are indicated. It remains to be tested whether the symbolized regulatory proteins play a role in all three pathovars studied.

(Bender *et al.*, 1987; Mittal and Davis, 1995; Tamura *et al.*, 1998). Analysis of the accumulation of mRNAs encoded by defense-related genes in *Arabidopsis* demonstrated that COR is required for successful infection under natural conditions and that COR plays a critical role during the early stages of infection by suppressing the activation of defense-related plant genes (Mittal and Davis, 1995). These findings were confirmed by Budde and Ullrich (2000), who investigated the impact of COR production by *P. syringae* PG4180 during compatible and incompatible interactions with soybean and tobacco plants, respectively. After spray inoculation, PG4180 caused typical bacterial blight symptoms on soybean plants only when the bacteria produced COR at 18°C. This effect was not observed when PG4180 was infiltrated into soybean leaves, indicating that COR synthesis is important for bacterial invasion via natural plant openings. In the incompatible interaction, PG4180 elicited a rapid localized cell death termed the hypersensitive response (HR) on tobacco plants regardless of COR synthesis. However, the HR was significantly delayed when tobacco plants were treated with cells of a CFA-overproducing derivative of PG4180. Results of Budde and Ullrich (2000) indicated that the amount of synthesized CFA but not that of COR influenced the outcome of the HR, shedding further light on CFA's putative role as a molecular mimic of jasmonic acid.

Genetics and Biosynthesis of COR

Naturally COR$^-$ strains of *P. syringae* acquired the ability to produce COR when transformed with p4180A, a 90-kb indigenous plasmid from PG4180 (Bender *et al.*, 1993). Tn*5* mutagenesis indicated that the COR biosynthetic genes are clustered within a 32.8-kb region on p4180A (Figure 1A). The phenotype of COR-defective mutants was determined by supplying them with CFA and CMA and by complementation studies with cloned DNA from the COR biosynthetic cluster. The regions encoding CFA and CMA synthesis and coupling activity were localized to 18.8-, 6.9- and 2.3-kb regions of the cluster, respectively. Mutants in a 3.4-kb region required the addition of both CFA and CMA for COR synthesis indicating a regulatory role for this part of the cluster (Bender *et al.*, 1993). The COR gene cluster in *P. syringae* is normally plasmid-encoded in pvs. atropurpurea, glycinea, morsprunorum, and tomato but chromosomally encoded in pv. maculicola (Bender *et al.*, 1991; Ullrich *et al.*, 1993). Alarcón-Chaidez *et al.* (1999) demonstrated that some COR plasmids were self-transmissible, and that all COR plasmids shared a related *oriV/par* region. Sequence analysis of the region adjacent to the COR gene cluster in p4180A indicated the presence of IS elements which may have contributed to the horizontal transfer of the COR gene cluster.

The biosynthetic regions within the COR gene cluster have been analyzed in detail. The 6.9-kb DNA region required for CMA biosynthesis was sequenced, revealing four distinct open reading frames, which shared a common orientation for transcription (Ullrich *et al.*, 1994; Ullrich and Bender, 1994; Budde *et al.*, 1998) (Figure 1A). The deduced amino acid sequence of *cmaA* contained six core sequences, which are present in amino acid-activating enzymes, including nonribosomal peptide synthetases. The *cmaB* gene showed high similarity to a gene required for syringomycin production in *P. syringae*. The deduced amino acid sequence of *cmaT* was related to thioesterases. These data suggest that CMA assembly is similar to the thiotemplate mechanism of nonribosomal peptide synthesis. No significant similarities were found between *cmaU* and other database entries (Ullrich and Bender, 1994). Furthermore, we localized and studied the promoter region upstream of *cmaABT*. Data of transcriptional fusions with a promoterless β-glucuronidase gene (*uidA*) indicated that CMA biosynthesis is regulated by temperature at the transcriptional level (see below).

Liyanage *et al.* (1995b) analyzed the nucleotide sequence of the coronafacate ligase-encoding gene (*cfl*), which is required for the amide linkage of CFA and CMA. The deduced amino acid sequence of *cfl* showed homology to a variety of adenylate-forming enzymes, which bind and hydrolyze ATP in order to activate their substrates for further ligation. Liyanage *et al.*, (1995a) showed that *cfl* and the CFA biosynthetic genes are located on a single transcript and that the *cfl* promoter directs its transcription. The *cfl* promoter was shown to be activated in a thermoresponsive manner.

Subsequently, nine genes (*cfa1* to *cfa9*) were identified that are required for CFA biosynthesis (Figure 1A). While no meaningful sequence similarity was observed for *cfa4*, the predicted translation products of *cfa*1, *cfa*2 *cfa*3, and *cfa5* were related to acyl carrier proteins, fatty acid dehydrases, β-ketoacyl synthases, and adenylating enzymes, respectively (Penfold *et al.*, 1996). Two large genes, *cfa*6 and *cfa*7, encode polyketide synthases (PKS) that are structurally and functionally similar to multifunctional modular PKS (Rangaswamy *et al.*, 1998a). Additionally, the predicted translation products of *cfa8*, and *cfa9* showed relatedness to oxidoreductases and thioesterases, respectively (Rangaswamy *et al.*, 1998b). Mutagenesis indicated that *cfa*8 is required for the production of CFA whereas *cfa*9 is dispensable for CFA synthesis but may accelerate the release of enzyme-bound products. Rangaswamy *et al.*, (1998a) presented a scheme for CFA biosynthesis that incorporates the activities of all proteins encoded by the *cfa1–9* region.

Thermoregulation of COR Biosynthesis

Transcription of fusions of the *cmaABT* and *cfl*/*cfa1–9* operons with *uidA* were influenced by temperature and showed maximal reporter gene activity at 18°C (Ullrich and Bender, 1994; Liyanage *et al.*, 1995a; Budde *et al.*, 1998). Furthermore, transcription of both biosynthetic operons was dependent on a modified two-component regulatory system located within the COR biosynthetic gene cluster (Ullrich *et al.*, 1995). A 3.4-kb DNA fragment from the COR biosynthetic gene cluster restored temperature-regulated phytotoxin production to Tn*5* mutants that were defective in production of both CFA and CMA. Nucleotide sequence analysis of this fragment revealed three genes, *corS*, *corP*, and *corR*, which encode an unconventional two-component regulatory system consisting of one sensor protein, CorS, and two response regulator proteins, CorP and CorR (Figure 1C). Although only one response regulator, CorR, has a DNA-binding domain, the phosphate-receiving domains of both response regulator proteins are highly conserved.

Transcriptional fusions of *corP* and *corR* to *uidA* indicated that these two genes are expressed constitutively. In contrast, a *corS*::*uidA* fusion exhibited the temperature dependence previously observed for COR biosynthetic promoters and exhibited maximal transcriptional activity at 18°C and low activity at 28°C suggesting a positive auto-regulation of the *corS* gene (Ullrich *et al.*, 1995). Furthermore, *uidA* activity for *corS*::*uidA*, *cmaABT*::*uidA*, and *cfl*::*uidA* was decreased in *corP*, *corR*, and *corS* mutants as compared to wild type levels. This difference was not found for *corP*::*uidA* and *corR*::*uidA* transcriptional fusions. The three regulatory genes functioned in a *P. syringae* strain lacking the COR gene cluster to achieve thermoresponsive activation of an introduced COR biosynthetic promoter, indicating that this triad of genes has the primary control over COR biosynthesis and is responsible for thermoregulation.

Later, we furthered our understanding of the transcriptional and post-translational effects of temperature on *cmaB*, which encodes an enzyme involved in CMA biosynthesis (Budde *et al.*, 1998). Transcriptional *cmaABT*::*uidA* fusions and Northern blot analyses were used to monitor promoter activities and transcript abundance, respectively, during bacterial growth at 18 and 28°C. Promoter activity and transcript abundance were maximal when cells were incubated at 18°C and sampled at mid-exponential phase. Western blot analysis indicated that CmaB accumulated in *P. syringae* cultures grown at 18°C but not in cultures incubated at 28°C. Temperature shift experiments indicated that CmaB was more stable at 18°C than at 28°C (Budde *et al.*, 1998). Rangaswamy *et al.*, (1997) used Anti-Cfl antibodies and a transcriptional *cfl*::*uidA* fusion to find that transcription of *cfl* is thermoresponsive. In summary, we propose that thermoregulation of both, transcription and protein stability, might control COR synthesis.

Using phenotypic determination of COR synthesis, a transcriptional *cmaABT*::*uidA* fusion, and Western blot analysis, we screened a representative number of natural isolates of *P. syringae* for effects of temperature on expression of *cmaABT* (Rohde *et al.*, 1998). Thermoregulation of *cmaABT* expression was frequent among the tested strains.

One important question currently addressed in our laboratory is whether the *in vitro* thermoregulation of COR gene expression also plays a role *in planta*. Transcriptional fusions of *cmaABT* with a *gfp* derivative encoding the enhanced green fluorescent protein were generated, introduced to PG4180, and shown to be thermoresponsive *in vitro* as well as *in planta* (Weingart and Ullrich, unpublished).

It was investigated whether CorR binds specifically to the DNA region upstream of *cfl* (Peñaloza-Vázquez and Bender, 1998). Complementation analysis with a *corR* mutant and transcriptional fusions indicated that CorR functions as a positive regulator of COR gene expression. Deletion analysis of the 5'-end of the *cfl* upstream region was used to define the minimal region required for COR gene expression. An area extending from -704 to -650 with respect to the *cfl* transcriptional start site was protected from DNase I in footprinting experiments, indicating a rather large binding area. This area was also conserved in the promoter region for *cmaABT*.

To substantiate this, fusions of CorR and CorP to the maltose-binding protein (MBP) were overproduced and tested for functionality by complementation of *corR* and *corP* mutants of PG4180 (Wang *et al.*, 1999). The *cmaABT* promoter region was defined by deletion mapping and the DNA-binding capability of CorR and CorP was examined by gel retardation assays. Once overproduced in *P. syringae* at 18°C and purified, MBP-CorR was shown to bind specifically to a DNA fragment corresponding to positions −841 to −623 bp upstream of the transcriptional start of *cmaABT*. In contrast, MBP-CorP and MBP did not bind to this or to any other DNA fragment analyzed. The MBP-CorR activity was completely abolished when the fusion protein was overproduced at 28°C or in a *corS* mutant, indicating that the DNA binding of CorR depended on the growth temperature at which it was produced, and was controlled by CorS. DNA-binding assays with CorR were also conducted at different temperatures. However, the actual temperature at which DNA-binding assays were carried out had no effect suggesting that thermally-induced changes in the DNA structure might not play a role in this system (Wang *et al.*, 1999). In addition, overproduction of MBP-CorR in a *corP* mutant of PG4180 also yielded inactive protein, highlighting the importance of CorP for CorR activation. Rangaswamy and Bender (2000) demonstrated the *in vitro* phosphorylation of both, CorR and an N-terminally truncated form of CorS, thereby biochemically proving their function in phospho-transfer. We propose that CorR is activated by CorS at low temperature and that CorP is required for this activation before CorR can bind to its target DNA sequences (Figure 1C).

The proportion of unsaturated fatty acids in bacterial membranes is increased when temperature decreases to maintain membrane fluidity. This common feature was also observed for *P. syringae* grown at 18 or 28°C using gas chromatography analysis of membrane fatty acid methyl esters (Smirnova *et al.*, unpublished). Since CorS is a membrane-bound sensor kinase, it was tempting to speculate that CorS might sense changes in the fatty acid composition of the inner membrane. Respective experiments to test this are currently underway in our laboratory. Translational fusions of CorS to alkaline phosphatase (AP) and β-galactosidase were generated to biochemically confirm the computer prediction of membrane topology for this enzyme. Six potential membrane-spanning helices could be confirmed by this approach (Smirnova *et al.*, unpublished). Interestingly, CorS-AP fusions downstream of one of these helices had either a cytoplasmic or a periplasmic location depending on the bacterial incubation temperature, suggesting that CorS might modify its conformation in a thermoresponsive manner.

Additional Regulatory Factors Influencing COR Synthesis

Palmer *et al.*, (1997) demonstrated that selected carbon and amino acid sources significantly decreased COR biosynthesis of PG4180 at the transcriptional level. Interestingly, changes in osmolarity and the addition of complex carbon and nitrogen sources to the growth medium did not significantly affect COR transcription. These results indicate that multiple factors impact COR production. The influence of plant-borne signals on COR gene expression in *P. syringae* pv. tomato DC3000 was studied by Li *et al.*, (1998). Crude leaf extracts and plant intercellular fluids strongly induced expression of COR genes in this organism.

The co-ordinated regulation of different systems required for pathogenicity and virulence was recently addressed by analyzing COR production in defined *hrp* mutants of *P. syringae* pv. tomato DC3000 (Peñaloza-Vázquez *et al.*, 2000). COR was produced *in vitro* by mutants of DC3000 defective in *hrcC*, which encodes an outer-membrane protein required for type III-mediated secretion. Furthermore, a DC3000 mutant containing a polar mutation in *hrcC*, which inactivates *hrcC*, *hrpT*, and *hrpV*, produced significantly higher amounts of COR than the wild type suggesting that the regulatory protein HrpV might negatively control COR synthesis (Figure 1C). The results indicated that the *hrp* secretion system is not required for COR production, but mutations in this system may have regulatory effects on COR production.

Hendrickson *et al.*, (2000) constructed a mutant defective in the *rpoN* gene encoding σ^{54} from *P. syringae* pv. maculicola and showed that *rpoN* was essential for production of COR and for expression of the *cmaABT* operon (Figure 1C). In summary, it remains to be tested whether the additional regulatory principles play a role in all COR-producing pathovars of *P. syringae*.

Perspectives

Several intriguing questions regarding the CorRPS triad remain to be answered in future studies. Most importantly, the molecular mechanism by which CorS senses temperature changes needs to be addressed.

The postulated role of CorP as a modulator of CorR/CorS activity will be studied using phosphorylation assays and two-hybrid protein-protein interactions systems. Another important task will be to dissect the regulatory cascades or networks governing the concerted regulation of different pathogenicity and virulence determinants, with special emphasis on COR, *hrpV*, and *rpoN*. Ultimately, our studies aim at a better understanding of the molecular processes taking place in *P. syringae* during the onset of virulence in response to micro-climatic temperature changes *in planta*.

Acknowledgements

We thank Carol L. Bender and her coworkers for continuing support and fruitful cooperation. Critical comments by Alexander Schenk, Honqiao Li, Hilary King, Erhard Bremer, and Uwe Völker are gratefully acknowledged. This work was supported by grants from the Deutscher Akademischer Austauschdienst, Deutsche Forschungsgemeinschaft, and the Max-Planck-Gesellschaft.

References

Alarcón-Chaidez, F.J., Peñaloza-Vázquez, A., Ullrich, M., and Bender, C.L. 1999. Characterization of plasmids encoding the phytotoxin coronatine in *Pseudomonas syringae*. Plasmid 42: 210–220.

Alfano, J.R., and Collmer, A. 1996. The type III (Hrp) secretion pathway of plant pathogenic bacteria: Trafficking harpins, Avr proteins, and death. J. Bacteriol. 179: 5655–5662.

Banta, L.M., Bohne, J., Lovejoy, S.D., and Dostal, K. 1998. Stability of the *Agrobacterium tumefaciens* VirB10 protein is modulated by growth temperature and periplasmic osmoadaptation. J. Bacteriol. 180: 6597–6606.

Bender, C.L., Liyanage, H., Palmer, D.A., Ullrich, M., Young, S.A., and Mitchell, R.E. 1993. Characterization of the genes controlling biosynthesis of the polyketide phytotoxin coronatine including conjugation between coronafacic and coronamic acid. Gene 133: 31–38.

Bender, C.L., Stone, H.E., Sims, J.J., and Cooksey, D.A. 1987. Reduces pathogen fitness of *Pseudomonas syringae* pv. tomato Tn*5* mutants defective in coronatine production. Physiol. Mol. Plant Pathol. 30: 273–283.

Bender, C.L., Young, S.A., and Mitchell, R.E. 1991. Conservation of plasmid DNA sequences in coronatineproducing pathovars of *Pseudomonas syringae*. Appl. Environm. Microbiol. 57: 993999.

Budde, I.P., Rohde, B.H., Bender, C.L., and Ullrich, M.S. 1998. Growth phase and temperature influence promoter activity, transcript abundance, and protein stability during biosynthesis of the *Pseudomonas syringae* phytotoxin coronatine. J. Bacteriol. 180: 1360–1367.

Budde, I.P., and Ullrich, M.S. 2000. Interactions of *Pseudomonas syringae* pv. glycinea with host and non-host plants in relation to temperature and phytotoxin synthesis. Mol. Plant-Microbe. Interact. 13: 951–961.

Feys, B.J., Benedetti, C.E., Penfold, C.N., Turner, J.G. 1994. *Arabidopsis* mutants selected for resistance to the phytotoxin coronatine are male sterile, insensitive to methyl jasmonate, and resistant to a bacterial pathogen. Plant Cell 6: 751–759.

Hendrickson, E.L., Guevera, P., Peñaloza-Vázquez, A., Shao, J., Bender, C., and Ausubel, F.M. 2000. Virulence of the phytopathogen *Pseudomonas syringae* pv. maculicola is *rpoN* dependent. J. Bacteriol. 182: 3498–507.

Hugovieux-Cotte-Pattat, N., Dominguez, H., and Robert-Baudouy, J. 1992. Environmental conditions affect transcription of the pectinase genes of *Erwinia chrysanthemi* 3937. J. Bacteriol. 174: 7807–7818.

Hurme, R. and Rhen, M. 1998. Temperature sensing in bacterial gene regulation – what it all boils down to. Mol. Microbiol. 30: 1–6.

Krumm, T., Bandemer, K., and Boland, W. 1995. Induction of volatile biosynthesis in the Lima bean (*Phaseolus lunatus*)leucine- and isoleucine conjugates of 1-oxo-and 1-hydroxyindan-4-carboxylic acid: evidence for amino acid conjugates of jasmonic acid as intermediates in the octadecanoid signaling pathway. FEBS Lett. 377: 523–529.

Li, X.Z., Starratt, A.N., and Cuppels, D.A. 1998. Identification of tomato leaf factors that activate toxin gene expression in *Pseudomonas syringae* pv. tomato DC3000. Phytopathol. 88: 1094–1100.

Liyanage, H., Palmer, D.A., Ullrich, M., and Bender, C.L. 1995a. Characterization and transcriptional analysis of the gene cluster for coronafacic acid, the polyketide component of the phytotoxin coronatine. Appl. Environ. Microbiol. 61: 3843–3848.

Liyanage, H., Penfold, C., Turner, J., and Bender, C.L. 1995b. Sequence, expression and transcriptional analysis of the coronafacate ligase-encoding gene required for coronatine biosynthesis by *Pseudomonas syringae*. Gene 153: 17–23.

Mitchell, R.E. 1982. Coronatine production by some phytopathogenic pseudomonads. Physiol. Plant Pathol. 220: 83–89.

Mittal, S. and Davis, K.R. 1995. Role of the phytotoxin coronatine in the infection of *Arabidopsis thaliana* by *Pseudomonas syringae*. Mol. Plant-Microbe. Interact. 1: 165–171.

Palmer, D.A., and Bender, C.L. 1995. Ultrastructure of tomato leaf tissue treated with the pseudomonad phytotoxin coronatine and comparison with methyl jasmonate. Mol. Plant Microbe. Interact. 8: 683–692.

Palmer, D.A., Bender, C.L., and Sharma, S.B. 1997. Use of Tn*5*-gus*A5* to investigate environmental and nutritional effects on gene expression in the coronatine biosynthetic gene cluster of *Pseudomonas syringae* pv. *glycinea*. Can. J. Microbiol. 43: 517–525.

Peñaloza-Vázquez, A. and Bender, C.L. 1998. Characterization of CorR, a transcriptional activator which is required for biosynthesis of the phytotoxin coronatine. J. Bacteriol. 180: 6252–6259.

Peñaloza-Vázquez, A., Preston, G.M., Collmer, A., and Bender, C.L. 2000. Regulatory interactions between the Hrp type III protein secretion system and coronatine biosynthesis in *Pseudomonas syringae* pv. tomato DC3000. Microbiol. 146: 2447–2456.

Penfold, C.N., Bender, C.L., and Turner, J.G. 1996. Characterization of genes involved in biosynthesis of coronafacic acid, the polyketide component of the phytotoxin coronatine. Gene 183: 167–173.

Rangaswamy, V. and Bender, C.L. 2000. Phosphorylation of CorS and CorR, regulatory proteins that modulate production of the phytotoxin coronatine in *Pseudomonas syringae*. FEMS Microbiol. Lett. 193: 13–18.

Rangaswamy, V., Jiralerspong, S., Parry, R., and Bender, C.L. 1998a. Biosynthesis of the *Pseudomonas* polyketide coronafacic acid requires monofunctional and multifunctional polyketide synthase proteins. Proc. Natl. Acad. Sci. USA 95: 15469–15474.

Rangaswamy, V., Mitchell, R., Ullrich, M., and Bender, C.L. 1998b. Analysis of genes involved in the synthesis of the polyketide phytotoxin coronatine. J. Bacteriol. 180: 3330–3338.

Rangaswamy, V., Ullrich, M., Jones, W., Mitchell, R., Parry, R., Reynolds, P., and Bender, C.L. 1997. Expression and analysis of coronafacate ligase, a thermoregulated gene required for production of the phytotoxin coronatine in *P. syringae*. FEMS Microbiol. Lett. 154: 65–72.

Rohde, B.H., Pohlack, B., and Ullrich, M.S. 1998. Occurrence of thermoregulation of genes involved in coronatine biosynthesis among various *Pseudomonas syringae* strains. J. Basic Microbiol. 38: 41–50.

Rowley, K.B., Xu, R., and Patil, S.S. 2000. Molecular analysis of thermoregulation of phaseolotoxin-resistant ornithine carbamoyltransferase (*argK*) from *Pseudomonas syringae* pv. *phaseolicola*. Mol. Plant-Microbe. Interact. 13: 1071–1080.

Sakai, R., Nishiyama, K., Ichihara, A., Shiraishi, K., and Sakamura, S. 1979. Studies on the mechanism of physiological activity of coronatine: effect of coronatine on cell wall extensibility and expansion of potato tuber tissue. Ann. Phytopathol. Soc. Japan 45: 645–653.

Smirnova, A.V., Aufhammer, S., and Ullrich, M.S. Unpublished results.

Tamura, K., Zhu, Y., Sato, M., Teraoka, T., Hosokawa, D., and Watanabe M. 1998. Roles of coronatine production by *Pseudomonas syringae* pv. *maculicola* for pathogenicity. Ann. Phytopathol. Soc. Jpn. 64: 299–302.

Ullrich, M., Bereswill, S., Völksch, B., Fritsche, W., and Geider, K. 1993. Molecular characterization of field isolates of *Pseudomonas syringae* pv. glycinea differing in coronatine production. J. Gen. Microbiol. 139: 1927–1937.

Ullrich, M. and Bender, C.L. 1994. The biosynthetic gene cluster for coronamic acid, an ethylcyclopropyl amino acid, contains genes homologous to amino acid-activating enzymes and thioesterases. J. Bacteriol. 176: 7574 –7586.

Ullrich, M., Guenzi, A.C., Mitchell, R.E., and Bender, C.L. 1994. Cloning and expression of genes required for coronamic acid

(2-ethyl-1-aminocyclo-propane-1-carboxylic acid), an intermediate in the biosynthesis of the phytotoxin coronatine. Appl. Environ. Microbiol. 60: 2890–2897.

Ullrich, M., Peñaloza-Vázquez, A., Bailey, A.M., and Bender, C.L. 1995. A modified two-component regulatory system is involved in temperature-dependent biosynthesis of the *Pseudomonas syringae* phytotoxin coronatine. J. Bacteriol. 177: 6160–6169.

Van Dijk, K., Fouts, D.E., Rehm, A.H., Hill, R.A., Collmer, A., and Alfano, J.R. 1999. The Avr (effector) proteins HrmA (HopPsyA) and AvrPto are secreted in culture from *Pseudomonas syringae* pathovars via the Hrp (type III) protein secretion system in a temperature- and pH-sensitive manner. J. Bacteriol. 181: 4790–4797.

Wang, L., Bender, C., and Ullrich, M. 1999. The transcriptional activator CorR is involved in biosynthesis of the phytotoxin coronatine and binds to the *cmaABT* promoter region in a temperature-dependent manner. Mol. Gen. Genet. 262: 250–260.

Wei, Z.M., Sneath, B.J., and Beer, S.V. 1992. Expression of *Erwinia amylovora hrp* genes in response to environmental stimuli. J. Bacteriol. 174: 1875–1882.

Weiler, E.W., Kutchan, T.M., Gorba, T., Brodschelm, W., Niesel, U., and Bublitz. F. 1994. The *Pseudomonas* phytotoxin coronatine mimics octadecanoid signaling molecules of higher plants. FEBS Lett. 345: 9–13

Weingart, H. and Ullrich, M. Unpublished results.

Chapter 6

Virulence Gene Regulation in *Bordetella pertussis*

Jochen König, Andreas Bock, Anne-Laure Perraud, Thilo M. Fuchs, Dagmar Beier, and Roy Gross

Abstract

Most pathogenic bacteria encounter changing growth conditions during their infectious cycle and, accordingly, have to modulate gene expression to enable the efficient colonization of different environments outside or within their host organisms. In *Bordetella pertussis* the transcription of most virulence factors including several toxins and adhesins is regulated coordinately by the BvgAS two-component system. The molecular characterization of the BvgAS system revealed that it belongs to the small group of unorthodox two-component systems applying an obligate multistep phosphorelay. Moreover, despite the coordinated control of the virulence regulon, subtle differences in the regulation of individual virulence genes were observed which led to the identification of sophisticated mechanisms possibly engaged in fine tuning of virulence gene expression.

Coordinate Regulation of the Bvg-Regulon

It has been known for a long time that the expression of virulence properties of *B. pertussis*, the etiological agent of whooping cough, is unstable. In fact, avirulent so-called *phase variants* may arise with high frequency. Moreover, the virulent phenotype depends on environmental conditions and is reversibly affected by changes in the temperature and by several chemical compounds, a phenomenon termed *phenotypic modulation*. For example, the virulence regulon is only expressed at body but not at room temperature. Both phenomena – *phase variation* and *phenotypic modulation* – were shown to involve the BvgAS two-component system (Arico *et al.*, 1989; Cotter and DiRita, 2000; Gross and Rappuoli, 1988,1989; Weiss and Falkow, 1984). *Phase variation* leads to a spontaneous and usually irreversible loss of virulence gene expression due to mutations, frequently short deletions, in the *bvgAS* gene locus (Monack *et al.*, 1989; Stibitz *et al.*, 1989). *Phenotypic modulation* depends on the activity of the histidine kinase BvgS which is an environmental sensor located in the cytoplasmic membrane. Under the appropriate conditions, the BvgS protein autophosphorylates at a histidine residue in its transmitter domain and, subsequently, the phosphate is transferred to an aspartic acid in the receiver domain of the BvgA response regulator. BvgA-P is then able to activate transcription from the virulence gene promoters leading to expression of several adhesins and toxins (Table 1) (Rappuoli, 1994).

The BvgAS system differs from typical two-component systems, because BvgS contains additional phosphorylation sites outside of its transmitter: In fact, an obligate multistep His-Asp-His-Asp phosphorelay occurs that involves the BvgS transmitter, receiver and HPt domains (Figure 1) (Arico *et al.*, 1989; Perraud *et al.*, 1999; Uhl and Miller, 1994,1996). The high energy phosphohistidine present in the C-terminal HPt domain is the exclusive phosphate source for BvgA (Perraud *et al.*, 1998). Furthermore, transcomplementation of mutant sensor proteins by the separate expression of individual signalling domains demonstrated that the phosphorelay in the BvgS histidine kinase involves BvgS homodimers (Beier *et al.*, 1995, 1996). BvgA also forms homodimers, but, in contrast to several other response regulators such as FixJ, there is no evidence for any influence of phosphorylation on its oligomerisation state (Bock, Rippe and Gross, unpublished; Perraud *et al.*, 2000). This shows that activation of BvgA by phosphorylation does not involve its dimerization (Perraud *et al.*, 2000; Bock *et al.*, 2001).

Little is known about the mechanisms of signal perception and the control of the histidine kinase activity present in the transmitter domain of two-component sensor proteins. However, like in other sensor proteins the linker region of BvgS connecting the transmitter domain with the membrane spanning region and the periplasmic domain appears to be crucial, because point mutations in this linker can either cause the inactivation of BvgS or constitutive kinase activity (Beier *et al.*, 1996; Manetti *et al.*, 1994; Miller *et al.*, 1992). Interestingly, the BvgS linker region was recently shown by sequence similarity to contain a PAS domain which should be affected by these point mutations (Taylor and Zhulin, 1999). This may indicate that the BvgS protein via its PAS domain is also able to perceive oxygen and/or the energy state of the bacteria. The perception of such stimuli is in agreement with the fact, that also housekeeping functions including several dehydrogenases and the cytochrome composition of the terminal oxidase of the respiratory chain are Bvg-regulated (Cotter and DiRita, 2000; Schneider *et al.*, 2002). In fact, we could recently show that the autophosphorylation activity of purified BvgS protein *in vitro* is responsive to the oxidation status of ubiquinone providing first evidence that BvgS is able to sense cellular signals in addition to environmental stimuli (Bock and Gross, 2002).

Table 1. Selected virulence factors of *B. pertussis* encoded by genes activated by the BvgAS two-component system.

Factors	Activity/Function
Pertussis Toxin (PTX)	ADP-ribosylation of G-proteins
Adenylate cyclase toxin (CYA)	Invasive adenylate cyclase and hemolysin
Dermonecrotic toxin (DNT)	Transglutaminase
Filamentous hemagglutinin (FHA)	Adhesion and colonization
Pertactin (PRN)	Adhesion and colonization
Tracheal colonizing factor (TCF)	Adhesion and colonization
BrkA	Serum resistance

Structurally, the BvgAS system is highly related to several other bacterial phosphorelay systems including the ArcAB and EvgAS systems of *E. coli.* The ArcAB system controls the aerobic modulation regulon and the EvgAS system controls expression of a drug efflux pump (Kato *et al.*, 2000; Sawers 1999). In general, the interaction of the histidine kinase and its cognate response regulator occurs with high specificity, although in some particular two-component systems or under artificial conditions a cross-talk between non-cognate proteins could be observed (Wanner, 1992). However, for the ArcB histidine kinase it was shown that its HPt domain is quite promiscuous being able to act as a quite efficient phosphodonor even for non-cognate response regulators such as CheY or OmpR (Yaku *et al.*, 1997). Accordingly, it was proposed that HPt domains may be signalling devices which may link different two-component systems thus creating a regulatory network based on cross-talk phenomena. To investigate signalling specificity mediated by

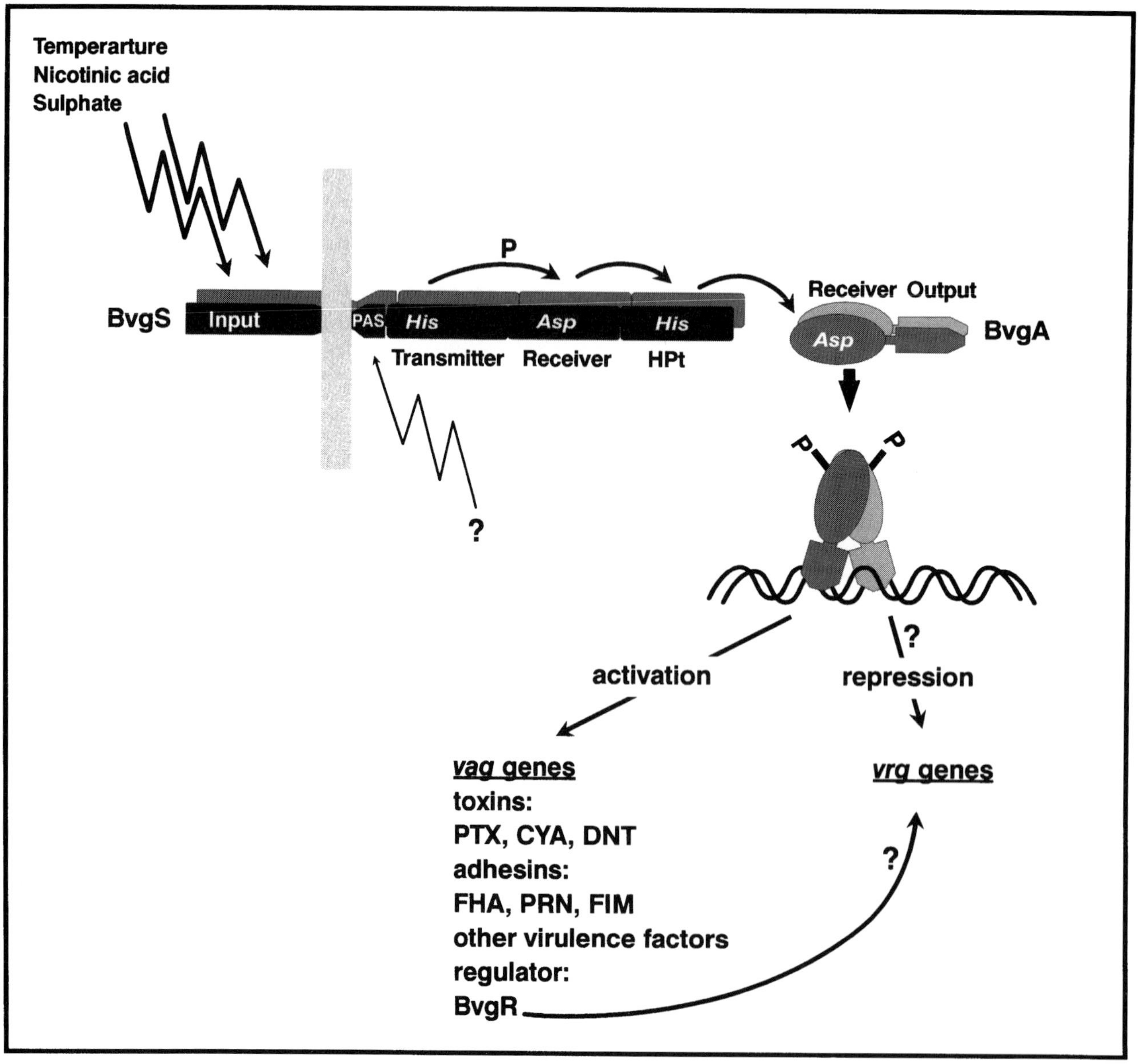

Figure 1. The BvgAS two-component system and its regulon.

the BvgS HPt domain we performed domain swapping experiments and constructed chimaeric histidine kinase proteins containing the highly related BvgS and EvgS receiver and HPt domains fused to the BvgS transmitter. Interestingly, the chimaeric proteins harbouring either BvgS or EvgS receiver and HPt domains, respectively, were as active as the wild type proteins regarding the intramolecular phosphorelay of the sensor protein. However, these experiments demonstrated that, despite their high level of sequence conservation, BvgA and EvgA could only be phosphorylated by their respective BvgS and EvgS HPt domains (Perraud *et al.*, 1998). Therefore, these data show that signal transduction via HPt domains does occur in a highly specific manner and does not support a general concept of HPt mediated cross-talk phenomena (Perraud *et al.*, 1999).

The BvgAS system does not only control expression of the BvgA activated virulence genes (also termed *vag* = **v**irulence **a**ctivated **g**enes), but also negatively controls expression of an additional set of genes (also termed *vrg* = **v**irulence **r**epressed **g**enes) (Figure 1) (Akerley and Miller, 1996; Beattie *et al.*, 1993; Knapp and Mekalanos, 1988; Martinez de Tejada *et al.*, 1998). Neither the functions of the *vrg* genes nor their regulation are yet well characterized. However, a Bvg-activated repressor protein, BvgR, was identified, which very likely controls expression of at least several of the *vrg* genes (Merkel *et al.*, 1998).

Differential Regulation within the Bvg-Regulon

The *bvgAS* gene locus is autoregulated and a weak constitutive promoter as well as two strong BvgA dependent promoters control its transcription (Scarlato *et al.*, 1990). Therefore, a shift of the bacteria from non-permissive to permissive growth conditions is followed by a strong and long-lasting increase in BvgA concentration. This autoregulation is a prerequisite for differential gene activation phenomena observed within the Bvg-regulon: There are different subsets of *vag* genes which are characterized by fast or very slow kinetics of transcriptional induction that were classified as "early" or "late" genes, respectively (Gross and Rappuoli, 1989; Scarlato *et al.*, 1991). Whereas the major adhesin FHA is an early factor, the toxins PTX and CYA will be transcribed only several hours after the switch in the environmental conditions. These differences in gene expression mainly depend on the concentration of phosphorylated BvgA (BvgA-P) as the promoters have different affinities for the transcriptional regulator and their maximal expression may either be achieved with low amount of BvgA-P (FHA) or may require high concentrations of the activator (PTX and CYA) (Steffen *et al.*, 1996; Zu *et al.*, 1996). Recently, a third class of *vag* genes was identified, which show an "intermediate" kinetics of expression and are switched off again when BvgA-P concentration reaches levels required for the activation of the "late" genes (Cotter and DiRita, 2000; Deora *et al.*, 2001). Among these intermediate factors is the *bipA* gene encoding a high molecular weight protein that shares amino acid sequence similarity at its N-terminus with the proposed outer membrane localization domains of intimin of enteropathogenic and enterohaemorrhagic *Escherichia coli* and invasin of *Yersinia* spp. (Stockbauer *et al.*, 2001). The promoter regions of the different classes of virulence genes show interesting structural differences implying that additional regulatory elements may be involved in their control. In fact, there is evidence that DNA supercoiling affects virulence gene expression in *B. pertussis* as well as a protein termed Baf which under certain conditions contributes to *ptx* expression (DeShazer *et al.*, 1994; Graeff-Wohlleben *et al.*, 1995; Scarlato *et al.*, 1993).

Interestingly, spontaneous phase variants were identified which showed a very peculiar phenotype: In contrast to typical phase variants which harbour inactivating mutations in the *bvgAS* gene locus and do not express the entire virulence regulon (Monack *et al.*, 1989; Stibitz *et al.*, 1989), in these variants only the expression of the "late" virulence genes encoding PTX and CYA was abolished (Carbonetti *et al.*, 1993; Cookson *et al.*, 1988). The subsequent characterization of these "partial" phase variants revealed a novel regulatory mechanism termed *phenotypic variation*. In these variants, single point mutations in the translational control region of the *rpoA* gene encoding the RNA polymerase α subunit caused an up to three-fold overproduction of the α subunit (Carbonetti *et al.*, 1994). Overproduction of α most likely caused the lack of expression of the toxin promoters by a direct interaction of excess α with BvgA, thereby reducing the amount of available BvgA below the threshold concentration required for expression of the "late" toxin promoters (Boucher *et al.*, 1997; Carbonetti *et al.*, 2000). So far, only two of these phenotypic variants have been identified, which may indicate that *phenotypic variation* is a relatively rare event. However, no extensive survey has been carried out yet to characterize the molecular basis of *phase variation* in a sufficient number of different phase variants and *phenotypic variation* may well contribute to the gradual disappearance of virulence traits in *B. pertussis* and the closely related organism *B. bronchiseptica* observed previously (Goldman *et al.*, 1984; Gueirard *et al.*, 1995).

As explained above the *phenotypic variation* is caused by a disequilibrium of factors building up the transcription machinery resulting in the lack of expression of PTX and CYA. These non-hemolytic variants were the basis for a novel strategy to identify regulatory genes possibly involved in toxin expression (Fuchs *et al.*, 1996). It was assumed that the phenotypic variants were more sensitive to further perturbations in their transcription apparatus than the wild type strains. Mutations affecting toxin gene expression were generated by chemical mutagenesis of the phenotypic variants and screening was performed for colonies with a reconstituted hemolytic phenotype. In most cases the resulting mutants carried either reversions or suppressor mutations in the *rpoA* gene itself or in the *bvgA* gene (Fuchs and Gross, unpublished). However, several mutants carried suppressor mutations in unknown gene

Table 2. Distribution of Tex protein homologs in *Eubacteria*. The names of those species that according to their complete genome sequences do not contain the *tex* gene are underlined.

Bacteria	
Aquificales	**Aquificaceae**
	Aquifex aeolicus
Chlamydiales	**Chlamydiaceae**
	Chlamydia trachomatis, *C. pneumoniae*
Cyanobacteria	**Chroococcales**
	Synechocystis PCC6803
Firmicutes	**Bacillus/Clostridium Gruppe (low G/C gram⁺)**
	Bacillus/Lactobacillus/Streptococcus Gruppe
	Bacillus anthracis, *B. halodurans*,
	B. subtilis, *B. stearothermophilus*
	Staphylococcus aureus
	Streptococcus pneumoniae, *S. pyogenes*,
	S. mutans, *S. equii*
	Mycoplasmataceae
	Mycoplasma genitalium, *M. pneumoniae*
	Ureaplasma urealyticum
	Clostridiaceae
	Clostridium acetobotylicum, *C. difficile*
	Enterococcaceae
	Enterococcus faecalis
	Actinobacteria (high G/C gram⁺)
	Corynebacterineae
	Corynebacterium diphteriae
	Mycobacterium tuberculosis
	Streptomycineae
	Streptomyces coeliclor
Proteobacteria	**α Gruppe**
	Rickettsiaceae
	Rickettsia prowazekii
	β Gruppe
	Alcaligenacea
	Bordetella bronchiseptica, *B. pertussis*
	Neisseriaceae
	Neisseria gonorrhoeae, *N. meningitidis*
	γ Gruppe
	Enterobacteriaceae
	Escherichia coli
	Salmonella typhi, *S. typhimurium*,
	S. paratyphi A
	Yersinia pestis
	Buchnera aphidicola
	Pasteurellaceae
	Pasteurella multocida
	Haemophilus influenzae, *H. ducreyi*
	Actinobacillus actinomycetemcomitans
	Pseudomonaceae
	Pseudomonas aeruginosa, *P. putida*
	Alteromonadaceae
	Shewanella putrefaciens
	Vibrionaceae Gruppe
	Vibrio cholerae
	Xanthomonas Gruppe
	Xylella fastidiosa
	Legionellaceae ε
	Legionella pneumophila
	δ Gruppe
	Geobacter sulfurreducens
	Desulfuromonas Gruppe
	Desulfovibrio vulgaris
	ε Gruppe
	Helicobacter pylori
	Campylobacter jejuni
Spirochaetales	Spirochaetaceae
	Treponema pallidum, *T. denticola*
Thermotogales	*Thermotoga maritima*

loci. These mutants were used for the search of antisuppressor loci by the introduction of a genomic library and allowed the identification of a gene termed *tex* (= **t**oxin **ex**pression) which, when slightly over-expressed, exerted a negative effect on transcription of PTX and CYA in the mutant background.

The deduced aminoacid sequence of the Tex protein revealed that it is strongly conserved in most of the eubacteria sequenced so far with amino acid similarities ranging from 50 to 80% (Table 2). Such a degree of sequence conservation is also observed with essential factors including RpoA and GyrB. However, in some phylogenetic lineages no Tex homologue is found, e.g. in the ε group of the *Proteobacteria*, or in several obligate parasites or symbionts with extremely reduced genomes such as *Mycoplasma* spp. or the endosymbiont of aphids *Buchnera aphidicola*. Nevertheless, the high degree of sequence conservation of the Tex protein in most other eubacteria indicates a basic role of this factor. In fact, the *tex* gene could not be deleted from *B. pertussis*, but, apparently, it is not essential for other bacteria including *E. coli* and *Neisseria gonorrhoea* (Fuchs *et al.*, 1996; König and Gross, unpublished; Petering *et al.*, 1996). The protein shows interesting sequence similarities with the mannitol repressor (MtlR) of *E. coli* in its N-terminal domain and harbours a S1 domain at its C-terminus (Figure 2) (Bycroft *et al.*, 1997; Fuchs *et al.*, 1996). The presence of the S1 domain suggested that the protein may be a nucleic acid binding protein, because in most cases other proteins carrying a S1 domain were reported to interact with RNA (Bycroft *et al.*, 1997). In fact, in solid phase binding assays using the purified Tex protein of *E. coli* linked to magnetic beads, RNA but not DNA could be found as a specific ligand (Figure 3). Apparently, the specific binding of Tex to RNA requires its N-terminus, because the deletion of several N-terminal amino acids abolished any preference for RNA and resulted in highly efficient binding of DNA as well as RNA (König and Gross, unpublished). Several proteins harbouring the S1 domain are involved in stress response pathways. For example, under cold-shock conditions polynucleotide phosphorylase (PNPase), containing a S1 domain at its C-terminus, proved to be essential and is one of the key enzymes for mRNA turnover at low temperatures (Jones *et al.*, 1987; Luttinger *et al.*, 1996). However, as shown by Western-blot analysis and 2-D gel electrophoresis, in *E. coli* expression of Tex occurs at a very low level (Fountoulakis *et al.*, 1999; König and Gross, unpublished), and no growth conditions could be identified yet leading to a significant increase in its expression. Attempts to identify a specific RNA target for the purified *E. coli* Tex protein were carried out once more using a solid phase binding assay. This approach identified 16S rRNA and CsrB as preferential binding partners. However, the functional relevance of these findings remains to be investigated. CsrB is a regulatory RNA, which by interaction with the CsrA protein controls message turnover in *E. coli*.

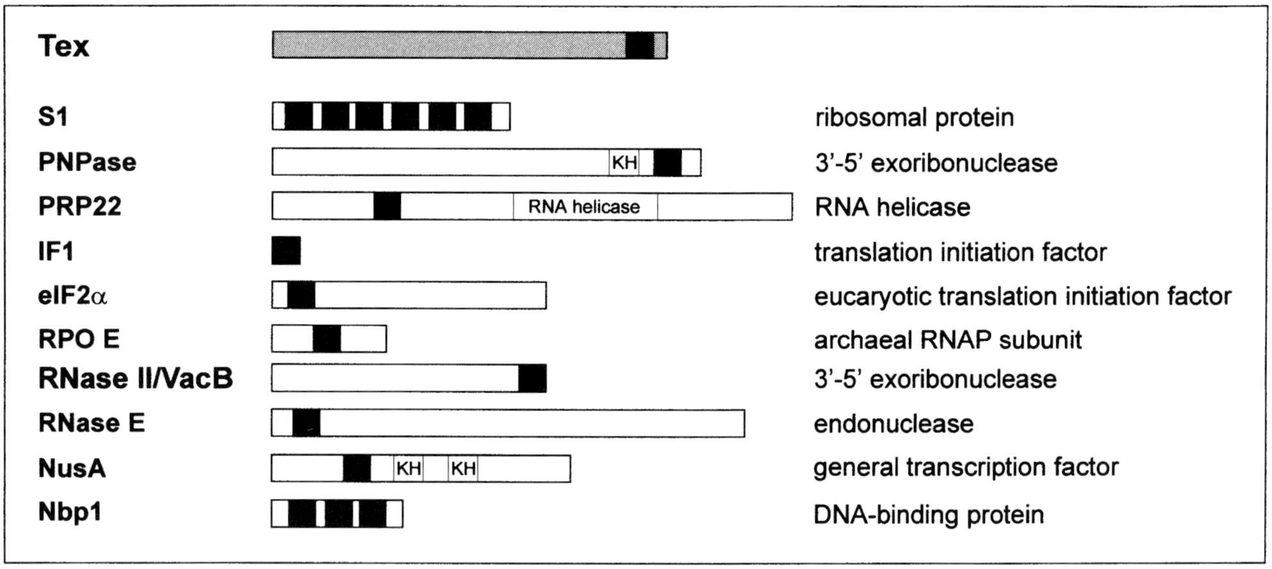

Figure 2. Schematic presentation of several members of the S1 protein family. The black boxes represent S1 domains, KH indicates the presence of KH-domains which are independent nucleic-acid-binding units. The figure has been adapted from Bycroft *et al.*, 1997 and Sugita *et al.*, 2000.

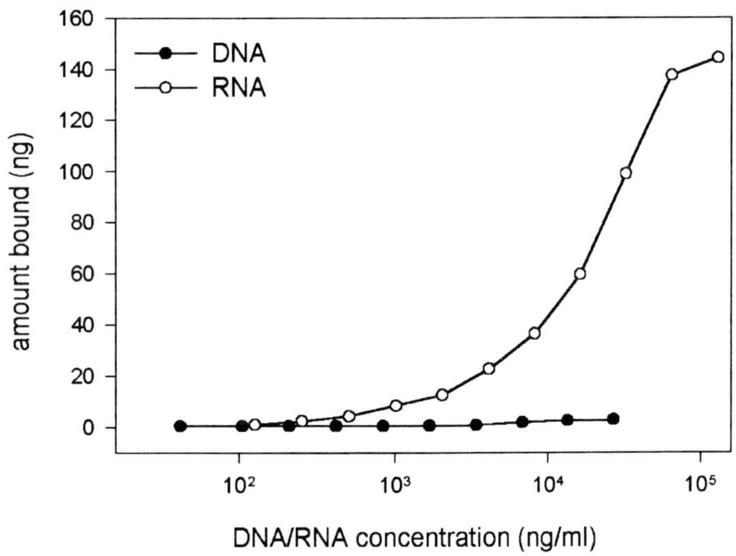

Figure 3. Binding of ^{3}H-labelled RNA (open symbols) and DNA (closed symbols) to the purified *E. coli* Tex protein.

Interestingly, the homologous regulator pairs of *Salmonella typhimurium* and *Erwinia carotovora* were found to be involved in virulence gene expression (Altier *et al.*, 2000; Cui *et al.*, 1999).

Acknowledgements

We thank Verena Weiss, Kirsten Jung, Karsten Rippe, Marcus Bantscheff, Michael Glocker, Nick Carbonetti and Vincenzo Scarlato for advice and fruitful collaboration throughout this project. This work was supported by grants from the Human Frontier Science Program Organization, the Priority Program "Regulatory Networks in Bacteria" of the Deutsche Forschungsgemeinschaft and by the Fonds der Chemischen Industrie.

References

Akerley, B.J., and Miller, J.F. 1996. Understanding signal transduction during bacterial infection. Trends Microbiol. 4: 141–146.

Altier, C., Suyemoto, M., Ruiz, A.I., Burnham, K.D., and Maurer, R. 2000. Characterization of two novel regulatory genes affecting *Salmonella* invasion gene expression. Mol. Microbiol. 35: 635–646.

Aricó, B., Miller, J., Roy, C., Stibitz, S., Monack, D., Falkow, S., Gross, R., Rappuoli, R. 1989. Sequences required for the expression of *Bordetella pertussis* virulence factors share homology with prokaryotic signal transduction proteins. Proc. Natl. Acad. Sci. USA 86: 6671–6675.

Beattie, D.T., Mahan, M.J., and Mekalanos, J.J. 1993. Repressor binding to a regulatory site in the DNA coding sequence is sufficient to confer transcriptional regulation of the *vir*-repressed genes (*vrg* genes) in *Bordetella pertussis*. J. Bacteriol. 175: 519–527.

Beier, D., Deppisch, H., and Gross, R. 1996. Conserved sequence motifs in the unorthodox BvgS two-component sensor protein of *Bordetella pertussis*. Mol. Gen. Genet. 252: 169–176.

Beier, D., Schwarz, B., Fuchs, T.M., and Gross, R. 1995. *In vivo* characterization of the unorthodox two-component BvgS sensor protein of *Bordetella pertussis*. J. Mol. Biol. 248: 596–610.

Bock, A., Bantscheff, M., Perraud, A.L., Rippe, K., Weiss, V., Glocker, M.O., and Gross, R. 2001. Rational design and molecular characterization of a chimaeric response regulator protein. J. Mol. Biol. 310: 283–290.

Bock, A., and Gross, R. 2002. The unorthodox histidine kinases BvgS and EvgS are responsive to the oxidation status of a quinone electron carrier. Eur. J. Biochem. 269: 3479–3484.

Boucher, P.E., Murakami, K., Ishihama, A., and Stibitz, S. 1997. Nature of DNA binding and RNA polymerase interaction of the *Bordetella pertussis* BvgA transcriptional activator at the *fha* promoter. J. Bacteriol. 179: 1755–1763.

Bycroft, M., Hubbard, T.J., Proctor, M., Freund, S.M., and Murzin, A.G. 1997. The solution structure of the S1 RNA binding domain: a member of an ancient nucleic acid-binding fold. Cell 88: 235–242.

Carbonetti, N.H., Romashko, A., and Irish, T.J. 2000. Overexpression of the RNA polymerase alpha subunit reduces transcription of Bvg-activated virulence genes in *Bordetella pertussis*. J. Bacteriol. 182: 529–531.

Carbonetti, N.H., Fuchs, T.M., Patamawenu, A.A., Irish, T., Deppisch, H., Gross, R. 1994. Effect of mutations causing overexpression of RNA polymerase subunit on regulation of virulence factors in *Bordetella pertussis*. J. Bacteriol. 176: 7267–7273.

Carbonetti, N.H., Khelef, N., Guiso, N., and Gross, R. 1993. A phase variant of *Bordetella pertussis* with a mutation in a new locus involved in the regulation of pertussis toxin and adenylate cyclase toxin expression. J. Bacteriol. 175: 6679–6688.

Cookson, B.T., Berg, D.E., and Goldman, W.E. 1990. Mutagenesis of *Bordetella pertussis* with transposon Tn*5tac1*: conditional expression of virulence-associated genes. J. Bacteriol. 172: 1681–1687.

Cotter, P.A., and DiRita, V.J. 2000. Bacterial virulence gene regulation: an evolutionary perspective. Annu. Rev. Microbiol. 54: 519–565.

Cui, Y., Mukherjee, A., Dumenyo, C.K., Liu, Y., and Chatterjee, A.K. 1999. *rsmC* of the soft-rotting bacterium *Erwinia carotovora* subsp. *carotovora* negatively controls extracellular enzyme and harpin (Ecc) production and virulence by modulating levels of regulatory RNA (*rsmB*) and RNA-binding protein (RsmA). J. Bacteriol. 181: 6042–6052.

DeShazer, D., Wood, G.E., and Friedman, R.L. 1995. Identification of a *Bordetella pertussis* regulatory factor required for transcription of the pertussis toxin operon in *Escherichia coli*. J. Bacteriol. 177: 3801–3807.

Deora, R., Bootsma, H.J., Miller, J.F., and Cotter, P.A. 2001. Diversity in the *Bordetella* virulence regulon: transcriptional control of a Bvg-intermediate phase gene. Mol. Microbiol. 40: 669–683.

Fountoulakis, M., Takacs, M.F., Berndt, P., Langen, H., and Takacs, B. 1999. Enrichment of low abundance proteins of *Escherichia coli* by hydroxyapatite chromatography. Electrophoresis 20: 2181–2195.

Fuchs, T.M., Deppisch, H., Scarlato, V., and Gross, R. 1996. A new gene locus of *Bordetella pertussis* defines a novel family of prokaryotic transcriptional accessory proteins. J. Bacteriol. 178: 4445–4452.

Goldman, S., Hanski, E., and Fish, F. 1984. Spontaneous phase variation in *Bordetella pertussis* is a multistep non-random process. EMBO J. 3: 1353–1356.

Graeff-Wohlleben, H., Deppisch, H., and Gross, R. 1995. Global regulatory mechanisms affect virulence gene expression in *Bordetella pertussis*. Mol. Gen. Genet. 247: 86–94.

Gross, R., and Rappuoli, R. 1988. Positive regulation of pertussis toxin expression. Proc. Natl. Acad. Sci. USA 85: 3913–3917.

Gross, R., and Rappuoli, R. 1989. Pertussis toxin promoter sequences involved in modulation. J. Bacteriol. 171: 4026– 4030.

Gueirard, P., Weber, C., LeCoustumier, A., and Guiso, N. 1995. Human *Bordetella bronchiseptica* infection related to contact with infected animals: persistence of bacteria in host. J. Clin. Microbiol, 33: 2002–2006.

Jones, P.G., VanBogelen, R.A., and Neidhardt, F.C. 1987. Induction of proteins in response to low temperature in *Escherichia coli*. J. Bacteriol. 169: 2092–2095.

Kato, A., Ohnishi, H., Yamamoto, K., Furuta, E., Tanabe, H., and Utsumi, R. 2000. Transcription of *emrKY* is regulated by the EvgA-EvgS two-component system in *Escherichia coli* K-12. Biosci. Biotechnol. Biochem. 64: 1203–1209.

Knapp, S., and Mekalanos, J.J. 1988. Two trans-acting regulatory genes (*vir* and *mod*) control antigenic modulation in *Bordetella pertussis*. J. Bacteriol. 170: 5059–5066.

Luttinger, A., Hahn, J., and Dubnau, D. 1996. Polynucleotide phosphorylase is necessary for competence development in *Bacillus subtilis*. Mol. Microbiol. 19: 343–356.

Manetti, R., Arico, B., Rappuoli, R., and Scarlato, V. 1994. Mutations in the linker region of BvgS abolish response to environmental signals for the regulation of the virulence factors in *Bordetella pertussis*. Gene 150: 123–127.

Martinez de Tejada, G., Cotter, P.A., Heininger, U., Camilli, A., Akerley, B.J., Mekalanos, J.J., and Miller, J.F. 1998. Neither the Bvg- phase nor the *vrg6* locus of *Bordetella pertussis* is required for respiratory infection in mice. Infect. Immun. 66: 2762–2768.

Merkel, T.J., Barros, C., and Stibitz, S. 1998. Characterization of the *bvgR* locus of *Bordetella pertussis*. J. Bacteriol. 180: 1682–1690.

Miller, J.F., Johnson, S.A., Black, W.J., Beattie, D.T., Mekalanos, J.J., and Falkow, S. 1992. Constitutive sensory transduction mutations in the *Bordetella pertussis bvgS* gene. J. Bacteriol. 174: 970–979.

Monack, D.M., Arico, B., Rappuoli, R., and Falkow, S. 1989. Phase variants of *Bordetella bronchiseptica* arise by spontaneous deletions in the *vir* locus. Mol. Microbiol. 3: 1719–1728.

Perraud, A.-L., Kimmel, B., Weiss, V., and Gross, R. 1998. Specificity of the BvgAS and EvgAS phosphorelay is mediated by the C-terminal HPt domains of the sensor proteins. Mol. Microbiol. 27: 875–888.

Perraud, A.-L., Rippe, K., Bantscheff, M., Glocker, M., Lucassen, M., Jung, K., Sebald, W., Weiss, V., and Gross, R. 2000. Dimerization of signalling domains of the EvgAS and BvgAS phosphorelay systems. Biochim. Biophys. Acta 1478: 341–354.

Perraud, A.L., Weiss, V., and Gross, R. 1999. Signalling pathways in two-component phosphorelay systems. Trends Microbiol. 7: 115–120.

Petering, H., Hammerschmidt, S., Frosch, M., van Putten, J.P., Ison, C.A., and Robertson, B.D. 1996. Genes associated with meningococcal capsule complex are also found in *Neisseria gonorrhoeae*. J. Bacteriol. 178: 3342–3345.

Rappuoli, R. 1994. Pathogenicity mechanisms of *Bordetella*. Curr. Top. Microbiol. Immunol. 192: 319–336.

Sawers, G. 1999. The aerobic/anaerobic interface. Curr. Opin. Microbiol. 2: 181–187.

Scarlato, V., Arico, B., Prugnola, A., and Rappuoli, R. 1991. Sequential activation and environmental regulation of virulence genes in *Bordetella pertussis*. EMBO J. 10: 3971–3975.

Scarlato, V., Arico, B., and Rappuoli, R. 1993. DNA topology affects transcriptional regulation of the pertussis toxin gene of *Bordetella pertussis* in *Escherichia coli* and *in vitro*. J. Bacteriol. 175: 4764–4771.

Scarlato, V., Prugnola, A., Arico, B., and Rappuoli, R. 1990. Positive transcriptional feedback at the *bvg* locus controls expression of virulence factors in *Bordetella pertussis*. Proc. Natl. Acad. Sci. USA 87: 10067–10071.

Schneider, B., Stübs, D., and Gross, R. 2002. Identification and genomic organization of gene loci negatively controlled by the virulence regulatory BvgAS two-component system in *Bordetella bronchiseptica*. Mol. Genet. Genomics 267: 526–535.

Steffen, P., Goyard, S., and Ullmann, A. 1996. Phosphorylated BvgA is sufficient for transcriptional activation of virulence-regulated genes in *Bordetella pertussis*. EMBO J. 15: 102–109.

Stibitz, S., Aaronson, W., Monack, D., and Falkow, S. 1989. Phase variation in *Bordetella pertussis* by frameshift mutation in a gene for a novel two-component system. Nature 338: 266–269.

Stockbauer, K.E., Fuchslocher, B., Miller, J.F., and Cotter, P.A. 2001. Identification and characterization of BipA, a *Bordetella* Bvg-intermediate phase protein. Mol. Microbiol. 39: 65–78.

Sugita, C., Sugiura, M., and Sugita, M. 2000. A novel nucleic acid-binding protein in the cyanobacterium *Synechococcus* sp. PCC6301: a soluble 33-kDa polypeptide with high sequence similarity to ribosomal protein S1. Mol. Gen. Genet. 263: 655–663.

Taylor, B.L., and Zhulin, I.B. 1999. PAS domains: internal sensors of oxygen, redox potential, and light. Microbiol. Mol. Biol. Rev. 63: 479–506.

Uhl, M.A., and Miller, J.F. 1994. Autophosphorylation and phosphotransfer in the *Bordetella pertussis* BvgAS signal transduction cascade. Proc. Natl. Acad. Sci. USA 91: 1163–1167.

Uhl, M.A., and Miller, J.F. 1996. Central role of the BvgS receiver as a phosphorylated intermediate in a complex two-component phosphorelay. J. Biol. Chem. 271: 33176–33180.

Wanner, B.L. 1992. Is cross regulation by phosphorylation of two-component response regulator proteins important in bacteria? J. Bacteriol. 174: 2053–2058.

Weiss, A.A., and Falkow, S. 1984. Genetic analysis of phase change in *Bordetella pertussis*. Infect. Immun. 43: 263–269.

Yaku, H., Kato, M., Hakoshima, T., Tsuzuki, M., and Mizuno, T. 1997. Interaction between the CheY response regulator and the histidine-containing phosphotransfer (HPt) domain of the ArcB sensory kinase in *Escherichia coli*. FEBS Lett. 408: 337–440.

Zu, T., Manetti, R., Rappuoli, R., and Scarlato, V. 1996. Differential binding of BvgA to two classes of virulence genes of *Bordetella pertussis* directs promoter selectivity by RNA polymerase. Mol. Microbiol. 21: 557–565.

Chapter 7

Influence of the *leuX*-Encoded $tRNA_5^{Leu}$ on the Regulation of Gene Expression in Pathogenic *Escherichia coli*

Ulrich Dobrindt, Katharine Piechaczek, Angelika Schierhorn, Gunter Fischer, Michael Hecker, and Jörg Hacker

Abstract

The *leuX* gene encoding the minor $tRNA_5^{Leu}$ is important for the expression of several virulence factors of pathogenic *Escherichia coli* strains. The differential usage of minor codons to control the expression of specialized genes has been proposed to be a general mechanism of bacteria to regulate gene expression at the posttranscriptional level. The minor codon usage theory foots on the biased codon usage of bacterial genes and the selective availability of tRNA isoacceptors. We aimed at the further investigation of the regulatory role of the $tRNA_5^{Leu}$ for gene expression in pathogenic *E. coli*. For this purpose, the molecular mechanism underlying the $tRNA_5^{Leu}$-dependent regulation of different virulence-associated genes of pathogenic *E. coli* as well as the regulation of *leuX* transcription under various growth conditions were investigated in detail. The global impact of the presence or absence of the *leuX*-encoded tRNA on gene expression of the uropathogenic *E. coli* strain 536 was studied by proteome analysis. The obtained results argue for a general importance of the $tRNA_5^{Leu}$ for gene expression of *E. coli* and the involvement of this tRNA in global regulatory networks.

Introduction

The expression of bacterial virulence-associated genes is often subject to coordinate global regulation and determined by a few common regulators in response to environmental conditions (Andersson *et al.*, 1999; Harel and Martin, 1999; Hengge-Aronis, 1999; Ibanez-Ruiz *et al.*, 2000; Wang and Kim, 2000; Winzer *et al.*, 2000). In bacteria, a broad variety of mechanisms exist to regulate gene expression resulting in an adaptation to environmental changes. Transfer RNAs and their encoding genes are important for several aspects of bacterial life. tRNAs not only function as amino acid donors during translation, several biosynthetic processes and during the determination of metabolic half lifes of proteins (Lennarz *et al.*, 1966; Katz *et al.*, 1967; Roberts *et al.*, 1968a, b; Petit *et al.*, 1968; Stewart *et al.*, 1971; Tobias *et al.*, 1991; Jahn *et al.*, 1992), they also seem to be of general importance for the regulation of gene expression in prokaryotes. It is well known that tRNAs are involved in the regulation of transcriptional termination (Williamson and Oxender, 1992; Yanofsky, 2000). In addition, certain transfer RNAs are considered as possible regulatory molecules participating in the regulation of gene expression at the posttranscriptional level (Saier Jr., 1995; Björk *et al.*, 1999). It has been speculated that the availability of tRNA isoacceptors together with the codon bias are part of an evolutionary strategy to control the expression rate of certain genes and more recent studies in various organisms imply that differential codon usage may ensure the proper temporal expression of specialized genes by a posttranscriptional mechanism (Leskiw *et al.*, 1991; Saier Jr., 1995). Several examples of tRNA-dependent gene expression in bacteria which underline this hypothesis are summarized in Table 1.

There is increasing evidence that transfer RNA-encoding genes frequently serve as target sites for the integration of horizontally transferred genetic elements, such as bacteriophages, plasmids or pathogenicity islands (PAIs), into the bacterial chromosome and that they are therefore important for the genetic diversity of bacteria. One characteristic feature of PAIs is their association with tRNA genes. (Cheetham and Katz, 1995; Kaper and Hacker, 1999; Kiewitz *et al.*, 2000; Jackson *et al.*, 2000; Perna, 2001). In the uropathogenic *Escherichia coli* strain 536 (O6:K15:H31), the *leuX* gene, which codes for the minor $tRNA_5^{Leu}$, is associated with the PAI II_{536} and is required for the expression of a broad variety of virulence traits. The lack of a functional $tRNA_5^{Leu}$ results in mutant strains with attenuated virulence (Blum *et al.*, 1994; Ritter *et al.*, 1995; Susa *et al.*, 1996; Dobrindt *et al.*, 1998; Hacker *et al.*, 1999). In order to understand the importance of the $tRNA_5^{Leu}$, which recognizes the codon UUG, for the expression of virulence-associated genes, the influence of this minor leucyl-tRNA on virulence gene expression as well as the regulation of *leuX* transcription were investigated in detail in the *E. coli* strain 536.

Influence of the $tRNA_5^{Leu}$ on Gene Expression in Pathogenic *E. coli* Strains

The differential usage of minor codons to control the expression of specialized genes has been proposed to be a general mechanism of bacteria to regulate gene

Table 1. Examples of tRNA-dependent gene expression in bacteria

Function	Organism	Codon	Amino acid specificity	Reference
Photosynthesis	*Rhodobacter capsulatus*	GCU GUC	Ala Val	Wu and Saier Jr., 1991
Fructose utilization	*R. capsulatus*	AAU UGU	Asn Cys	Wu and Saier Jr., 1991
Production of acetone and butanol	*Clostridium acetobutylicum*	ACG	Thr	Sauer and Dürre, 1992
Production of aerial mycelium, conidiospores and antibiotics	*Streptomyces coelicolor* A3(2), *S. lividans, S. alboniger*	UUA	Leu	Lawlor *et al.*, 1987; Guthrie and Chater, 1990; Tercero *et al.*, 1998
Expression of type 1-fimbriae	*Salmonella typhimurium, S. enteritidis*	UCU AGA AGG	Ser Arg Arg	Clouthier *et al.*, 1998
Global gene expression	*Escherichia coli*	AGA AGG	Arg Arg	Chen *et al.*, 1990; Chen and Inouye, 1994
Expression of type 1-fimbriae-, enterobactin and flagella, serum resistence, virulence	*E. coli* 536 (uropathogenic)	UUG	Leu	Ritter *et al.*, 1995, 1997
Persistence in stationary phase in cecal or bladder mucus of CD-1 mice	*E. coli* F-18, *E. coli* 536	UUG	Leu	Newman *et al.*, 1994; Dobrindt *et al.*, 1998
Cell division	*E. coli*	UCA UCG CUA UUG	Ser Ser Leu Leu	Tamura *et al.*, 1984; Leclerc *et al.*, 1989; Chen *et al.*, 1991; Nakayashiki and Inokuchi, 1998
DNA-Replication	*E. coli*	AGA	Arg	Garcia *et al.*, 1986
Resistence to calmodulin-inhibitor	*E. coli*	CUA	Leu	Chen *et al.*, 1991; Bouquin *et al.*, 1996
Temperature sensitivity	*E. coli*	CUA	Leu	Chen *et al.*, 1991; Bouquin *et al.*, 1996
Production of protease and syringomycin	*Pseudomonas syringae*	AUG	Met	Rich and Willis, 1997
		CUU CUC	Leu Leu	
Virulence	*Shigella flexneri*	CUG UAC UAC	Leu Tyr	Hromockyj *et al.*, 1992

expression at the posttranscriptional level. The minor codon usage theory foots on the biased codon usage of bacterial genes and the selective availability of tRNA isoacceptors (Saier Jr., 1995).

In the uropathogenic *E. coli* strain 536, a functional $tRNA_5^{Leu}$ is required for expression of type 1 fimbriae, enterobactin, motility, serum resistance and *in vivo* virulence (Ritter *et al.*, 1995). In order to understand how the $tRNA_5^{Leu}$ influences the expression of these virulence factors, we focussed on the molecular mechanism underlying the influence of this minor leucyl-tRNA on the expression of different virulence factors. The availability of the *leuX*-encoded tRNA affects expression of type 1 fimbriae by facilitated translation of *fimB* transcripts which code for the FimB recombinase. Transcription of the *fimA* gene encoding the major type 1 fimbrial subunit is directed by a promoter located on an invertible DNA switch. The orientation of this switch is due to the interplay between two recombinases, FimB and FimE. FimB is able to turn the switch on, whereas FimE can only turn it off. *fimB* transcripts of *E. coli* 536 contain five *leuX*-specific codons (UUG), but only two UUG codons are present in the *fimE*-specific mRNA. By the exchange of the five *leuX*-specific codons of *fimB* transcripts into codons recognized by the major leucyl-tRNA (CUG), it has been demonstrated that the loss of the functional $tRNA_5^{Leu}$ limits translation of *fimB*- relative to that of *fimE* transcripts, in turn influencing *fimA* transcription (Ritter *et al.*, 1997). The lack of the *leuX* gene also results in a reduced expression of the α-haemolysin determinant due to a reduced transcription of the encoding *hly* operon. Increased expression of the *hha* gene whose encoded product acts as a repressor of *hly* transcription may be at least partially responsible for this effect (Dobrindt *et al.*, 2000; Dobrindt, *et al.*, 2002). By the use *leuX*-specific antisense constructs in different pathogenic *E. coli* strains causing urinary tract infections, newborn meningitis or diarrhea, the expression of type 1 fimbriae, flagella and siderophore systems has been reduced in analogy to the phenotype of *leuX*-negative mutants of the uropathogenic *E. coli* strain 536 (Dobrindt *et al.*, 2000; Dobrindt, unpublished). These results support the model that the $tRNA_5^{Leu}$ is required for the expression of several virulence factors of *E. coli*.

The influence of the $tRNA_5^{Leu}$ on global gene expression of the *E. coli* strain 536 was studied by comparative analysis of proteome patterns of the wild type strain 536 and different *leuX*-negative mutants.

The proteome comparison revealed that the expression of various genes, including many house keeping genes, is affected by the presence of the $tRNA_5^{Leu}$. Among them are also several genes which are involved in iron acquisition. The fact that the presence or absence of the *leuX*-encoded $tRNA_5^{Leu}$ influences the expression of 31 different genes coding for intracellular proteins supports the view that this minor tRNA plays an important role for the expression of a broad variety of different genes in *E. coli* which is not restricted to virulence-associated genes and that it may be part of a regulatory network (Piechaczek *et al.*, 2000).

Regulation of *leuX* Transcription

If the *leuX*-encoded tRNA was important for gene expression as a part of a regulatory network, its transcriptional regulation should be affected by environmental stimuli and should differ from that of other tRNA genes, i.e. that of the *leuV* encoded major leucyl-tRNA. In order to investigate whether the $tRNA_5^{Leu}$ has a regulatory function as a part of a regulatory network which coordinately controls the expression of different virulence associated genes in the *E. coli* strain 536, *leuX* and *leuV* transcription as a response to several stress stimuli and growth conditions was compared in order to gain further insights into the regulation of *leuX* transcription (Dobrindt and Hacker, 2001).

The comparison of *leuX*- and *leuV* transcript levels in response to different environmental stimuli and growth conditions demonstrated that *leuX*- and *leuV* transcription are differently regulated and that the *leuX* promoter sequence is important for these regulatory characteristics. The ratio of $tRNA_5^{Leu}/tRNA_1^{Leu}$ within the tRNA pool increased under several stress conditions (high growth temperature or high osmolarity or ethanol concentrations). Under these conditions *leuX* transcription was stable or slightly enhanced concomitant whereas that of *leuV* was reduced. *leuX* transcription has been shown to be specifically affected by the alternative sigma factor RpoH (Dobrindt and Hacker, 2001). These results were also supported by the fact that the *leuX* promoter sequence exhibits similarity to heat shock-dependent promoter sequences. In stationary phase, *leuX*- but not *leuV* primary transcript levels were also increased in a *rpoS* mutant of the *E. coli* strain 536 in comparison to the wild type strain. The described alterations of the $tRNA_5^{Leu}$-

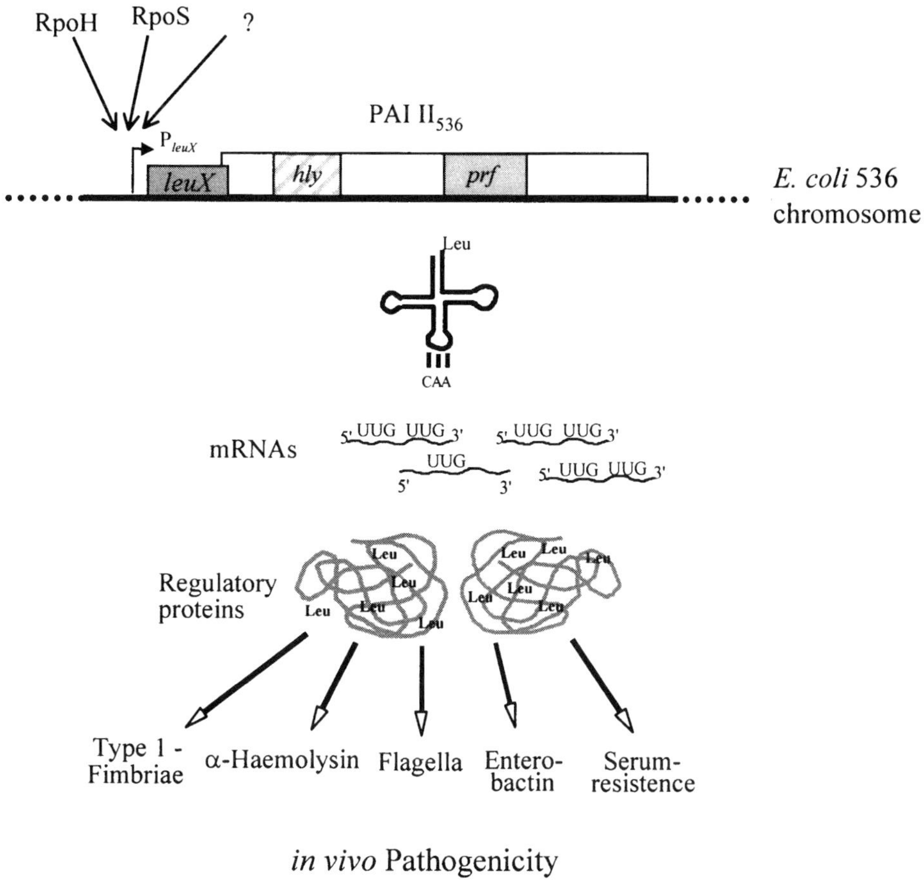

Figure 1. Influence of the *leuX*-encoded $tRNA_5^{Leu}$ on gene expression in the uropathogenic *E. coli* strain 536. Global regulatory networks affect *leuX* transcription in response to environmental stimuli. The availability of the $tRNA_5^{Leu}$ influences the expression of genes with a high percentage of *leuX*-specific codons. The expression of a broad variety of different subordinate genes can be affected by $tRNA_5^{Leu}$-dependent facilitated translation of genes encoding regulatory proteins. PAI, pathogenicity island; *hly*, α-haemolysin determinant; *prf*, P-related fimbriae determinant

as well as of the $tRNA_1^{Leu}$ transcript levels result in an increase of the availability of the *leuX*-encoded minor $tRNA_5^{Leu}$ within the tRNA isoacceptor pool. This implies a possible regulatory function of the $tRNA_5^{Leu}$ for gene expression of *E. coli* (Dobrindt and Hacker, 2001). Additionally, *leuX*- and *leuV* promoter activities were analyzed upon growth in human urine. The corresponding ratio of *leuX*- and *leuV* promoter activities resembled those obtained under different stress conditions (Dobrindt and Hacker, 2001). This underlines that under conditions similar to the natural environment of uropathogenic *E. coli* the availability of the $tRNA_5^{Leu}$ can be increased in relation to that of the $tRNA_1^{Leu}$ and argues for a regulatory role of the $tRNA_5^{Leu}$ under these conditions according to the minor codon usage theory.

Conclusions

Many studies dealing with the importance of minor tRNAs for the expression of certain genes have been published (see Table 1). A well characterized example of tRNA-dependent gene expression is the *bldA*-dependent sporulation and production of antibiotics and aerial mycelium of *Streptomyces coelicolor* A3(2) which shows a remarkable similarity to characteristics of *leuX*-dependent expression of genes in *E. coli*. The presence of *bldA*-specific codons is limited to specialized genes, the *bldA*-encoded tRNA accumulates in the stationary phase simultaneously with the occurrence of *bldA*-dependent gene expression (Leskiw *et al.*, 1991; Kataoka *et al.*, 1999) and *bldA* transcription is affected by an unknown factor (Leskiw and Mah 1995). In case of *bldA*-dependent production of the antibiotic undecylprodigiosin, the *bldA*-encoded tRNA is essential for the efficient translation of the *redZ* transcript coding for a superimposed regulator of the undecylprodigiosin biosynthesis pathway (White and Bibb, 1997). This suggests that the availability of the *bldA*-encoded tRNA seems to affect gene expression at the translational level. It has also been published that in *E. coli* the transcription of some tRNA genes is enhanced under special conditions (Espinosa-Urgel and Kolter, 1998). These results suggest that the transcription of certain tRNA genes can be specifically regulated, although in these cases the regulatory function of a tRNA has not been proven. Our results demonstrate that the *leuX*-encoded tRNA is of great importance in pathogenic *E. coli* as it influences the expression of a broad variety of genes including virulence determinants. The fact that the transcription of the tRNA gene *leuX* is specifically affected in response to environmental conditions is consistent with the postulated participation of certain minor tRNAs as regulatory components within global regulatory networks (see Figure 1). The availability of this tRNA ensures the expression of genes with a high amount of *leuX*-specific codons by facilitated translation. If these genes code for regulatory proteins, their efficient expression in turn affects that of other subordinate genes within these networks.

Acknowledgements

Our work was supported by the DFG (Ha 1434/8-2, SFB 479) and by the Fonds der Chemischen Industrie. UD was supported by a grant from the Boehringer Ingelheim Fonds.

References

Andersson, R.A., Koiv, V., Norman-Setterblad, C., and Pirhonen, 1999. M. Role of RpoS in virulence and stress tolerance of the plant pathogen *Erwinia carotovora* subsp. *carotovora*. Microbiology 145: 3547–3556.

Björk, G.R., Durand, J.M., Hagervall, T.G., Leipuviene, R., Lundgren, H.K., Nilsson, K., Chen, P., Qian, Q., and Urbonavicius, J. 1999. Transfer RNA modification: influence on translational frameshifting and metabolism. FEBS Lett. 452: 47–51.

Blum, G., Ott, M., Lischewski, A., Ritter, A., Imrich, H., Tschäpe, H., and Hacker, J. 1994. Excision of large DNA regions termed pathogenicity islands from tRNA-specific loci in the chromosome of an *Escherichia coli* wild-type pathogen. Infect. Immun. 62: 606–614.

Bouquin, N., Chen, M.X., Kim, S., Vannier, F., Bernard, S., Holland, I.B., and Seror S.J. 1996. Characterization of an *Escherichia coli* mutant, *feeA*, displaying resistance to the calmodulin inhibitor 48/80 and reduced expression of the rare $tRNA_3^{Leu}$. Mol. Microbiol. 20: 853–865.

Cheetham, B.F., and Katz, M.E. 1995. A role for bacteriophages in the evolution and transfer of bacterial virulence determinants. Mol. Microbiol. 18: 201–208.

Chen, G.F., and Inouye, M. 1994. Role of the AGA/AGG codons, the rarest codons in global gene expression in *Escherichia coli*. Genes and Development 8: 2641–2652.

Chen, K.S., Peters, T.C., and Walker, J.R. 1990. A minor arginine tRNA mutant limits translation preferentially of a protein dependent on the cognate codon. J. Bacteriol. 172: 2504–2510.

Chen, M.X., Bouquin, N., Norris, V., Casaregola, S., Seror, S.J., and Holland, I.B. 1991. A single base change in the acceptor stem of tRNA(3Leu) confers resistance upon *Escherichia coli* to the calmodulin inhibitor, 48/80. EMBO J. 10: 3113–3122.

Clouthier, S.C., Collinson, S.K., White, A.P., Banser, P.A., and Kay, W.W. 1998. $tRNA^{Arg}$ (*fimU*) and expression of SEF14 and SEF21 in *Salmonella enteritidis*. J. Bacteriol. 180: 840–845.

Dobrindt, U., and Hacker, J. 2001. Regulation of the pathogenicity island associated $tRNA_5^{Leu}$-encoding gene *leuX* in the uropathogenic *Escherichia coli* strain 536. Mol. Gen. Genomics 265: 895–904.

Dobrindt, U., Cohen, P.S., Utley, M., Mühldorfer, I., and Hacker, J. 1998. The *leuX*-encoded $tRNA_5^{Leu}$ but not the pathogenicity islands I and II influence the survival of the uropathogenic *Escherichia coli* strain 536 in CD-1 mouse bladder mucus in the stationary phase. FEMS Microbiol. Lett. 162: 135–141.

Dobrindt, U., Janke, B., Piechaczek, K., Nagy, G., Ziebuhr, W., Fischer, G., Schierhorn, A., Hecker, M., Blum-Oehler, G., and Hacker, J. 2000. Toxin genes on pathogenicity islands: impact for microbial evolution. Int. J. Med. Microbiol. 290: 307–311.

Dobrindt, U., Emödy, L., Gentschev, J., Goebel, W., Hacker, J. 2002. Efficient expression of the alpha-haemolysin determinant in the Uropathogenic *Escherichia coli* Strain 536 requires the *leuX*-encoded $tRNA_5^{Leu}$. Mol. Genet. Genomics 267: 370–379.

Espinosa-Urgel, M., and Kolter, R. 1998. *Escherichia coli* genes expressed preferentially in an aquatic environment. Mol. Microbiol. 28: 325–332.

Garcia, G.M., Mar, P.K., Mullin, D.A., Walker, J.R., and Prather, N.E. 1986. The *E. coli dnaY* gene encodes an arginine transfer RNA. Cell 45: 453–459.

Guthrie, E.P., and Chater, K.F. 1990. The level of a transcript required for production of a *Streptomyces coelicolor* antibiotic is conditionally dependent on a tRNA gene. J. Bacteriol. 172: 6189–6193.

Hacker, J., Blum-Oehler, G., Janke, B., Nagy, G., and Goebel, W. 1999. Pathogenicity islands of extraintestinal *Escherichia coli*. In: Kaper, J.B., and Hacker, J. (eds.), Pathogenicity islands, plasmids and other mobile elements. ASM Press, Washington, DC, pp. 59–76.

Harel, J., and Martin, C. 1999. Virulence gene regulation in pathogenic *Escherichia coli*. Vet. Res. 30:131–155.

Hengge-Aronis, R. 1999. Interplay of global regulators and cell physiology in the general stress response of *Escherichia coli*. Curr. Opin. Microbiol. 2: 148–152.

Hromockyj, A.E., Tucker, S.C., and Maurelli, A.T. 1992. Temperature regulation of *Shigella* virulence: identification of the repressor gene *virR*, an analogue of *hns*, and partial complementation by tyrosyl transfer RNA ($tRNA_1^{Tyr}$). Mol. Microbiol. 6: 2113–2124.

Ibanez-Ruiz, M., Robbe-Saule, V., Hermant, D., Labrude, S., and Norel, F. 2000. Identification of RpoS (sigma(S))-regulated genes in *Salmonella enterica* serovar *typhimurium*. J. Bacteriol. 182: 5749–5756.

Jackson, R.W., Mansfield, J.W., Arnold, D.L., Sesma, A., Paynter, C.D., Murillo, J., Taylor, J.D., and Vivian, A. 2000. Excision from tRNA genes of a large chromosomal region, carrying *avrPphB*, associated with race change in the bean pathogen, *Pseudomonas syringae* pv. *phaseolicola*. Mol. Microbiol. 38: 186–197.

Jahn, D., Verkamp, E., and Söll, D. 1992. Glutamyl-transfer RNA: a precursor of heme and chlorophyll biosynthesis. TIBS 17: 215–218.

Kaper, J.B., and Hacker, J. 1999. Pathogenicity islands, plasmids and other mobile elements, ASM Press, Washington, D.C.

Kataoka, M., Kosono, S., and Tsujimoto, G. 1999. Spatial and temporal regulation of protein expression by *bldA* within a *Streptomyces lividans* colony. FEBS Lett. 462: 425–429.

Katz, W., Matsuhashi, M., Dietrich, C.P., and Strominger, J.L. 1967. Biosynthesis of the peptidoglycan of bacterial cell walls. IV. Incorporation of glycine in *Micrococcus lysodeikticus*. J. Biol. Chem. 242: 3207–3217.

Kiewitz, C., Larbig, K., Klockgether, J., Weinel, C., and Tümmler, B. 2000. Monitoring genome evolution *ex vivo*: reversible chromosomal integration of a 106 kb plasmid at two tRNA(Lys) gene loci in sequential *Pseudomonas aeruginosa* airway isolates. Microbiology 146: 2365–2373.

Lawlor, E.J., Baylis, H.A., and Chater, K.F. 1987. Pleiotropic morphological and antibiotic deficiencies result from mutations in a gene encoding a tRNA-like product in *Streptomyces coelicolor* A3(2). Gene and Development 1: 1305–1310.

Leclerc, G., Sirard, C., and Drapeau, G.R. 1989. The *Escherichia coli* cell division mutation *ftsM*1 is in *serU*. J. Bacteriol. 171: 2090–2095.

Lennarz, W.J., Nesbitt,J.A. 3rd, and Reiss, J. 1966. The participation of sRNA in the enzymatic synthesis of O-L-lysyl phosphatidylgylcerol in *Staphylococcus aureus*. Proc. Natl. Acad. Sci. USA 55: 934–941.

Leskiw, B.K., and Mah, R. 1995. The *bldA*-encoded tRNA is poorly expressed in the *bldI* mutant of *Streptomyces coelicolor* A3(2). Microbiology 141: 1921–1926.

Leskiw, B.K., Bibb, M.J., and Chater, K.F. 1991. The use of a rare codon specifically during development? Mol. Microbiol. 5: 2861–2867.

Nakayashiki, T., and Inokuchi, H. 1998. Novel temperature-sensitive mutants of *Escherichia coli* that are unable to grow in the absence of wild-type $tRNA_6^{Leu}$. J. Bacteriol. 180: 2931–2935.

Newman, J.V., Kolter, R., Laux, D.C., and Cohen, P.S. 1994. Role of *leuX* in *Escherichia coli* colonization of the streptomycin-treated mouse large intestine. Microb. Pathog. 17: 301–311.

Ochman, H., Lawrence, J.G., and Groisman, E.A. 2000. Lateral gene transfer and the nature of bacterial innovation. Nature 405: 299–304.

Perna, N.T., Plunkett, G. 3rd, Burland, V., Mau, B., Glasner, J.D., Rose, D.J., Mayhew, G.F., Evans, P.S., Gregor, J., Kirkpatrick, H.A., Posfai, G., Hackett, J., Klink, S., Boutin, A., Shao, Y., Miller, L., Grotbeck, E.J., Davis, N.W., Lim, A., Dimalanta, E.T., Potamousis, K.D., Apodaca, J., Anantharaman, T.S., Lin, J., Yen, G., Schwartz, D.C., Welch, R.A., Blattner, F.R. 2001. Genome sequence of enterohemorrhagic *Escherichia coli* 0157:H7, Nature 409: 529–533.

Petit, J.F., Strominger, J.L., and Söll, D. 1968. Biosynthesis of the peptidoglycan of bacterial cell walls. VII. Incorporation of serine and glycine into interpeptide bridges in *Staphylococcus epidermidis*. J. Biol. Chem. 243: 757–767.

Piechaczek, K., Dobrindt, U., Schierhorn, A., Fischer, G., Hecker, M., and Hacker, J. 2000. Influence of pathogenicity islands and the minor *leuX*-encoded $tRNA_5^{Leu}$ on the proteome pattern of the uropathogenic *Escherichia coli* strain 536. Int. J. Med. Microbiol. 290: 75–84.

Rich, J.J., and Willis, D.K. 1997. Multiple loci of *Pseudomonas syringae* pv. *syringae* are involved in pathogenicity on bean: restoration of one lesion-deficient mutant requires two tRNA genes. J. Bacteriol. 179: 2247–2258.

Ritter, A., Blum, G., Emödy, L., Kerenyi, M., Böck, A., Neuhierl, B., Rabsch, W., Scheutz, F., and Hacker, J. 1995. tRNA genes and pathogenicity islands: influence on virulence and metabolic properties of uropathogenic *Escherichia coli*. Mol. Microbiol. 17: 109–121.

Ritter, A., Gally, D.L., Olsen, P.B., Dobrindt, U., Friedrich, A., Klemm, P., and Hacker, J. 1997. The Pai-associated *leuX* specific $tRNA_5^{Leu}$ affects type 1 fimbriation in pathogenic *Escherichia coli* by control of the FimB recombinase expression. Mol. Microbiol. 25: 871–882.

Roberts, W.S., Petit, J.F., and Strominger, J.L. 1968b. Biosynthesis of the peptidoglycan of bacterial cell walls. IIX. Specificity in the utilization of L-alanyl transfer ribonucleic acid for interpeptide bridge synthesis in *Arthrobacter crystallopoietes*. J. Biol. Chem. 243: 768–772.

Roberts, W.S., Strominger, J.L., and Söll, D. 1968a. Biosynthesis of the peptidoglycan of bacterial cell walls. VI. Incorporation of L-threonine into interpeptide bridges in *Micrococcus roseus*. J. Biol. Chem. 243: 749–756.

Saier, M.H., Jr. 1995. Differential codon usage: a safeguard against inappropriate expression of specialized genes? FEBS Lett. 362: 1–4.

Sauer, U., and Dürre, P. 1992. Possible function of $tRNA_{ACG}^{Tyr}$ in regulation of solvent formation in *Clostridium acetobutylicum*. FEMS Microbiol. Lett. 100: 147–154.

Stewart, T.S., Roberts, R.J., and Strominger, J.L. 1971. Novel species of tRNA. Nature 230: 36–38.

Susa, M., Kreft, B., Wasenauer, G., Ritter, A., Hacker, J., and Marre, R. 1996. Influence of cloned tRNA genes from a uropathogenic *Escherichia coli* strain on adherence to primary human renal tubular epithelial cells and nephropathogenicity in rats. Infect. Immun. 64: 5390–5394.

Tamura, F., Nishimura, S., and Ohki, M. 1984. The *E. coli divE* mutation, which differentially inhibits synthesis of certain proteins, is in tRNASer1. EMBO J. 3: 1103–1107.

Tercero, J.A., Espinosa, J.C., and Jimenez, A. 1998. Expression of the *Streptomyces alboniger pur* cluster in *Streptomyces lividans* is dependent on the *bldA*-encoded tRNALeu. FEBS Lett. 421: 221–223.

Tobias, J.W., Shrader, T.E., Rocap, G., and Varshavsky, A. 1991. The N-end rule in bacteria. Science 254: 1374–1377.

Williamson, R.M., and Oxender, D.L. 1992. Premature termination of *in vivo* transcription of a gene encoding a branched-chain amino acid transport protein in *Escherichia coli*. J. Bacteriol. 174: 1777–1782.

Winzer, K., Falconer, C., Garber, N.C., Diggle, S.P., Camara, M., and Williams, P. 2000. The *Pseudomonas aeruginosa* lectins PA-IL and PA-IIL are controlled by quorum sensing and by RpoS. J. Bacteriol. 182: 6401–6411.

Wang, Y., and Kim, K.S. 2000. Effect of *rpoS* mutations on stress-resistance and invasion of brain microvascular endothelial cells in *Escherichia coli* K1. FEMS Microbiol. Lett. 182: 241–247.

White, J., and Bibb, M. 1997. *bldA* dependence of undecylprodigiosin production in *Streptomyces coelicolor* A3(2) involves a pathway-specific regulatory cascade. J. Bacteriol. 179: 627–633.

Wu, L.F., and Saier, M.H., Jr. 1991. Differences in codon usage among genes encoding proteins of different function in *Rhodobacter capsulatus*. Res. Microbiol. 142: 943–949.

Yanofsky, C. 2000. Transcription attenuation: once viewed as a novel regulatory strategy. J. Bacteriol. 182: 1–8.

Chapter 8

The *ciaR/ciaH* Regulatory System of *Streptococcus pneumoniae* is Involved in Beta-Lactam Resistance and Genetic Competence

Dorothea Zähner, Kristina Kaminski, Mark van der Linden, Thorsten Mascher, Michelle Merai, and Regine Hakenbeck

Abstract

New mechanisms for β-lactam resistance independent on the target penicillin-binding proteins were detected in β-lactam-resistant laboratory mutants of *Streptococcus pneumoniae*. The link between mutations in the histidine protein kinase CiaH and phenotypic expression of cefotaxime resistance suggests that the cell is able to monitor the integrity of the cell wall and in emergency cases such as during the action of β-lactams can counteract such danger. At least one *ciaH* mutation Thr230 > Pro is likely to affect its phosphatase activity resulting in elevated phosphorylation of CiaR, the cognate response regulator, but other CiaH-independent signaling pathways may also result in CiaR phosphorylation. Mutants in CiaH, either alone or in combination with a mutated penicillin-binding protein 2x(PBP2x) fail to develop genetic competence. In all cases complementation of this phenotype was observed upon addition of the competence inducing pheromone peptide CSP, the processed product of the *comC* gene. This indicates that the *cia* system is part of a regulatory network that includes another two component system *comDE*. The DNA binding property of CiaR and ComE were exploited to isolate specifically interacting DNA fragments as a first step to identify genes targeted by individual response regulators.

New Mechanisms for β-lactam Resistance in *Streptococcus pneumoniae*

In the late 80s we isolated spontaneous mutants of *S. pneumoniae* resisting to increasing concentrations of β-lactams in order to investigate the nature of point mutations in penicillin binding proteins (PBPs), the penicillin-target proteins, that contribute to the resistance phenotype by reducing the PBP affinity to the antibiotic (Laible and Hakenbeck, 1987). Two different β-lactams were used for selection of the resistant mutants: piperacillin, a penicillin that induces cellular lysis at concentrations as low as 0.01 μg/ml; and cefotaxime, a third generation cephalosporin that curiously does not interact with PBP2b and does not cause rapid lysis even though pneumococci are also highly susceptible to this drug (Hakenbeck *et al.*, 1987). At this time, alterations in PBPs were believed to be the only exclusive resistance mechanism in the non β-lactamase producing pneumococcus. However, in addition to identifying a series of PBP mutations, other, non-PBP genes were also affected and contributed to increased resistance in the mutants. In fact, mutations in these genes conferred resistance without mutations in PBPs, i.e. they functioned as primary resistance determinants (for review, see (Hakenbeck *et al.*, 1999)).

Not only were different PBPs affected in piperacillin-resistant mutants (P-mutants) versus cefotaxime resistant mutants (C-mutants), i.e. primarily PBP2b versus PBP2x, but also different non-PBP genes were mutated in P- and C-mutants: a putative glycosyl-transferase gene *cpoA* was altered in the P-mutants, and *ciaH*, encoding a histidine protein kinase of a two component system in the C-mutants (Grebe *et al.*, 1997; Guenzi *et al.*, 1994). Both, *ciaH* and *cpoA* alleles, could act as primary resistance determinants. In other words, depending on the class of β-lactam used for selection, distinct mutational pathways were involved in the evolution of resistance. That implies that the damages induced by piperacillin, a highly lytic antibiotic, differ from those induced by the non-lytic cefotaxime, and thus required a different set of genes for counteracting these effects. In the following sections, results obtained with *cia* mutants are summarized.

Phenotypes Associated with *ciaH* Mutations

Every C-mutant lineage contained at least one *ciaH* mutation (Table 1). Some mutations are located in the sensor domain, which is placed between two putative membrane spanning regions and therefore presumably facing the outside of the cell, or in the kinase domain (Figure 1A). Each of the *ciaH* mutations is associated with an increase of resistance to β-lactams by approximately 2 fold, independent on the selection step it occurred. Disruption of *ciaH* also results in resistance (Table 1).

The *ciaH* gene is part of an operon and is located immediately downstream from the *ciaR* gene, the cognate response regulator (Figure 1B). The two genes are transcribed as a unique RNA (Giammarinaro *et al.*,

Table 1. Cefotaxime resistant mutant lineages.

lineage	Mutant	Protein	Mutation	Competence[1] /complementation	MIC CTX
	WT (R6)				0.02
1	C101	PBP2x	L600 W	+	0.09
	C201	nd		+	0.09
	C301	PBP2x	T550 A	+	0.2
	C401	CiaH	W9 stop	() yes	0.6
2	C102	CiaH	Q236 K	() yes	0.04
	C202	CiaH	A413 T	− yes	0.05
3	C103	CiaH	A203 V	+	0.04
	C203	PBP2x	G597 D	() yes	0.2
	C303	PBP2x	T550 A	− yes	0.9
4	C104	CiaH	A203 V	+	0.04
	C204	PBP2x	G601 E	+	0.08
	C304	nd		− yes	0.12
5	C105	nd		+	0.02
	C205	PBP2x	T526 S	+	0.07
	C305	CiaH	N95 D	− yes	0.12
6	C106	PBP2x	G601 V	+	0.09
	C206	PBP2x	G597 D	+	0.16
	C306	CiaH	T230 P	− yes	0.35
	R6-$T_{ciaHC306}$	CiaH	T230 P	− yes	0.04
	$R6_{ciaRnull}$[2]	CiaR		+	0.02
	$zR6_{ciaHnull}$[2]	CiaH		+	0.04

Only the first mutants of each of the lineages are described up to the step where competence deficiency occurred. The first digit indicates the selection step, the last digit specifies the lineage. Mutants in *cia* genes obtained in the R6 strain are also shown: R6-$T_{ciaHC306}$ is a transformant containing the *ciaH* allele of C306; null indicates mutants in *ciaR* and *ciaH*, respectively, generated by insertion duplication mutagenesis using a pJDC9 derivative (Chen and Morrison 1987) with a cloned internal *ciaR* (*ciaH*) gene fragment.

[1]Transformation efficiency: +, wild type; (): 10^{-2}–10^{-3} fold reduction; −: < 10^{-5} fold reduction. For complementation with CSP, 1 to 10 ng/ml synthetic peptide were added during the transformation assay.

[2] Disruption by plasmid integration of a pJDC9-derivatives.

CTX: cefotaxime; MIC values were determined after 24 h at 30°C on blood agar plates containing a narrow concentration range of the antibiotic; nd: not determined.

1999), and no CiaH protein is detectable in mutants with an insertional inactivated *ciaR* gene (Figure 1B). Disruption of *ciaR*, i.e. complete inactivation of the *cia* system, does not affect susceptibility to β-lactams in the parental strain R6 and abolishes the *ciaH*-mediated resistance in mutant strains (Guenzi *et al.*, 1994; Zähner *et al.*, 1996) (and unpublished results). This led to the assumption that a functional response regulator CiaR is required for the phenotype mediated by the *ciaH* mutations.

Curiously, another phenotype was associated with *ciaH* mutations: a decreased transformation efficiency (Guenzi *et al.*, 1994). This phenotype was differentially expressed depending on the individual *ciaH* mutation. The combination of *ciaH* point mutations with the *pbp2x* mutations in the C-mutants always resulted in loss of transformability, although *pbp2x* mutations alone had no significant effect on transformation (Table 1). Some *ciaH* alleles, such as those in C103 and C104 caused no significant decrease in transformation efficiency, whereas it was highly reduced in the mutant C102, and the $ciaH_{C306}$ mutation when introduced into the R6 strain conferred a complete transformation minus phenotype (Guenzi *et al.*, 1994; Zähner *et al.*, 1996). Disruption of *ciaR* in such mutants not only abolished the *ciaH* mediated resistance phenotype but also restored competence (Guenzi and Hakenbeck, 1995). Recently, a *ciaR* frame shift mutant that resulted in a truncated gene product was isolated that reversed the phenotype in the C306 mutant itself, in agreement with the notion that CiaR is required to mediate the effect of *ciaH* mutations (Giammarinaro *et al.*, 1999). The authors also noted a the *cia* operon is constitutively activated in strains carrying the $ciaH_{C306}$ allele suggesting that the *cia* system is positively autoregulated (Giammarinaro *et al.*, 1999).

The data can be explained if one assumes that it is the phosphorylated form of CiaR that results in β-lactam resistance as well as in competence deficiency. This in turn means that CiaR is phosphorylated in the *ciaH* missense mutations due to a defective phosphatase activity of the mutated CiaH. A mutation in the *E. coli* EnvZ histidine protein kinase which has been characterized as a phosphatase mutation occurs at a site homologous to the mutation in the $ciaH_{C306}$ allele, supporting the first assumption (Forst and

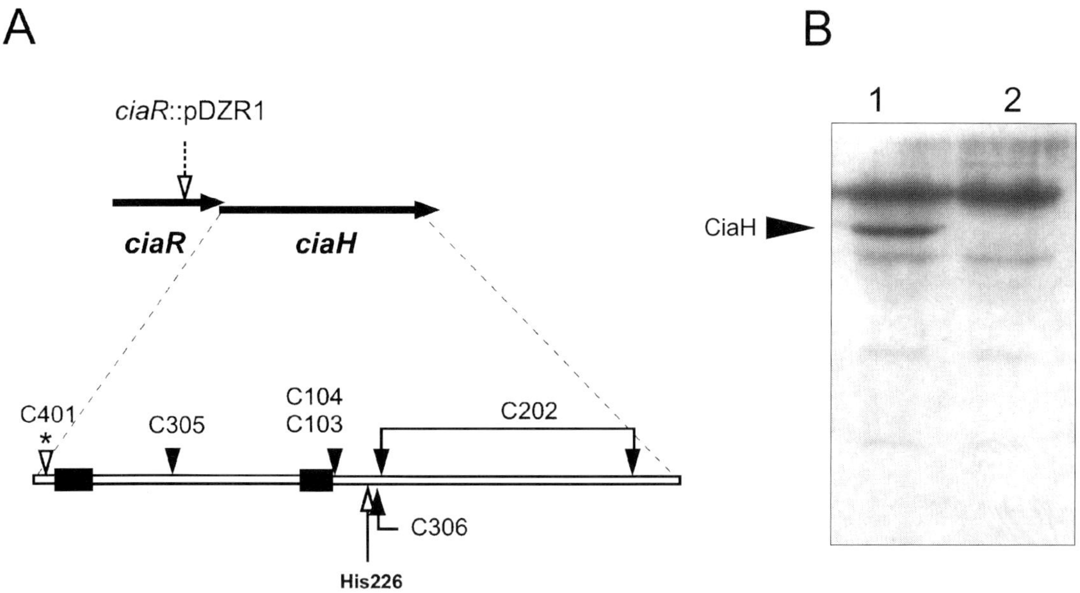

Figure 1. Mutations in the *cia* two component system. A. Insertion duplication mutagenesis of *ciaR* with the plasmid *ciaR*::pDZR1 results in a truncated *ciaR* gene as indicated by the open arrow (top). The positions of amino acid alterations in the histidine protein kinase CiaH in cefotaxime resistant laboratory mutants are indicated below (see Table 1). The star marks the stop codon introduced in mutant C401. Putative transmembrane regions are indicated by black boxes, the conserved His226 is marked. B. Western blot of cell lysates of the parental R6 strain and the mutant with a disrupted *ciaR* gene as depicted in A. A rabbit antiserum raised against a purified CiaH-derivative covering the N-terminal domain amino acids 48–182 was used at a dilution of 1:10.000. The arrow indicates the position of intact CiaH.

Roberts, 1994), and this conclusion was reached recently by comparing transcript profiles in wildtype and via mutants (Mascher *et al.*, in press). The data also suggest that CiaR can be phosphorylated via a CiaH-independent pathway, e.g. crosstalk with another protein kinase. This crosstalk mechanism becomes visible in mutants with disrupted *ciaH* gene, that are β-lactam resistant despite the fact that no signal can be transduced through the CiaH kinase; and that are non transformable in combination with a *pbp2x* mutation. In other words, signals are generated by the antibiotic, or via the PBP2x mutation, that result in increased CiaR phosphorylation in the absence of CiaH as shown in the model in Figure 2.

It has been reported previously that the competence defect in *ciaH* mutants could not be restored with competence factor (CF) (Giammarinaro *et al.*, 1999; Zähner *et al.*, 1996). In these papers, "CF" referred to a crude preparation of solubilized material from competent R6 cells prepared as described by Tomasz and Mosser (Tomasz and Mosser, 1966). Meanwhile, CF has been identified as a small unmodified pheromone peptide termed CSP (competence stimulating peptide), and in vitro synthesized CSP is fully active in competence induction (Håvarstein *et al.*, 1995). Using CSP at concentrations of 1–10 ng/ml, the competence defect of every *ciaH* mutant could be complemented up to 30% in terms of transformation efficiency when compared to that of the R6 strain. Most likely, the concentration of peptide in the crude preparations used earlier was not sufficient to induce competence in the *ciaH* mutants. The data indicate that *ciaH* mutants are affected at an early step of the competence pathway, involving either the synthesis or the secretion of CSP (Hakenbeck *et al.*, 1999). Similar conclusions were reached recently where it was shown that *comCDE* expression paralleled derepression of competence in mutants with disrupted *cia* genes, convincingly confirming the link between the *cia* system and competence (Echenique *et al.*, 2000; Martin *et al.*, 2000).

The *cia* system may thus be visualized as a "super-control" mechanism for transformation in *S. pneumoniae* which in itself is a complex regulation system. A two component regulatory system is required for CSP sensing, resulting in autocatalytic activation of the *comCDE* operon by the response regulator ComE which also induces "late" competence genes that are involved in DNA uptake and recombination (for review, see (Morrison 1997)). The regulatory network between the *ciaRH* and the *comCDE* system is schematically shown in Figure 2.

What is the signal for the *cia* system? Curiously, the $ciaH_{C306}$ allele conferred not only a change in cefotaxime but also in vancomycin susceptibility, resulting in a twofold increase in resistance to both drugs. Giammarinaro *et al.* did not detect differences in vancomycin susceptibility in mutant versus wildtype, but it is possible that the concentration range tested did not allow detection of a twofold difference, and experimental details were not reported (Giammarinaro *et al.*, 1999). We therefore assume that the CiaH kinase plays a role as sensor for cell wall damage rather than in targeting the specific antibiotic. In addition to specific cell wall associated effects, the *cia* system apparently also responds to changes in the growth medium such as Ca^{++} and phosphate concentration (Giammarinaro

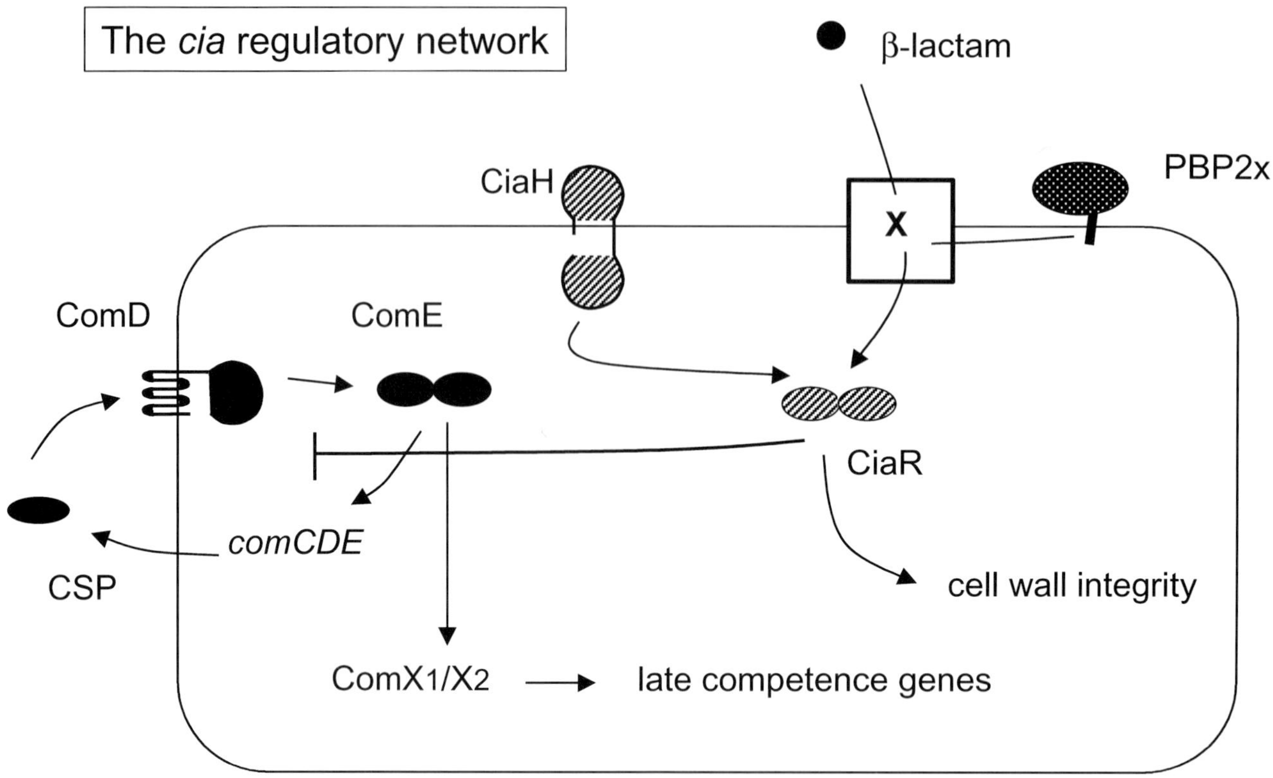

Figure 2. Model of the regulatory network between the *ciaR*/*ciaH* and *comD*/*comE* two component systems. CSP is the processed, exported *comC* gene product; X depicts an unknown link that is responsible for CiaH independent phosphorylation of CiaR in the presence of β-lactam antibiotics and in PBP2x mutants.

et al., 1999), oxygen (Echenique *et al.*, 2000), or is required for growth in complex medium (Lange *et al.*, 1999). It is possible that alterations in cell surface properties induced under these conditions, rather than specific molecules, activate the *cia* system. Direct sensing of phosphate or calcium seems unlikely to us, since BLAST searches conducted with the extracytoplasmic sensor domain alone did not reveal any significant similarities to PhoQ kinases except for the two putative transmembrane domains.

Toward Identification of the *cia* Regulon

In order to identify the genes controlled by the *ciaRH* regulatory system, the DNA binding property of the C-terminal domain of response regulators such as CiaR was exploited using the solid-phase DNA binding (SPDB) assay described by Roth and Messer for analyzing the DNA binding specificity of *E. coli* DnaA (Roth and Messer, 1995). In this assay, the gene encoding the DNA binding is fused to a gene encoding the subunit of the transcarboxylase complex of *Propionibacterium shermanii* that becomes biotinylated when expressed in *E. coli*. The fusion protein can easily be extracted from crude cell lysates using streptavidin coupled magnetic beads. Consecutively, DNA fragments that specifically interact with the DNA binding domain can be isolated after incubation of the fusion protein with restricted DNA (in the present case with restricted chromosomal DNA of *S. pneumoniae R6*), followed by washing of the DNA-protein complex in order to remove unspecifically bound DNA fragments, and elution of specifically bound DNA in buffer containing high salt concentration (1.3 M NaCl; Mascher, *et al.*, in press). Only small amounts of DNA could be obtained this way, and characterization of fragments required amplification, cloning of the fragments, and confirmation of specific interaction with the response regulator using the cloned DNA fragment in the presence of competitor DNA in the binding assay (Zähner, 1999).

In addition to CiaR, the ComE protein which belongs to a different class of response regulators (Grebe and Stock, 1999) was also used in this approach (Kaminski, 1999). Four and five DNA fragments were isolated from the pool of the SPDB-eluted fragments and were shown to specifically interact with CiaR and ComE, respectively (see Table 2).

ComE regulates the expression of the *comC* gene via interacting with a region containing a direct repeat upstream *comC* (Ween *et al.*, 1999). The fact that one of the five DNA fragments identified as specifically interacting with ComE was indeed a 1.4 kb *Sau*3AI DNA fragment that included the 5′-end of *comD*, *comC* and upstream regions confirms that the binding assay represents a valid method for identifying target sequences. However, it should be pointed out that binding also occurred to a subfragment located upstream *comD* that did not contain the direct repeat

Table 2. Putative target regions for CiaR and ComE.

Fusion protein[1]	Flanking gene (subfragment) homologue (organism)		Gene annotation[2]	Remarks[3]
CiaR[1]				
	uppS	(161 bp)	spr 0240	
	undecaprenyl-pyrophosphate synthetase			
	hypothetical	(474 bp)	spr 1281	
	membrane spanning protein (ABC transporter)			
	hypothetical	(249)	spr 1280	(–248 bp)
	membrane spanning protein (ABC transporter)			
	cylM	(385 bp)	spr 1767	
		(375 bp)	spr 1767	
	cytolysin (*E. faecalis*)			
	RUP	(281)	upstream tRNA-Pro2	
	repetitive element – tRNA genes			
	hypothetical[4]	(832)	spr 0703	(–780 bp)
ComE[1]				
	bglG	(453)	spr 0279	
	transcription antiterminator BlgG family			
	recN	(373)	spr 1084	(–60 bp)
	DNA repair and recombination			
	dnaE	(381)	spr 1556	
	chromosome replication initiation protein			
	clyB/nisP	(1233)	spr 1770/spr 1771	
	ABC transporter/subtilisin like serine protease (*E. faecalis*)			
	comC	(596)	spr 2043	
	competence stimulating peptide precursor			
	comD	(398)	spr 2042	
	histidine protein kinase			

[1] Fusion protein used in SPDB assay.
[2] Gene name as listed for the *S. pneumoniae* R6 genome (Hoskins *et al.*, 2001).
[3] Regions missing in the *S. pneumoniae* KNR_7/87 genome that are present in the R6 strain. Differences of single basepairs between the two strains are not listed.
[4] Identified on a 4.8 kb PCR fragment upstream *ciaR* in the R6 strain.

previously identified as the ComE specific DNA binding motif (Ween *et al.*, 1999), and none of the other three ComE interacting fragments contained the repeat, suggesting that the nonphosphorylated ComE might bind to other DNA motifs as well.

For all fragments, subfragments obtained either by restriction endonuclease digestion or by PCR amplification were tested to identify smaller regions that still retained binding specificity. A list of the loci associated with the smallest DNA fragments that specifically interact with either the CiaR or the ComE-fusion protein are shown in Table 2. These fragments included the 5'-end of a gene and upstream regions except for one region downstream of a tRNA cluster that specifically interacted with CiaR. Among the CiaR-interacting DNA fragments was also an intergenic region upstream of the *uppS* gene encoding the undecaprenyl-pyrophosphate synthetase, a key enzyme in cell wall polysaccharide biosynthesis (Apfel *et al.*, 1999). Other regions with both, CiaR and ComE binding sites, included a potential bacteriocin locus related to the *Enterococcus faecalis* cytolysin region which is located on a potential mobile element (Hakenbeck *et al.*, 2001). Of particular interest are also fragments in the vicinity of putative regulatory elements such as the *celR*-homologue in case of ComE. Mean-while, the *cia*-dependent regulation of specific genes has been confirmed by comparing global expression pattern in wild type and *cia*-mutant cells (Mascher, *et al.*, in press).

The genes located on the DNA fragments identified thus far represent only part of the potential CiaR respectively ComE regulated genes due to the methodology used for isolation of the DNA fragments. The ability to isolate a library of DNA fragments that specifically interact with DNA binding transcriptional regulators, combined with the availability of the *ciaH* mutants described above for transcriptome analyses, provide a powerful basis for future investigations. The age of genomics presents us with tools such as DNA microarrays suitable for accessing regulatory networks that will help us to understand the role of the *cia* regulatory system.

Acknowledgement

This work was supported by the DFG and the Schwerpunkt Biotechnologie Rheinland Pfalz.

References

Apfel, C.M., Takács, B., Fountoulakis, M., Stieger, M., and Keck, W. 1999. Use of genomics to identify bacterial undecaprenyl pyrophosphate synthetase: cloning, expression, and characterization of the essential *uppS* gene. J. Bacteriol. 181: 483–492.

Chen, J.-D., and Morrison, D.A. 1987. Cloning of *Streptococcus pneumoniae* DNA fragments in *Escherichia coli* requires vectors protected by strong transcriptional terminators. Gene 55: 179–187.

Echenique, J.R., Chapuy-Regaud, S., and Trombe, M.C. 2000. Competence regulation by oxygen in *Streptococcus pneumoniae*: involvement of *ciaRH* and *comCDE*. Mol. Microbiol. 36: 688–696.

Forst, S.A., and Roberts, D.L. 1994. Signal transduction by the EnvZ-OmpR phosphotransfer system in bacteria. Res. Microbiol. 145: 363–373.

Giammarinaro, P., Sicard, A.M., and Gasc, A.-M. 1999. Genetic and physiological studies of the CiaH-CiaR two-component signal-transducing system involved in cefotaxime resistance and competence of *Streptococcus pneumoniae*. Microbiology 145: 1859–1869.

Grebe, T., Paik, J., and Hakenbeck, R. 1997. A novel resistance mechanism for β-lactams in *Streptococcus pneumoniae* involves CpoA, a putative glycosyltransferase. J. Bacteriol. 179: 3342–3349.

Grebe, T., and Stock, J.B. 1999. The histidine protein kinase superfamily. Adv. Microb. Phys. 41: 140–227.

Guenzi, E., Gasc, A.M., Sicard, M.A., and Hakenbeck, R. 1994. A two-component signal-transducing system is involved in competence and penicillin susceptibility in laboratory mutants of *Streptococcus pneumoniae*. Mol. Microbiol. 12: 505–515.

Guenzi, E., and Hakenbeck, R. 1995. Genetic competence and susceptibility to β-lactam antibiotics in *Streptococcus pneumoniae* R6 are linked via a two-component signal-transducing system. pp. 125–128 in J.J. Ferretti, M.S. Gilmore, T.R. Klaenhammer, and F. Brown (eds.) Genetics of Streptococci, Enterococci and Lactococci. S. Karger AG, Basel.

Hakenbeck, R., Balmelle, N., Weber, B., Gardes, C., Keck, W., and de Saizieu, A. 2001. Mosaic genes and mosaic chromosomes: intra- and interspecies variation of *Streptococcus pneumoniae*. Infect. Immun. 69: 2477–2486.

Hakenbeck, R., Grebe, T., Zähner, D., and Stock, J.B. 1999. β-Lactam resistance in *Streptococcus pneumoniae*: penicillin-binding proteins and non penicillin-binding proteins. Mol. Microbiol. 33: 673–678.

Hakenbeck, R., Tornette, S., and Adkinson, N.F. 1987. Interaction of non-lytic β-lactams with penicillin-binding proteins in *Streptococcus pneumoniae*. J. Gen. Microbiol. 133: 755–760.

Håvarstein, L.S., Coomaraswamy, G., and Morrison, D.A. 1995. An unmodified heptadecapeptide pheromone induces competence for genetic transformation in *Streptococcus pneumoniae*. Proc. Natl. Acad. Sci. USA 92: 11140–11144.

Hoskins, J., Alborn, W.E. Jr., Arnold J., Blaszczak, L.C., Burgett, S., DeHoff, B.S., Estrem, S.T., Fritz, L., Fu, D.J., Fuller, W., Geringer, C., Gilmour, R., Glass, J.S., Khoja, H., Kraft, A.R., Lagace, R.E., LeBlanc, D.J., Lee, L.N., Lefkowitz, E.J., Lu, J., Matsushima, P., McAhren, S.M., McHenney, M. McLeaster, K., Mundy, C.W., Nicas, T.I., Norris, F.H., O'Gara, M., Peery, R.B., Robertson, G.T., Rockey, P., Sun, P.M., Winkler, M.E., Yang, Y., Young-Bellido, M., Zhao, G., Zook, C.A., Baltz, R.H., Jaskunas, S.R., Rosteck, P.R.J., Skatrud, P.L., and Glass, J.I. 2001. Genome of the bacterium *Streptococcus pneumoniae* strain R6. J. Bacteriol. 183: 5709–5717.

Kaminski, K. 1999. Genetische Kompetenz in *Streptococcus pneumoniae* und *Streptococcus* spp.: Charakterisierung des *comCDE* Regulons. PhD Thesis, Humboldt University.

Laible, G. and Hakenbeck, R. 1987. Penicillin-binding proteins in β-lactam-resistant laboratory mutants of *Streptococcus pneumoniae*. Mol. Microbiol. 1: 355–363.

Lange, R., Wagner, C., de Saizieu, A., Flint, N., Molnos, J., Stieger, M., Caspers, P., Kamber, M., Keck, W., and Amrein, K.E. 1999. Domain organization and molecular characterization of 13 two-component systems identified by genome sequencing of *Streptococcus pneumoniae*. Gene 237: 223–234.

Martin, B., Prudhomme, M., Alloing, G., Granadel, C., and Claverys, J.-P. 2000. Cross-regulation of competence pheromone production and export in the early control of transformation in *Streptococcus pneumoniae*. Mol. Microbiol. 38: 867–878.

Mascher, T., Zähner, D., Merai, M., Balmelle, N., de Saizieu, A., and Hakenbeck, R. 2002. The *Streptococcus pneumoniae cia* regulon: CiaR target sites and transcription profile analysis. J. Bacteriol.: in press.

Morrison, D.A. 1997. Streptococcal competence for genetic transformation: regulation by peptide pheromones. Microb. Drug Resist. 3: 27–37.

Roth, A., and Messer, W. 1995. The DNA binding domain of the initiator protein DnaA. EMBO J. 14: 2106–2111.

Tettelin, H., Nelson, K.E., Paulsen, I.T., Eisen, J.A., Read, T.D., Peterson, S., Heidelberg, J., DeBoy, R.T., Haft, D.H., Dodson, R.J., Durkin, A.S., Gwinn, M., Kolonay, J.F., Nelson, W.C., Peterson, J.D., Umayam, L.W., White, O., Salzberg, S.L., Lewis, M.R., Radune, D., Holtzapple, E., Khouri, H., Wolf, A.M., Utterback, T.R., Hansen, C.L., McDonald, L.A., Feldblyum, T.V., Angiuoli, S., Dickinson, T., Hickey, E.K., Holt, I.R., Loftus, B.J., Yang, F., Smith, H.O., Vetern, J.C., Dougherty, B.A., Morrison, D.A., Hollingshead, S.K., and Fraser, C.M. 2001. Complete genome sequence of a virulent isolate of *Streptococcus pneumoniae*. Science 293: 498–506.

Tomasz, A., and Mosser, J.L. 1966. On the nature of the pneumococcal activator substance. Proc. Natl. Acad. Sci. USA 55: 58–66.

Ween, O., Gaustad, P., and Håvarstein, L.S. 1999. Identification of DNA binding sites for ComE, a key regulator of natural competence in *Streptococcus pneumoniae*. Mol. Microbiol. 33: 817–827.

Zähner, D. 1999. Identifizierung von Zielgenen des signaltransduzierenden Zwei-Komponenten-Systems *cia* von *Streptococcus pneumoniae*. PhD Thesis, Universität Kaiserslautern.

Zähner, D., Grebe, T., Guenzi, E., Krauß, J., van der Linden, M., Terhune, K., Stock, J.B., and Hakenbeck, R. 1996. Resistance determinants for β-lactam antibiotics in laboratory mutants of *Streptococcus pneumoniae* that are involved in genetic competence. Microb. Drug Resist. 2: 187–191.

Chapter 9

Members of the Fur Protein Family Regulate Iron and Zinc Transport in *E. coli* and Characteristics of the Fur-Regulated FhuF Protein

Klaus Hantke

Abstract

The regulator Fur represses with Fe^{2+} as cofactor iron uptake genes. The *fhuF* gene reacts very sensitive to minor changes of Fe^{2+} and Fur. It is assumed that FhuF helps in the mobilisation of iron out of the hydroxamate siderophores transported into the cell. Analysis of the protein revealed an unusual [2Fe-2S] cluster bound to a Cys–Cys–X_{10}–Cys–X_2–Cys motif in FhuF. *suf* genes responsible for the synthesis of the iron sulfur center were identified.

The Zur protein shows 27% identity to the Fur protein of *E. coli*. It regulates as a repressor the high affinity uptake system *znuACB*. Only two additional Zur binding sites in the promoter region of genes with unknown function were found. Properties of Zur and Fur proteins from different bacteria are compared.

The Fur Protein

The Fur protein of *E. coli* was detected as a regulator of iron transport and siderophore biosynthesis. Later it was found out that Fur in addition regulated genes of the oxidative stress response, iron metabolism, and virulence determinants from pathogenic bacteria (Hantke and Braun, 2000). The protein was isolated and characterised in different laboratories. It acts mainly as a repressor with Fe^{2+} as corepressor. The N-terminal domain contains an unusual helix-turn-helix motif necessary for binding to a palindromic DNA sequence operator site also called Fur-box which is mostly found in the promoter region of iron regulated genes (Figure 1). The C-terminal half of Fur contains a little histidine cluster and four cysteines two of which are binding zinc. The regulatory Fe^{2+} is bound by histidine and aspartate or glutamate.

Fur-like proteins were detected in gram negative and gram positive bacteria – and most of them were regulating genes for iron uptake. However, in *E. coli* a second Fur-like protein was found which turned out to be Zur, a zinc dependent repressor (see below). In the genome sequence of *Bacillus subtilis* even three *fur* like genes were found, one called *fur* regulating iron uptake, the second one called *perR* regulating oxidative stress response and the third one called *zur* regulated zinc uptake genes (Gaballa and Helmann, 1998).

FhuF, a Gene Regulated by Fur

Important for the study of iron regulation was the development of the reporter gene technology by Casadaban and Cohen (1978). The *lac* gene expression is put under the control of the promoter of the gene of interest. The regulation of this gene under different growth conditions, for instance +/– iron, can be seen directly on MacConkey lactose plates as an indicator of β-galactosidase activity. The *lac* gene under control of a promoter of an iron uptake gene responds to low iron growth conditions by strong expression – the colonies are red on MacConkey lactose plates – with high iron the colonies are white. An iron regulated *lacZ* operon fusion on the chromosome allowed to test for Fur-boxes on plasmids. The high copy number of Fur binding sites on the plasmid titrates the Fur proteins in the cell and leads to a derepression of the *lac* gene and red colonies appear on MacCokey lactose iron plates (Stojiljkovic *et al.*, 1994). One gene fusion routinely used for the Fur titration assay was *fhuF-lacZ* since it was extremely sensitive to even small perturbations of the iron and Fur level in the cell.

The phenotype of *fhuF* mutants was a diminished utilisation of the siderophore ferrioxamine B on low iron plates. Ferrioxamine B is an inferior iron source for *E. coli* – iron regulated genes are derepressed when ferrioxamine B is the only iron source for *E. coli*. The reason may be that *E. coli* K-12 does not have an outer membrane receptor for ferrioxamine B like *Yersinia enterocolitica* (Bäumler and Hantke, 1992). In addition, it was shown that ferrioxamine B is a poor substrate for the hydroxamate binding protein FhuD which is part of the inner membrane ABC transporter for hydroxamates (Rohrbach *et al.*, 1995). From these observations and considerations FhuF was expected to be a minor component in the hydroxamate uptake system. So it was a surprise when the FhuF protein was found to be a cytoplasmic protein.

Characterization of FhuF

The *fhuF* gene was identified as open reading frame f262b at 99.2 min on the genome sequence map of *E. coli* K-12. The gene was cloned and the FhuF protein was labeled with a His-tag and then purified to

Structure predictions for the Fur protein

```
                                helix                                  helix
                          ΛΛΛΛΛΛΛΛΛΛΛ                           ΛΛΛΛΛΛΛΛΛΛ
MTDNNTALKKAGLKVTLPRLKILEVLQEPDNHHVSAEDLYKRLIDMGEE    49

           helix           beta-sheet      beta-sheet
TURN      ΛΛΛΛΛ            ⇒⇒⇒⇒            ⇐⇐⇐⇐
IGLATVYRVLNQFDDAGIVTRHNFEGGKSVFELTQQHHHDHLICLDCGK    98
                                                |  |
                                                 Zn

VIEFSDDSIEARQREIAAKHGIRLTNHSLYLYGHCAEGDCREDEHAHEGK 148
```

Fur-box: GATAATGATAATCATTATC

Figure 1. Structure predictions for the Fur protein according to Holm *et al.*, 1994 and the zinc binding site (Jaquamet *et al.*, 1998), the Fe^{2+} binding sites have not been defined. The Fur-box, the consensus sequence of Fur binding sites on the DNA is given. The larger the size of the letters the better conserved are the bases (Stojiljkovic *et al.*, 1994).

electrophoretic homogeneity. Based on sulfur determinations and UV/VIS, Mössbauer and EPR spectroscopy, FhuF was identified as a [2Fe-2S] protein. The g values ($g_x = 1.886$, $g_y = 1.961$, $g_z = 1.994$) and some of the Mössbauer parameters of FhuF obtained [oxidized protein as isolated: $\Delta E_{Q,4.2K} = 0.474$ mm s^{-1}; Fe^{3+} (reduced protein): $\Delta E_Q = 0.978$ mm s^{-1}] are not typical of common [2Fe-2S] proteins and indicate that FhuF has unusual structural properties (Müller *et al.*, 1998). The primary sequence of FhuF does not show any sequence similiarities to known [2Fe-2S] proteins. By site-directed mutagenesis, each of the six cysteines of FhuF was replaced by serine. EPR of the six reduced mutant proteins revealed that the terminal cysteine residues 244, 245, 256, and 259 form the [2Fe-2S]cys_4 cluster. Mutants having the Cys-to-Ser replacement at positions 244, 245, 256, or 259 did not complement a *fhuF* mutant (Müller *et al.*, 1998). The motif Cys–Cys–X_{10}–Cys–X_2–Cys in FhuF differs considerably from the motif Cys–X_2–Cys–X–$_{9-15}$–Cys–X_2–Cys found in other [2Fe-2S] proteins. The unusual Cys–Cys terminal group of the cluster may explain the atypical EPR and Mössbauer spectroscopic properties of the FhuF protein; possibly the tetrahedral symmetry at the ferric ion site is distorted.

This result leads to a new interpretation of the function of FhuF in ferrioxamine B uptake and utilisation. Transport of Fe^{3+} via iron binding siderophores has been studied in many bacteria down to the molecular level (Braun *et al.*, 1998). However, the release of iron from the siderophore is less well known. It is assumed that Fe^{3+} is reduced and indeed, in some cases the siderophore transported iron was found 30 minutes later in the Fe^{2+} pool of the cells (Matzanke *et al.*, 1991).

Many siderophore reductases have been found in *in vitro* assays but the purified proteins turned out to be flavin reductases with other, well defined functions. No specialised siderophore-iron reductases have been identified. It has been suggested that the reduced flavins from flavin oxidoreductases are the electron donors for ferric iron reduction (Fontecave *et al.*, 1994). The phenotype of *fhuF* mutants and the structural features of the FhuF protein suggest that FhuF is involved in the reduction and/or mobilisation of iron out of the claws of ferrioxamine B and other hydroxamate siderophores (Hantke, unpublished).

Synthesis of the Iron Cluster in FhuF

In an attempt of further defining the function of FhuF in ferrioxamine B uptake *E. coli* derivatives with the phenotype of *fhuF* mutants were selected. Two mutants were identified and characterised. The genes were named *sufD* and *sufS*. The two genes were part of the Fur regulated *suf* operon of six genes.

The protein SufS belongs to the family of NifS-like cysteine desulfurases, which liberate sulfur from cysteine and provide it for the [Fe-S] cluster synthesis (Zheng *et al.*, 1998; Agar *et al.*, 2000). In *E. coli* three NifS-like proteins are found, IscS, SufS and Csd. In many bacteria and even in mitochondria IscS and its homologs are part of a complex machinery for the synthesis of [Fe-S] centres and the maturation of [Fe-S] proteins (Lill and Kispal, 2000). The *isc* gene cluster in *E. coli* encodes IscU, a scaffold for the synthesis of an [Fe-S] cluster, heat shock proteins, a ferredoxin and some other proteines with less well characterised properties. The *suf* gene cluster may fulfil similar functions, however, their genes have not been characterised and only two of them are similar to proteins from other *isc* gene clusters. The T7 expression system and a His-tag allow the isolation in good yield of the FhuF protein from a wild-type strain. In contrast, overproduction of the protein in a Δ*sufD* strain failed. Radioactive labeling of N-His-FhuF with [^{35}S]methionine showed that the protein was unstable in the Δ*sufD* mutant (Patzer and Hantke, 1999). These observations indicate that the *isc* gene cluster is not the only responsible for [Fe–S] centre synthesis in *E. coli*. From our observations it seems that the mature FhuF protein can only be synthesised with the help of *suf* gene products. In addition it is interesting to note that the inability to mature FhuF labilised the protein in the cell.

High Affinity Zinc Uptake

During our studies on iron and Fur regulated genes *lacZ* operon fusions were isolated that were derepressed under iron repletion and repressed under iron depletion. Two fusions were localized in genes that formed an operon, whose gene products had characteristics of a binding-protein-dependent transport system, which had low similarity to a Mn^{2+} transporter in *Synechocystis* spc. (Bartsevich and Pakrasi, 1995). The growth defect of these mutants on TY medium containing 5 mM EGTA was compensated for by the addition of Zn^{2+}. In the presence of 0.5 mM EGTA, only the parental strain was able to take up $^{65}Zn^{2+}$. This high-affinity transport was energized by ATP. The genes were named *znuABC* (for *z*i*n*c *u*ptake; former name *yebLMI*) and localized at 42 min on the genetic map of *E. coli*. At high Zn^{2+} concentrations, the *znu* mutants took up more $^{65}Zn^{2+}$ than the parental strain (Patzer and Hantke, 1998).

Regulation of High Affinity Zinc Uptake by Zur

The high-affinity $^{65}Zn^{2+}$ uptake was repressed by growth in the presence of 10 μM Zn^{2+}. A *znuA-lacZ* operon fusion was repressed by 5 μM Zn^{2+} and showed a more than 20-fold increase of β-galactosidase activity when Zn^{2+} was bound to TPEN [tetrakis-(2-pyridylmethyl) ethylenediamine].

In order to identify the Zn^{2+}-dependent regulator, constitutive mutants were isolated and tested for complementation by a gene bank of *E. coli*. A complementing gene, *yjbK* of the *E. coli* genome, was identified and named *zur* (for *z*inc *u*ptake *r*egulation). The Zur protein showed 27 % sequence identity with the iron regulator Fur. High-affinity $^{65}Zn^{2+}$ transport of the constitutive *zur* mutant was tenfold higher than that of the uninduced parental strain (Patzer and Hantke, 1998). In additon to Zur regulating zinc uptake, there is a zinc dependent MerR type regulator for ZntA, the P-type ATPase, which exports zinc (Brocklehurst *et al.*, 1999). This system operates at concentrations above 0.5 mM Zn^{2+} while the Znu uptake system is induced below 10 μM Zn^{2+}.The two component sensor/regulator system ZraS and ZraR senses high zinc in the periplasm and regulates the periplasmic zinc binding protein ZraP (Leonhartsberger *et al.*, 2001).

By inactivation of the *zur* gene, it was demonstrated that Zur acts as a repressor and not as an activator. Eight chromosomal mutant *zur* alleles were sequenced to correlate the loss of Zur function with individual mutations. The purified proteins of wild-type Zur and of ZurΔ46–91 formed homo- and heterodimers. Dimerization was independent of metal ions since it also occurred in the presence of metal chelators (Patzer and Hantke, 2000). Using an *in vivo* titration assay, the site affording Zur regulation was narrowed down to a 31-bp region in the promoter region of *znuA* and *znuCB*. This location was confirmed by DNase I footprinting assays. A single region comprising a nearly perfect palindrome was protected, which indicated direct binding of Zur. Zinc chelators completely inhibited DNA binding of Zur, and addition of Zn^{2+} in low concentrations enhanced binding. Zur occupied its binding site only in the presence of zinc or other divalent metal cations at low concentrations, as shown by the DNase I footprinting data. Zur protected a 29-nt approximate palindrome on each strand of the *znu* operator with a 3′ stagger of 4 nt (Patzer and Hantke, 2000). This footprint resembles that of typical DNA binding dimers, such as classical helix-turn-helix proteins, e.g., the CI repressor from bacteriophage λ (Jordan and Pabo, 1988). The observed 3′ stagger is indicative for coverage of the minor groove at the ends, but provides no information about the protein–DNA recognition contacts. Analysis of the mutant Zur proteins suggested an amino-terminal DNA contact domain around residue 65 and a carboxy-terminal dimerization and Zn^{2+}-binding domain. Footprinting experiments indicated that although most of the mutant Zur proteins bound to the *znu* promoter *in vitro*, no protection was observed *in vivo* (Patzer and Hantke, 2000).

Characteristics of Zur in Comparison to Fur

Zur just as Fur seems to be widespread among bacteria, even in gram-positive bacteria and cyanobacteria, as indicated by sequence similarity searches. However, only in some cases a functional zinc dependent regulator was shown. Since small changes in the sequence may change the metal specificity one always has to check the prediction. In *B. subtilis,* apart from Fur and PerR (regulator of the peroxide stress response) (Bsat *et al.*, 1998), the third Fur-like homologue YqfV can act as Zur (Gaballa and Helmann, 1998). The sequence similarity to *E. coli* Zur is not strong (Figure 2). A *zur* gene has also been found in *L. monocytogenes* (Dalet *et al.*, 1999). Based on sequence similarity, we expect that the proposed Fur protein in *Staphylococcus epidermidis* (Heidrich *et al.*, 1996) is Zur rather than Fur. It is possible that other proteins designated as Fur homologues will turn out to be Zur proteins. In many of the partially sequenced genomes of various bacterial species, a Zur equivalent is found, e.g., in *Salmonella* strains, *Klebsiella pneumoniae*, *Yersinia pestis*, *Vibrio cholerae*, *Bordetella pertussis*, *Caulobacter crescentus*, *Pseudomonas aeruginosa*, and *Neisseria* strains. Since these organisms also possess a system homologous to Znu, we propose that these proteins likewise are regulators of zinc uptake. Some of these Zur-like proteins are aligned in Figure 2. In the dendrogram two groups of

```
ZUR_ECOLI    MEKTTT---Q-------------ELLAQAEKICAQRNVRLTPQRLEVLRLMSLQD    39
ZUR_YERPE    MNMMNPINQE-------------KLLAQAESLCQQRNVRLTPQRLEVLRLMAQQP    42
ZUR_VIBCH    MVM---VLDN-------------KLTKQIEEICAARGVRLTSQRKTVFELICASK    39
ZUR_PSEAE    MYKIAPKTPLAC-QPHDHSQCVSTALAEADALCARQGVRLTELRRRVLELVWQSH    54
ZUR_XYLFA    M-----SNDHGCTAPHHHVHDAPEFIHAVERACSERGLRLTPIRANVLQLIAHAG    50
ZUR_NEIGO    MK--------------------TNFKQKIIEQARREDLQVTALREQVLDIVLQQS    35
             *                      ^            ^^^^*  *  *^ ^^
Conserved    M                                   ^ ^ T  R  ^^ ^
ZURR_LISMO   MGLTAT---------------------EALMKMKEKGYKHTDKREFLINLLARKN    34
ZUR_BACSU    MNVQ-----------------------EALNLLKGNGYKYTNKREDMLQLFADSD    32
FUR_STAEP    MN--TN---------------------DAIKILKDNGLKYTDKRKDMLDIFVEED    32
FUR_ECOLI    M-------------------------TDNNTALKKAGLKVTLPRLKILEVLQEPD    30

ZUR_ECOLI    GAISAYDLLDLLREAE--PQAKPPTVYRALDFLLEQGFVHKVESTNSYVLCHLFD    92
ZUR_YERPE    GAISAYDLLDLLRVAE--PQAKPPTVYRALDFLLEQGFIHRVESANSYVLCHHFE    95
ZUR_VIBCH    RSSSAYELLEDLKKSE--PQAKPPTVYRALDFLLEQGFIHRVESTNSFISCCSCN    92
ZUR_PSEAE    KPLGAYDILA-VLSETDGRRAAPPTVYRALDFLQENGLVHRIASLNAFVGCNNPE   108
ZUR_XYLFA    KPVKAYELLEWVREGKDVGANAPPTVYRALDFLMTHGFIHKLESINAFIACHHPN   105
ZUR_NEIGO    GVIKAYNVLSQMQQQSEGVLA-PPTAYRALDFWADQGVLHKVAAVNGYILCSHAQ    89
                 **^^*^ ^   ^    ^ ***^******  ^^* ^*^^^^ *^^^ *   ^
Conserved        A ^^   ^   ^        T YR^L ^  ^ G  ^  ^  ^^
ZURR_LISMO   KYLTAKDVLENMK--DDFPGISFDTIYRNLSLFVELGIFEETDLSGE----RNFR    83
ZUR_BACSU    RYLTAKNVLSALN--DDYPGLSFDTIYRNLSLYEELGILETTELSGE----KLFR    81
FUR_STAEP    KYLNAKHIQQKMD--KDYPGISFDTVYRNLHLFKDLGIIESTELEGE----MKFR    81
FUR_ECOLI   NHHVSAEDLYKRLI--DMGEEIGLATVYRVLNQFDDAGIVTRHNFEGG----KSVF    80

ZUR_ECOLI    Q-------------PTHTSAMFICDRCGAVKEECAEGVEDIMHTLAAKMGFALRH   134
ZUR_YERPE    E-------------PTHTSALFICDRCKIVTERPTVGIEEALAQLAKQSGFTLRH   137
ZUR_VIBCH    A-------------HKHFSHLLICDQCGNVIESRDDSLVALLASNAEKHGFTISN   134
ZUR_PSEAE    H-------------SHQGQ-FLICRTCHTAIELEQPDISRAIVAGANSVGFAVES   149
ZUR_XYLFA    S-------------SRHSAPFLICDKCHNAVELEDKDVVAQLNACARALGFQPQM   147
ZUR_NEIGO    HECNDHCHDHEEAEAHHSAFILVCTECGAADEQTLSHEWAALRAGVAESGFALKE   144
                             ^   ^^^*  *  ^ *         ^   ^   **
Conserved                         ^^C  C  ^                   ^
ZURR_LISMO   LACTHEHHHHH----------FICMKCGKTKEIMMCPMDFLTEALP---GYQIDG   125
ZUR_BACSU    FKCSFTHHHHH----------FICLACGKTKEIESCPMDKLCDDLD---GYQVSG   123
FUR_STAEP    IACT-NHHHHH----------FICENCGDTKVIDFCPIEQIKQYLP---NVTIHT   122
FUR_ECOLI    -ELTQQHHHDH----------LICLDCGKVIEFSDDSIEARQREIAAKHGIRLTN   123

ZUR_ECOLI    NVIEAHGLCAA--CVEVEACRHPEQCQHDHSVQVKKKPR   171
ZUR_YERPE    SVVEAHGLCTE--CGVVEACESHDDCEHDHSIVVKKK--   172
ZUR_VIBCH    HVIESHGVCQT--CLSKE-----------------KKQQ   154
ZUR_PSEAE    QTVEVVGLCGT--CRD----------QKDA---------   167
ZUR_XYLFA    QTLEVHGICAR--CAA----------QHDPTATV-----   169
ZUR_NEIGO    EHVVLTGICKK--C-----------------------QK   158
               ^   *^*    *
Conserved          G C    C
ZURR_LISMO   HKFEVYGECPE--CLQ---------------------AS   141
ZUR_BACSU    HKFEIYGTCPD--CTA----ENQEN-----------TTA   145
FUR_STAEP    HKLEVYGVCES--CQ---------------------KNA   138
FUR_ECOLI    HSLYLYGHCAEGDCREDEHAHEGK                  148
```

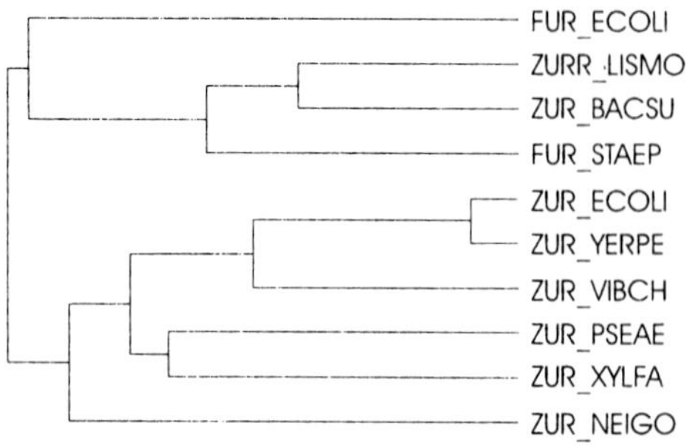

Figure 2. Alignment and dendrogram of selected members of the Zur protein family. For comparison the Fur protein of *E. coli* has been included in the dendrogram and is shown in italics below the aligned Zur sequences. The two groups of Zur proteins were aligned seperately and their consensus sequence is given in between the alignments. A * indicates an identical residue, a ^ indicates a similar residue. The abbreviation for the organisms are LISMO *Listeria monocytogenes*, BACSU *Bacillus subtilis*, STAEP *Staphylococcus epidermidis*, YERPE *Yersinia pestis*, VIBCH *Vibrio cholerae*, PSEAE *Pseudomonas aeruginosa*, XYLFA *Xylella fastidiosa*, NEIGO *Neisseria gonorrhoeae*. The ClustalW server at EBI was used (Thompson *et al.*, 1994).

Table 1. Comparison of the Fur protein of *E. coli* to Fur proteins of different bacteria and *E. coli* Zur to Zur proteins of the same bacteria.

Organism	Percent identity to the protein of *E. coli*	
	Fur	Zur
Escherichia coli	100	100
Salmonella enterica	99	94
Yersinia pestis	84	68
Vibrio cholerae	76	48
Pseudomonas aeruginosa	57	34
Neisseria gonorrhoeae	49	29

Zur proteins show up, one is found in Gram negative bacteria, the second one in Gram positive bacteria. Zur from Gram positive bacteria is more similar to the *E. coli* Fur than Zur from Gram negative ones.

It is generally observed in Gram negative bacteria that the sequence variation within the Zur proteins is larger than within the iron dependent Fur proteins (Table 1). In addition, it is astonishing that the Fur binding site in different organisms is very similar. Even in the promoter region of iron regulated genes of the gram positive *B. subtilis* a Fur-box is recognised by the *E. coli* Fur protein. This similarity is the basis for the use of the Fur titration test for promoters of iron regulated genes from very divers bacteria (Stojiljkovic *et al.*, 1994). In contrast, the Zur binding sites seem to differ distinctly in different organisms.

In footprinting experiments at high concentrations of Fur, polymerization and extension of the protected region occurred at several Fur-dependent promoters (de Lorenzo *et al.*, 1987). This was also observed by electron and atomic force microscopy (Le Cam *et al.*, 1994). Even at high concentrations of Zur no enlargement of the protected regions was observed, which is a clear-cut difference in the behaviour of the two repressors (Patzer and Hantke, 2000).

Zur is only active in the reduced form. As a cytoplasmic protein it has predominantly reduced thiols rather than oxidized disulfides due to the reducing conditions in the cytoplasm (Gilbert, 1990). *In vitro*, the cysteine residues of Zur are easily oxidised to disulfides, as judged by the slower migration in SDS-polyacrylamide gels under reducing conditions compared to non-reducing conditions. Oxidised Zur does not bind DNA or considerable amounts of Zn^{2+}. This may indicate that Zn^{2+} is mainly bound by some of the 9 cysteines found in Zur. In *E. coli* Fur only 4 cysteines are found which are all conserved in Zur. At least two to three zinc ions per dimer bind specifically to Zur (Patzer and Hantke, 2000). It remains to be seen if one of these zinc ions is bound to the same site in Zur as it was found in Fur at the conserved cysteines (positions 92 and 95, see Figure 1; Jacquamet *et al.*, 1998). These two cysteines are found in all Fur-like proteins except those from *Pseudomonas* and related organisms.

The Fur protein is not only auto-regulated and regulated by the cAMP-CAP catabolite repression system (De Lorenzo *et al.*, 1988), but also by OxyR and SoxRS (Zheng *et al.*, 1999). After treatment of cells with H_2O_2, OxyR induces a set of genes with clearly antioxidant activities. OxyR induction of Fur synthesis should diminish the formation of HO• radicals generated by H_2O_2 reacting with intracellular iron. SoxRS induces oxidative stress response genes after treatment of cells with O_2•-generating compounds. The induction of Fur by SoxRS should decrease the generation of HO• radicals (Zheng *et al.*, 1999). No evidence for autoregulation of Zur or for the influence of other regulators on Zur was found. This may indicate that Zur has the very limited function to regulate zinc uptake and metabolism in a zinc poor environment. Up to now besides the *znuA* promoter only two other Zur binding sites were identified in the *E. coli* chromosome; the genes or the encoded proteins had no obvious functions.

Acknowledgements

I thank Silke Patzer for her collaboration and her enthusiasm for the project, Christian Bayertz and Kerstin Haible for technical assistance, Volkmar Braun (this institute) for discussion and Karen A. Brune (Konstanz) for critical reading of the manuscript. This work was supported by the Deutsche Forschungsgemeinschaft (grant HA 1186/2-3).

References

Agar, J.N., Krebs, C., Frazzon, J., Huynh, B.H., Dean, D.R., and Johnson, M.K. 2000. IscU as a scaffold for iron-sulfur cluster biosynthesis: sequential assembly of [2Fe-2S] and [4Fe-4S] clusters in IscU. Biochemistry 39: 7856–7862.

Bartsevich V.V., and Pakrasi H.B. 1995. Molecular identification of an ABC transporter complex for manganese: analysis of a cyanobacterial mutant strain impaired in the photosynthetic oxygen evolution process. EMBO J. 14: 1845–1853.

Bäumler, A.J., and Hantke, K. 1992. Ferrioxamine uptake in *Yersinia enterocolitica*: characterization of the receptor protein FoxA. Mol. Microbiol. 6: 1309–1321.

Braun, V., Hantke, K., and Köster, W. 1998. Bacterial iron transport: mechanisms, genetics, and regulation. Met. Ions. Biol. Syst. 35: 67–145.

Brocklehurst, K.R., Hobman, J.L., Lawley, B., Blank, L., Marshall, S.J., Brown, N.L., and Morby, A.P. 1999. ZntR is a Zn(II)-responsive MerR-like transcriptional regulator of zntA in *Escherichia coli*. Mol. Microbiol. 31: 893–902.

Bsat, N., Herbig, A., Casillas-Martinez, L., Setlow, P., and Helmann, J.D. 1998. *Bacillus subtilis* contains multiple Fur homologues: identification of the iron uptake (Fur) and peroxide regulon (PerR) repressors. Mol. Microbiol. 29: 189–198.

Casadaban, M.J., and Cohen, S.N. 1979. Lactose genes fused to exogenous promoters in one step using a Mu*lac* bacteriophage: *In vitro* probe for transcriptional control sequences. Proc. Natl. Acad. Sci. USA 76: 4530–4533.

Dalet, K., Gouin, E., Cenatiempo, Y., Cossart, P., and Hechard, Y. 1-5-1999. Characterisation of a new operon encoding a Zur-like protein and an associated ABC zinc permease in Listeria monocytogenes. FEMS Microbiol Lett. 174: 111–116.

De Lorenzo, V., Wee, S., Herrero, M., and Neilands, J.B. 1987. Operator sequences of the aerobactin operon of plasmid ColV-K30 binding the ferric uptake regulation (*fur*) repressor. J. Bacteriol. 169: 2624–2630.

De Lorenzo, V., Herrero, M., Giovannini, F., and Neilands, J.B. 1988. Fur (ferric uptake regulation) protein and CAP (catabolite-activator protein) modulate transcription of *fur* gene in *Escherichia coli*. Eur. J. Biochem. 173: 537–546.

Fontecave, M., Coves, J., and Pierre, J.L. 1994. Ferric reductases or flavin reductases? Biometals. 7: 3–8.

Gaballa, A., and Helmann, J.D. 1998. Identification of a zinc-specific metalloregulatory protein, Zur, controlling zinc transport operons in *Bacillus subtilis*. J. Bacteriol. 180: 5815–5821.

Gilbert, H.F. 1990. Molecular and cellular aspects of thiol-disulfide exchange. Adv. Enzymol Relat. Areas. Mol. Biol. 63: 69–172.

Hantke, K., and Braun, V. (2000) The art of keeping low and high iron concentrations in balance. In: Bacterial stress responses. Storz, G. and Hengge-Aronis, R. ASM Press, Washington, D.C., 275–288.

Heidrich, C., Hantke, K., Bierbaum, G., and Sahl, H.G. 1-7-1996. Identification and analysis of a gene encoding a Fur-like protein of *Staphylococcus epidermidis*. FEMS Microbiol Lett. 140: 253–259.

Holm, L., Sander, C., Rüterjans, H., Schnarr, M., Fogh, R., Boelens, R., and Kaptein, R. 1994. LexA repressor and iron uptake regulator from *Escherichia coli*: new members of the CAP-like DNA binding domain superfamily. Protein Eng. 7: 1449–1453.

Jacquamet, L., Aberdam, D., Adrait, A., Hazemann, J.L., Latour, J.M., and Michaud-Soret, I. 1998. X-ray absorption spectroscopy of a new zinc site in the fur protein from *Escherichia coli*. Biochemistry 37: 2564–2571.

Jordan, S.R., and Pabo, C.O. 1988. Structure of the lambda complex at 2.5 A resolution: details of the repressor-operator interactions. Science 242: 893–899.

Le Cam, E., Frechon, D., Barray, M., Fourcade, A., and Delain, E. 1994. Observation of binding and polymerization of Fur repressor onto operator-containing DNA with electron and atomic force microscopes. Proc. Natl. Acad. Sci. USA 91: 11816–11820.

Leonhartsberger, S., Huber, A., Lottspeich, F., and Böck, A. 2001 The hydH/G Genes from *Escherichia coli* code for a zinc and lead responsive two-component regulatory system. J. Mol. Biol. 307: 93–105.

Lill, R., and Kispal, G. 2000. Maturation of cellular Fe-S proteins: an essential function of mitochondria. Trends Biochem. Sci. 25: 352–356.

Matzanke, B.F., Berner, I., Bill, E., Trautwein, A.X., and Winkelmann, G. 1991. Transport and utilization of ferrioxamine-E-bound iron in *Erwinia herbicola* (*Pantoea agglomerans*). Biol. Met. 4: 181–185.

Müller, K., Matzanke, B.F., Schünemann, V., Trautwein, A.X., and Hantke, K. 1998. FhuF, an iron-regulated protein of *Escherichia coli* with a new type of [2Fe-2S] center. Eur. J. Biochem. 258: 1001–1008.

Patzer, S.I., and Hantke, K. 1998. The ZnuABC high-affinity zinc uptake system and its regulator Zur in *Escherichia coli*. Mol. Microbiol. 28: 1199–1210.

Patzer, S.I., and Hantke, K. 1999. SufS is a NifS-like protein, and SufD is necessary for stability of the [2Fe-2S] FhuF protein in *Escherichia coli*. J. Bacteriol. 181: 3307–3309.

Patzer, S.I., and Hantke, K. 2000. The zinc-responsive regulator Zur and its control of the znu gene cluster encoding the ZnuABC zinc uptake system in *Escherichia coli*. J Biol Chem 275: 24321–24332.

Rohrbach, M.R., Braun, V., and Köster, W. 1995. Ferrichrome transport in *Escherichia coli* K-12: altered substrate specificity of mutated periplasmic FhuD and interaction of FhuD with the integral membrane protein FhuB. J. Bacteriol. 177: 7186–7193.

Stojiljkovic, I., Bäumler, A.J., and Hantke, K. 1994. Fur regulon in Gram-negative bacteria: Identification and characterization of new iron-regulated *Escherichia coli* genes by a fur titration assay. J. Mol. Biol. 236: 531–545.

Thompson J.D., Higgins D.G., and Gibson T.J. 1994. CLUSTAL W: improving the sensitivity of progressive multiple sequence alignment through sequence weighting, position-specific gap penalties and weight matrix choice. Nucleic Acids Res. 22: 4673–4680.

Zheng, M., Doan, B., Schneider, T.D., and Storz, G. 1999. OxyR and SoxRS regulation of fur. J. Bacteriol. 181: 4639–4643.

Chapter 10

Stimulus Perception and Signal Transduction by the KdpD/KdpE System of *Escherichia coli*

Kirsten Jung and Karlheinz Altendorf

Abstract

The membrane–bound histidine kinase KdpD is a putative turgor sensor that regulates, together with the response regulator KdpE, expression of the *kdpFABC* operon. This operon encodes the high affinity K^+-uptake system KdpFABC of *Escherichia coli*. Expression of *kdpFABC* is induced under K^+-limiting growth conditions and in response to an osmotic upshift. Various structural features of KdpD and KdpE, which are important for stimulus perception and/or signal transduction were identified and are described here. Furthermore, various studies undertaken to elucidate the nature of the stimulus for KdpD result in a new model for KdpD stimulus perception. According to this, autophosphorylation activity of KdpD is not a result of changes in turgor *per se*. Instead, various – mainly intracellular parameters – that are related to changes of environmental conditions influence the activities of KdpD.

Introduction

Maintenance of turgor is a fundamental regulatory process in microorganisms. When *Escherichia coli* is exposed to an osmotic upshift, which is accompanied by a loss of turgor, then the primary response is an accumulation of K^+ (Csonka, 1989; Csonka and Hanson, 1991). *E. coli* uses several K^+ transport systems to adjust the intracellular K^+ concentration (Altendorf and Epstein, 1996). Under physiological conditions the constitutive K^+ uptake systems TrkG, TrkH and Kup are operating. Upon osmotic upshift and under K^+-limiting growth conditions ($[K^+] < 2$ mM) the high affinity K^+-transport complex KdpFABC is synthesized. Expression of the *kdpFABC* operon is under control of the regulatory proteins KdpD and KdpE, which constitute a typical sensor kinase/response regulator system (Polarek *et al.*, 1992; Walderhaug *et al.*, 1992). The sensor kinase KdpD is an integral protein of the cytoplasmic membrane consisting of a large cytoplasmic N-terminal domain, four putative transmembrane domains and an extended cytoplasmic C-terminal domain (Zimmann *et al.*, 1995) (Figure 1). KdpD undergoes autophosphorylation (probably at His-673) (Voelkner *et al.*, 1993) and subsequently, the phosphoryl group is transferred to the response regulator KdpE (probably at Asp-52). Phosphorylated KdpE exhibits an increased affinity for a 23 bp sequence immediately upstream of the canonical −35 and −10 regions of the *kdpFABC* promoter (Sugiura *et al.*, 1992), thereby triggering *kdpFABC* transcription.

The stimulus, which KdpD senses, is postulated to be a decrease in turgor or some effect thereof reflecting the role of K^+ as an important cytoplasmic osmotic solute (Laimins *et al.*, 1981). This model has been challenged by the finding that under some conditions expression of *kdpFABC* is only significantly induced when the osmolality of the medium is increased by salt and not in the case of sugar (Asha and Gowrishankar, 1993; Jung and Altendorf, 1998a; Sugiura *et al.*, 1994). Analysis of *kdpD* mutants, which constitutively express *kdpFABC* independent of the K^+ concentration of the medium, but retain the ability to respond to changes in medium osmolality, led to the suggestion that the sensing mechanisms of KdpD for K^+ limitation and osmotic upshift may be mechanistically different (Sugiura *et al.*, 1994).

Characterization of the Enzymatic Activities of KdpD and KdpE *In Vitro*

To address the question of the stimulus for KdpD under *in vitro* conditions and to characterize this protein biochemically, it was necessary to develop efficient and rapid purification and reconstitution methods for KdpD. KdpD can be solubilized from membrane preparations enriched in KdpD-6His (KdpD accounts for about 10% of the total membrane proteins) by the detergent lauryldimethylamine oxide (LDAO) in the presence of high salt. Solubilized KdpD is enzymatically inactive, which is in contrast to e.g., solubilized EnvZ (Jung *et al.*, 2001). However, reconstitution of the protein into *E. coli* liposomes restores full activity of the sensor kinase. Purified and reconstituted KdpD-6His exhibits autokinase activity and the phosphoryl group is rapidly transferred to KdpE. Furthermore, KdpD-6His is the only protein that mediates dephosphorylation of KdpE~P (Jung *et al.*, 1997) (Figure 1). The phosphatase activity of KdpD is stimulated in the presence of ATP, whereas GTP, ITP, CTP, ADP, GDP have no effect, although only ATP-binding is required, since non-hydrolyzable analogs (ATP-γ-S, AMP-PNP) work as well (Jung and Altendorf, 1998b).

The transfer of the phosphoryl group from KdpD~P to KdpE has been demonstrated (Nakashima *et al.*, 1992; Voelkner *et al.* 1993, Lucassen, 1998). KdpE can also be overproduced and an efficient purification protocol was established (Puppe *et al.*, 1996). However, the phosphorylation efficiency *in vitro* is very low (3–4%). A much higher value (50–60%) has been obtained using purified KdpE and acetyl phosphate as the phosphoryl donor (Lucassen, 1998). It is important to note that in the latter case complete dephosphorylation of KdpE~P by KdpD occurs indicating that phosphorylation of KdpE by acetyl phosphate leads to a functional product. Gel permeation chromatography and native gel electrophoresis revealed that KdpE exists as a monomer whereas KdpE~P forms a dimer (Lucassen, 1998).

The replacement of the conserved Asp52, which has been identified in other response regulators as being the phosphorylation site, with Asn leads to a KdpE derivative which can no longer be phosphorylated (Lucassen, 1998). Together with the pH stability profile of KdpE~P (Zimmann, 1995) these data argue that Asp52 is the phosphorylation site in KdpE. In contrast, the replacement of Asp9 with Asn leads to a KdpE derivative which still can be phosphorylated although with lower efficiency. This lends support to the notion that Asp9 participates in the catalysis of phosphorylation and plays a similar role in the "acid pocket", as has already been postulated for the corresponding Asp residues of other response regulators (Lukat *et al.*, 1991).

Identification of Structural Features Important for Stimulus Perception by the Sensor KdpD

1. KdpD Forms a Dimer

Co-production of two derivatives of KdpD (KdpD-N788D and KdpD-H673Q), which cannot be autophosphorylated by ATP, results in active KdpD heterooligomers. This result suggests that the functional state of KdpD is a dimer or an oligomer and that the phosphorylation of KdpD is a reaction between two monomers (Heermann *et al.*, 1998). Similar observations were obtained for other sensor kinases (Yang and Inouye, 1991; Ninfa

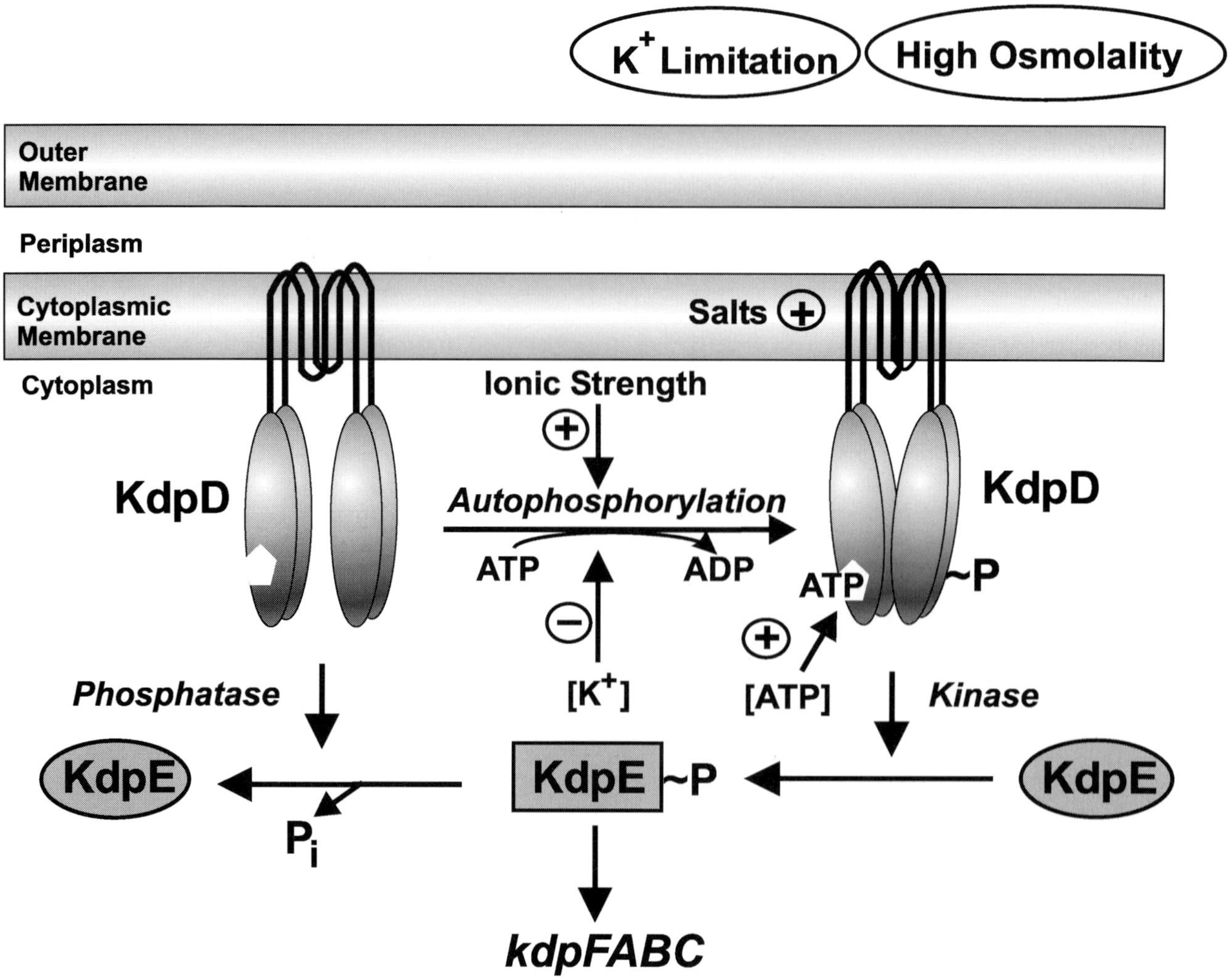

Figure 1. Model for stimulus perception and signal transduction of the KdpD/KdpE system. The model depicts structural features of KdpD and KdpE, which are important for stimulus perception and/or signal transduction. In addition, the effects of primary stimuli on the enzymatic activities of KdpD are marked.

et al., 1993; Swanson *et al.*, 1993; Wolfe and Stewart, 1993). To solve the question whether the switch between autophosphorylation and phosphatase activity of KdpD involves a dimer to monomer transition, we characterized the oligomeric state of the phosphorylated and non-phosphorylated form of KdpD. Results obtained by various techniques revealed that KdpD does not undergo a monomer-to-dimer transition upon phosphorylation. Crosslinking experiments with copper phenanthroline demonstrate that cysteine residues located in the transmitter domain or transmembrane domain I are easily oxidized resulting in dimers or higher order KdpD oligomers. From this it is concluded that the transmembrane domains and the transmitter domains, respectively, of the monomers are in close contact (Jung *et al.*, 1998).

The role of the 6 native cysteine residues for KdpD dimer formation was analyzed and the results indicate that cysteine residues at positions 852 and 874 are able to form a disulfide bridge (Jung *et al.*, 1998). However, a protein devoid of cysteine residues still appears as a dimer concluding that cysteine residues do not participate in dimer formation (Heermann *et al.*, 1998).

2. The Input Domain of KdpD Contains a Regulatory ATP-Binding Site

KdpD is characterized by a large, about 400 amino acids comprising N-terminal domain, which is not found in any other known sensor kinase. Comparison of 50 KdpD sequences from various microorganisms revealed that this part of the sensor kinase is highly conserved and includes two motifs (Walker A and Walker B) which are very similar to the classical ATP-binding sites of ATP-requiring enzymes. Interestingly, in *Synechocystis* sp. (Kaneko *et al.*, 1996), in *Anabaena* sp. L-31 (GenBank™ accession number AF213466), and some other bacteria short versions of KdpD, which comprise only the N-terminal domain and lack the transmembrane as well as the transmitter domains have been found.

Photoaffinity labeling with 8-azido-[α-^{32}P]ATP confirmed the existence of an ATP-binding site located in the N-terminal domain of KdpD of *E. coli* (Heermann *et al.*, 2000) (Figure 1). KdpD proteins missing this ATP-binding site either by truncation (KdpD/Δ12-395, KdpD/Δ12-228) or by amino acid replacements (KdpD/G37A, K38A,T39C) are characterized by a high phosphatase activity, which in case of wild-type KdpD is only detectable in the presence of ATP. Furthermore, these modified proteins are responsible for a drastically lower expression level of the *kdpFABC* operon under K^+-limiting growth conditions and do not induce expression when cells are shifted to high osmolality. These results imply that ATP-binding to this regulatory site modulates the phosphatase activity of KdpD (Jung and Altendorf, 1998b).

We have shown that a hybrid protein consisting of the *Anabaena* KdpD and a truncated KdpD of *E. coli* (Anacoli KdpD) mediates *kdpFABC* expression like full-length KdpD of *E. coli* indicating that the *Anabaena* part functionally interacts with the *E. coli* part (Ballal *et al.*, 2002). Moreover, the N-terminal domain of *E. coli*, KdpD/1-395, was overproduced and purified. Although predicted to be hydrophilic, it was found to be membrane-associated. The membrane-associated form, but not the solubilized one, retains the ability to bind 8-azido-[α-^{32}P]ATP (Heermann *et al.*, 2000). Whether membrane association is a prerequisite of ATP-binding and hence part of the signal transduction mechanism is currently under investigation.

Whereas in a strain that carries a truncated form of KdpD/Δ12-395 *kdpFABC* expression is drastically reduced, co-expression of *kdpD*/Δ12-395 and *kdpD*/1-395 significantly increases *kdpFABC* expression under inducing conditions. These results provide first evidence for an interaction of the N-terminal input domain and the C-terminal transmitter domain of KdpD (Heermann *et al.*, 2000).

3. Role of the transmembrane domains in KdpD

Replacement of various amino acid residues located within the transmembrane domains (TM's) led to altered sensing properties of KdpD (Jung *et al.*, 1998; Sugiura *et al.*, 1994; Brandon *et al.*, 2000). Furthermore, truncated KdpD proteins missing the membrane-spanning segments and 32 amino acids of the adjacent C-terminal cytoplasmic domain including a cluster of positively charged amino acid residues (see next paragraph) were characterized by a loss of all enzymatic activities and were unable to induce *kdpFABC* expression *in vivo* (Puppe *et al.*, 1996). From these results it was concluded that the TM's play an essential role for KdpD activity. However, recent studies shed new light on the role of the transmembrane domains for the activity of the sensor kinase KdpD of *E. coli* (Heermann *et al.*, 2003). Importantly, a truncated KdpD lacking again all four transmembrane domains but retaining the cluster of positively charged residues supported *kdpFABC* expression in a K^+-dependent manner at K^+ concentrations below 1 mM. Furthermore, results obtained with truncated KdpD proteins lacking only two transmembrane domains or derivatives in which a linker peptide or two transmembrane domains of PutP, the Na^+/proline transporter of *Escherichia coli*, replaced the missing part indicated that the transmembrane domains are not essential for sensing K^+ limitation, but may be important for the correct positioning of the large N- and C-terminal cytoplasmic domains to each other (Heermann *et al.*, 2003).

There are some indications that the transmembrane domains of KdpD are involved in osmosensing. A truncated KdpD lacking TM1 and TM2 was unable to sense an increase of medium osmolality (Heermann *et al.*, 2003). Furthermore, the systematic replacement of each single amino acid of TM1 revealed that amino acids of this transmembrane domain are involved in osmosensing but not K^+-sensing (Stallkamp *et al.*, 2002). Mutational analysis of TM3, TM4, and the adjacent C-terminal hydrophilic region identified a number of KdpD derivatives which were insensitive towards the K^+ signal but sensitive towards osmotic shifts (Sugiura *et al.*, 1994). Finally, experiments with right-side-out membrane vesicles have shown that salts from the outside stimulate the autophosphorylation activity of KdpD (Jung *et al.*, 2000).

4. Role of a Cluster of Positively Charged Residues in KdpD

It is known that a cluster of at least five arginine or lysine residues with interspersed electrically neutral amino acids within a soluble protein is sufficient to contribute to its temporary binding to a phospholipid bilayer (Ben Tal *et al.*, 1996). A cluster of arginine residues following the transmembrane domains (Figure 1) is found in all known KdpD sequences. Therefore, we were interested if that cluster is involved in the interaction with negatively charged phospholipids and whether changes of this interaction may be one of the key players in the activation mechanism of KdpD upon stimulus perception. To study the role of these residues, arginine at positions 503, 506, 508, 511 or 513 was replaced by glutamine (Jung and Altendorf, 1998a). Three groups of KdpD proteins emerged from this study. Removal of the positive charge at position 511 led to the loss of phosphatase activity. On the other hand, KdpD proteins with Arg replacements at positions 503, 506 or 508 are characterized by an increase in the phosphatase activity and a decrease in the autophosphorylation activity. Furthermore, when Arg513 was replaced with Gln the amount of KdpD detected in the membrane was drastically reduced. Thus, this substitution does probably not affect the enzymatic activities of KdpD, but rather the insertion and/or stability of the protein in the membrane. These results do not favor the hypothesis of an interaction of these residues with the acidic phospholipids of the membrane. However, they indicate that these Arg residues are important for maintaining the equilibrium between the autophosphorylation and phosphatase activities of KdpD. Since the removal of one positively charged amino acid residue is sufficient to disturb the ratio of the KdpD activities, we propose that an electrostatic switch mechanism within the protein exists that regulates the ratio of phosphatase to autophosphorylation activity in KdpD. According to this, stimulus perception leads to conformational changes thereby altering electrostatic interactions which in turn shift the enzymatic activities of KdpD. Regulation of activity *via* a protein domain that has a net positive charge is already well characterized for the Shaker K^+ channel. The so called "ball and chain" model has been shown to be suitable to explain the rapid inactivation of voltage-dependent K^+ channels (Kondoh *et al.*, 1997; Hoshi *et al.*, 1990).

It is interesting to note that various KdpD derivatives causing constitutive *kdpFABC* expression turned out to be single amino acid replacements, which are in close proximity to the cluster of positively charged residues (Brandon *et al.*, 2000; Sugiura *et al.*, 1994). The enzymatic activities of two of these proteins were determined (Brandon *et al.*, 2000). Since the phosphatase activities were severely affected, the idea of a switch mechanism between autophosphorylation and phosphatase activity is further supported.

Conformational Changes within KdpE upon Phosphorylation

Limited proteolysis with trypsin led to distinct differences in the fragmentation pattern when purified KdpE or KdpE~P were used (Lucassen, 1998). A detailed analysis revealed that in response to phosphorylation at the receiver domain a conformational change occurs at the DNA binding domain. This also argues in favor of a change in communication between both domains upon phosphorylation.

DNase I footprinting analysis and gel retardation experiments demonstrated that KdpE as well as KdpE~P bind to a [Tn]-rich region upstream of the *kdpFABC* promotor (Sugiura *et al.*, 1992; 1993). Furthermore, it has been shown that KdpE~P binds with higher affinity compared to KdpE (Nakashima *et al.*, 1993). This has also been confirmed using acetyl phosphate-modified KdpE (Lucassen, 1998). In addition, it has been demonstrated that stimulation of transcription of the *kdpFABC* operon *in vitro* depends on the presence of KdpE~P (Nakashima *et al.*, 1993).

The Stimulus for KdpD

It has been puzzling for a long time to which stimulus (stimuli) the membrane-bound sensor kinase KdpD is responding to. Epstein and coworkers have put forward the hypothesis that KdpD is a turgor sensor (Malli and Epstein, 1998, Laimins *et al.*, 1981). In contrast, the model of Mizuno *et al.* describes two mechanisms for KdpD activation: K^+ limitation and osmotic upshift (Sugiura *et al.*, 1994). Other groups argue that the K^+ signal is related to the internal K^+ level and/or the processes of K^+ transport (Asha and Gowrishankar, 1993; Frymier *et al.*, 1997) or the external K^+ concentration (Roe *et al.*, 2000).

In vitro phosphorylation assays with inverted membrane vesicles (Voelkner *et al.*, 1993) or proteoliposomes (Nakashima *et al.*, 1993) revealed that KdpD autophosphorylation activity is stimulated by salts, whereby both NaCl and KCl are effective. Recently, we developed a new *in vitro* test system based on right-side-out membrane vesicles (RSO-MV) (Jung *et al.*, 2000). Using the RSO-MV test system, in which KdpD has the same orientation as in whole cells, an inhibitory effect of K^+ (provided from the inside of the vesicles) on KdpD autophosphorylation activity is detected. For the first time these data indicate a correlation between K^+ concentration and KdpD autophosphorylation activity. Furthermore, these results reveal that the domains, which are exposed to the cytoplasmic side, are affected by K^+. Based on these results it is proposed that the intracellular K^+ concentration directly influences KdpD by down-regulating the autophosphorylation activity (Figure 1). This is supported by the fact that extracellular Cs^+ significantly induces *kdpFABC* expression by lowering the intracellular K^+ content (Jung *et al.*, 2001). It is conceivable that under K^+-limiting growth conditions the intracellular K^+ concentration falls below a certain threshold thereby suspending the inhibitory effect of K^+ on KdpD autophosphorylation activity.

Furthermore, an increase of the ionic strength inside the vesicles, imposed by either NaCl, RbCl or HEPES-Na buffer, was accompanied by an increase of the autophosphorylation activity of KdpD (Figure 1) (Jung *et al.*, 2000). *In vivo*, due to the loss of K^+ or due to an osmotic upshift cells lose water leading to an increase of the concentration of all molecules (Record *et al.*, 1998) which in turn increases the ionic strength. Thus, it is conceivable that the autophosphorylation activity of KdpD is stimulated under these conditions.

Although turgor cannot be established across the membrane of RSO-MV, they still behave like osmometers. For example, it has been shown that the transporter and osmosensor ProP of *Escherichia coli* is activated by hyperosmotic shifts imposed by NaCl or sucrose in RSO-MV (Milner *et al.*, 1988). An increased osmolality outside of the RSO-MV also stimulates KdpD autophosphorylation activity, however, the effects with salts as osmolytes are much higher compared to sugars. Moreover, the maximal stimulatory effect from the outside is smaller compared to that observed by an increase of the salt concentration inside the RSO-MV. Furthermore, KdpD in RSO-MV cannot be activated in the presence of amphiphilic compounds (Jung *et al.*, 2000). These results are in favor of a special effect of salts rather than changes of membrane strain on KdpD activation. *In vivo,* salts are also more effective to induce *kdpFABC* expression than non-polar compounds at the same osmolality. This can be explained either by a specific effect of salts on phospholipids or by an effect on the interaction between phospholipids and protein. In accord with this is the finding that KdpD autophosphorylation activity is dependent on negatively charged phospholipids, whereas the structure of the phospholipids plays a minor role (Stallkamp *et al.*, 1999).

Model for Stimulus Perception and Signal Transduction via KdpD/KdpE

Based on the results described above the following model for stimulus perception and signal transduction via the sensor kinase KdpD and the response regulator KdpE is proposed (Figure 1). KdpD catalyzes the autophosphorylation by ATP and the dephosphorylation of the corresponding phosphorylated response regulator KdpE. KdpD is a dimer and the autophosphorylation reaction is a result of a *trans*-phosphorylation between two monomers. There is no change of the oligomeric state of KdpD upon phosphorylation. An equilibrium between autophosphorylation and phosphatase activities is proposed which is shifted via an electrostatic switch mechanism. The phosphoryl group from KdpD is rapidly transferred to KdpE. KdpE undergoes dimerization upon phosphorylation. The intracellular K^+ concentration as well as ionic strength affect KdpD autophosphorylation activity of KdpD directly whereby K^+ has an inhibitory and ionic strength a stimulatory effect. Furthermore, an increase of the intracellular ATP concentration, which is observed after an osmotic upshift (Ohwada and Sagisaka, 1987) and also when cells are exposed to K^+ limitation (K. Jung, unpublished observation), modulates the phosphatase activity of KdpD via ATP-binding to the regulatory ATP-binding site. Salts from the outside affect the autophosphorylation activity of KdpD. It is proposed that salts influence the interaction of domains of KdpD with the phospholipid bilayer. Therefore, various mainly intracellular parameters are integrated by KdpD and via phosphorylation reactions transmitted to KdpE to initiate expression of the *kdpFABC* operon. Although KdpD is special in such a way that it is able to respond to high osmolality AND K^+ limitation, the obtained results should be compared to other proteins with osmosensing properties. The characterization of purified and reconstituted EnvZ, a putative osmosensor of *E. coli*, revealed that the autophosphorylation activity of this protein is stimulated by monovalent cations, specifically K^+, suggesting that the increased intracellular K^+ concentration upon an osmotic upshock is a trigger for the activation of EnvZ (Jung *et al.*, 2001). The transport activity of the glycine betaine carrier BetP of *Corynebacterium glutamicum* has also been shown to be triggered by the intracellular K^+ concentration (Rübenhagen *et al.*, 2000, 2001). In case of PutP, the osmoregulatory proline transporter of *E. coli*, it was found that the protein responds to osmotically induced changes in both cytoplasmic K^+ levels and the concentration of cytoplasmic macromolecules (Racher *et al.*, 1999; J. Wood, personal communication). The osmotic activity of the ATP-binding cassette (ABC) transporter OpuA of *Lactococcus lactis* was found to be dependent on the nature of the surrounding lipids, and the activity was stimulated by a high concentration of ionic osmolytes indicating that the intracellular ionic strength and the membrane signal osmotic stress to the transporter (van der Heide *et al.*, 2001; Poolmann *et al.*, 2002).

The integration of positive and negative stimuli is unique for KdpD and has not been found for other osmoresponsive proteins. However, it is a well-known principle for tight regulation allowing in case of KdpD optimal adaptation of *E. coli* to altered environmental conditions.

References

Altendorf, K., and Epstein, W. 1996. The Kdp-ATPase of *Escherichia coli*. In: Biomembranes. A.G. Lee, (ed.), JAI Press Inc, London. 5: 403–420.

Asha, H., and Gowrishankar, J. 1993. Regulation of *kdp* operon expression in *Escherichia coli*: evidence against turgor as signal for transcriptional control. J. Bacteriol. 175: 4528–4537.

Ballal, A., Heermann, R., Jung, K., Gassel, M., Apte, S.K. and Altendorf, K. 2002. A chimeric Anabaena/*Escherichia coli* KdpD protein (Anacoli KdpD) functionally interacts with *E. coli* KdpE and activates *kdp* expression in *E. coli*. Arch. Microbiol. 178: 141–148.

Ben Tal, N., Honig, B., Peitzsch, R.M., Denisov, G., and McLaughlin, S. 1996. Binding of small basic peptides to membranes containing acidic lipids: theoretical models and experimental results. Biophys. J. 71: 561–575.

Brandon, L., Dorus, S., Epstein, W., Altendorf, K., and Jung, K. 2000. Modulation of KdpD phosphatase implicated in the physiological expression of the Kdp ATPase of *Escherichia coli*. Mol. Microbiol. 38: 1086–1092.

Csonka, L.N. 1989. Physiological and genetic responses of bacteria to osmotic stress. Microbiol. Rev. 53: 121–147.

Csonka, L.N., and Hanson, A.D. 1991. Prokaryotic osmoregulation: genetics and physiology. Annu. Rev. Microbiol. 45: 569–606.

Frymier, J.S., Reed, T.D., Fletcher, S.A., and Csonka, L.N. 1997. Characterization of transcriptional regulation of the *kdp* operon of *Salmonella typhimurium*. J. Bacteriol. 179: 3061–3063.

Heermann, R., Altendorf, K., and Jung, K. 2000. The hydrophilic N-terminal domain complements the membrane-anchored C-terminal domain of the sensor kinase KdpD of *Escherichia coli*. J. Biol. Chem. 275: 17080–17085.

Heermann, R., Altendorf, K., and Jung, K. 1998. The turgor sensor KdpD of *Escherichia coli* is a homodimer. Biochim. Biophys. Acta 1415: 114–124.

Heermann, R., Fohrman, A., Altendorf, K. and Jung, K. 2003. The transmembrane domains of the sensor kinase KdpD of *E. coli* are not essential for sensing K^+ limitation. Mol. Microbiol. 47: 839–848.

Hoshi, T., Zagotta, W.N., and Aldrich, R.W. 1990. Biophysical and molecular mechanisms of Shaker potassium channel inactivation. Science 250: 533–538.

Jung, K., and Altendorf, K. 1998a. Individual substitutions of clustered arginine residues of the sensor kinase KdpD of *Escherichia coli* modulate the ratio of kinase to phosphatase activity. J. Biol. Chem. 273: 26415–26420.

Jung, K., and Altendorf, K. 1998b. Truncation of amino acids 12–128 causes deregulation of the phosphatase activity of the sensor kinase KdpD of *Escherichia coli*. J. Biol. Chem. 273: 17406–17410.

Jung, K., Hamann, K., and Revermann, A. 2001. K^+ stimulates specifically the autokinase activity of purified and reconstituted EnvZ of *Escherichia coli*. J. Biol. Chem. 276: 40896–40902.

Jung, K., Heermann, R., Meyer, M., and Altendorf, K. 1998. Effect of cysteine replacements on the properties of the turgor sensor KdpD of *Escherichia coli*. Biochim. Biophys. Acta 1372: 311–322.

Jung, K., Krabusch, M. and Altendorf, K. 2001. Cs^+ induces the *kdp* operon by lowering the intracellular K^+ concentration. J. Bacteriol. 183: 3800–3803.

Jung, K., Tjaden, B., and Altendorf, K. 1997. Purification, reconstitution, and characterization of KdpD, the turgor sensor of *Escherichia coli*. J. Biol. Chem. 272: 10847–10852.

Jung, K., Veen, M., and Altendorf, K. 2000. K^+ and ionic strength directly influence the autophosphorylation activity of the putative turgor sensor KdpD of *Escherichia coli*. J. Biol. Chem. 275: 40142–40147.

Kaneko, T., Sato, S., Kotani, H., Tanaka, A., Asamizu, E., Nakamura, Y., Miyajima, N., Hirosawa, M., Sugiura, M., Sasamoto, S., Kimura, T., Hosouchi, T., Matsuno, A., Muraki, A., Nakazaki, N., Naruo, K., Okumura, S., Shimpo, S., Takeuchi, C., Wada, T., Watanabe, A., Yamada, M., Yasuda, M., and Tabata, S. 1996. Sequence analysis of the genome of the unicellular cyanobacterium *Synechocystis* sp. strain PCC6803. II. Sequence determination of the entire genome and assignment of potential protein-coding regions. DNA Res. 3: 109–136.

Kondoh, S., Ishii, K., Nakamura, Y., and Taira, N. 1997. A mammalian transient type K^+ channel, rat Kv1.4, has two potential domains that could produce rapid inactivation. J. Biol. Chem. 272: 19333–19338.

Laimins, L.A., Rhoads, D.B., and Epstein, W. 1981. Osmotic control of *kdp* operon expression in *Escherichia coli*. Proc. Natl. Acad. Sci. U. S. A. 78: 464–468.

Lucassen, M. 1998. Regulation des Kdp-Systems aus *Escherichia coli*: Biochemische Charakterisierung des Antwortregulators KdpE und Nachweis von Konformationsänderungen im Zuge der Aktivierung, PhD Thesis, Universität Osnabrück.

Lukat, G.S., Lee, B.H., Mottonen, J.M., Stock, A.M., and Stock, J.B. 1991. Roles of the highly conserved aspartate and lysine residues in the response regulator of bacterial chemotaxis. J. Biol. Chem. 266: 8348–8354.

Malli, R., and Epstein, W. 1998. Expression of the Kdp ATPase is consistent with regulation by turgor pressure. J. Bacteriol. 180: 5102–5108.

Milner, J.L., Grothe, S., and Wood, J.M. 1988. Proline porter II is activated by a hyperosmotic shift in both whole cells and membrane vesicles of *Escherichia coli* K12. J. Biol. Chem. 263: 14900–14905.

Nakashima, K., Sugiura, A., Kanamaru, K., and Mizuno, T. 1993. Signal transduction between the two regulatory components involved in the regulation of the *kdpABC* operon in *Escherichia coli*: Phosphorylation-dependent functioning of the positive regulator, KdpE. Mol. Microbiol. 7: 109–116.

Nakashima, K., Sugiura, A., and Mizuno, T. 1993. Functional reconstitution of the putative *Escherichia coli* osmosensor, KdpD, into liposomes. J. Biochem. (Tokyo) 114: 615–621.

Nakashima, K., Sugiura, A., Momoi, H., and Mizuno, T. 1992. Phosphotransfer signal transduction between two regulatory factors involved in the osmoregulated *kdp* operon in *Escherichia coli*. Mol. Microbiol. 6: 1777–1784.

Ninfa, E.G., Atkinson, M.R., Kamberov, E.S., and Ninfa, A.J. 1993. Mechanism of autophosphorylation of *Escherichia coli* nitrogen regulator II (NRII or NtrB): trans-phosphorylation between subunits. J. Bacteriol. 175: 7024–7032.

Ohwada, T., and Sagisaka, S. 1987. An immediate and steep increase in ATP concentration in response to reduced turgor pressure in *Escherichia coli* B. Arch. Biochem. Biophys. 259: 157–163.

Polarek, J.W., Williams, G., and Epstein, W. 1992. The products of the *kdpDE* operon are required for expression of the Kdp ATPase of *Escherichia coli*. J. Bacteriol. 174: 2145–2151.

Poolman, B., Blount, P., Folgering, J.H.A., Friesen, R.H.E., Moe, P.C., and van der Heide, T. 2002. How do membrane proteins sense water stress? Mol. Microbiol. 44: 889–902.

Puppe, W., Zimmann, P., Jung, K., Lucassen, M., and Altendorf, K. 1996. Characterization of truncated forms of the KdpD protein, the sensor kinase of the K^+-translocating Kdp system of *Escherichia coli*. J. Biol. Chem. 271: 25027–25034.

Racher, K.I., Voegele, R.T., Marshall, E.V., Culham, D.E., Wood, J.M., Jung, H., Bacon, M., Cairns, M.T., Ferguson, S.M., Liang, W.J., Henderson, P.J., White, G., and Hallett, F.R. 1999. Purification and reconstitution of an osmosensor: transporter ProP of *Escherichia coli* senses and responds to osmotic shifts. Biochemistry 38: 1676–1684.

Record, M.T.J., Courtenay, E.S., Cayley, D.S., and Guttman, H.J. 1998. Responses of *E. coli* to osmotic stress: large changes in amounts of cytoplasmic solutes and water. Trends. Biochem. Sci. 23: 143–148.

Roe, A.J., McLaggan, D., O'Byrne, C.P., and Booth, I.R. 2000. Rapid inactivation of the *Escherichia coli* Kdp K^+ uptake system by high potassium concentrations. Mol. Microbiol. 35: 1235–1243.

Rübenhagen, R., Morbach, S., and Krämer, R. 2001. The osmoreactive betaine carrier BetP from Corynebacterium glutamicum is a sensor for cytoplasmic K^+. EMBO J. 20: 5412–5420.

Rübenhagen, R., Ronsch, H., Jung, H., Krämer, R., and Morbach, S. 2000. Osmosensor and osmoregulator properties of the betaine carrier BetP from *Corynebacterium glutamicum* in proteoliposomes. J. Biol. Chem. 275: 735–741.

Stallkamp, I., Altendorf, K. and Jung, K. 2002. Amino acid replacements in transmembrane domain 1 influence osmosensing but not K^+ sensing by the sensor kinase KdpD of *Escherichia coli*. Arch. Microbiol., 178:525–530.

Stallkamp, I., Dowhan, W., Altendorf, K., and Jung, K. 1999. Negatively charged phospholipids influence the activity of the sensor kinase KdpD of *Escherichia coli*. Arch. Microbiol. 172: 295–302.

Sugiura, A., Hirokawa, K., Nakashima, K., and Mizuno, T. 1994. Signal-sensing mechanisms of the putative osmosensor KdpD in *Escherichia coli*. Mol. Microbiol. 14: 929–938.

Sugiura, A., Nakashima, K., and Mizuno, T. 1993. Sequence-directed DNA curvature in activator-binding sequence in the *Escherichia coli kdpABC* promoter. Biosci. Biotech. Biochem. 57: 356–357.

Sugiura, A., Nakashima, K., Tanaka, K., and Mizuno, T. 1992. Clarification of the structural and functional features of the osmoregulated *kdp* operon of *Escherichia coli*. Mol. Microbiol. 6: 1769–1776.

Swanson, R.V., Bourret, R.B., and Simon, M.I. 1993. Intermolecular complementation of the kinase activity of CheA. Mol. Microbiol. 8: 435–441.

van der Heide, T., Stuart, M.C.A., and Poolman, B. 2001. On the osmotic signal and osmosensing mechanism of an ABC transport system for glycine betaine. EMBO J. 20: 7022–7032.

Voelkner, P., Puppe, W., and Altendorf K. 1993. Characterization of the KdpD protein, the sensor kinase of the K^+-translocating Kdp system of *Escherichia coli*. Eur. J. Biochem. 217: 1019–1026.

Walderhaug, M.O., Polarek, J.W., Voelkner, P., Daniel, J.M., Hesse, J.E., Altendorf, K., and Epstein, W. 1992. KdpD and KdpE, proteins that control expression of the *kdpABC* operon, are members of the two-component sensor-effector class of regulators. J. Bacteriol. 174: 2152–2159.

Wolfe, A.J., and Stewart, R.C. 1993. The short form of the CheA protein restores kinase activity and chemotactic ability to kinase-deficient mutants. Proc. Natl. Acad. Sci. USA 90: 1518–1522.

Yang, Y., and Inouye, M. 1991. Intermolecular complementation between two defective mutant signal-transducing receptors of *Escherichia coli*. Proc. Natl. Acad. Sci. USA 88: 11057–11061.

Zimmann, P. 1995. Molekulare Analyse des Regulationsproteins KdpD des Kdp-Systems von *Escherichia coli*, PhD thesis, Universität Osnabrück.

Zimmann, P., Puppe, W., and Altendorf, K. 1995. Membrane topology analysis of the sensor kinase KdpD of *Escherichia coli*. J. Biol. Chem. 270: 28282–28288.

Chapter 11

Mechanism of Regulation of the Bifunctional Histidine Kinase NtrB in *Escherichia coli*

Verena Weiss, Günter Kramer, Thomas Dünnebier, and Annette Flotho

Abstract

NtrB is the bifunctional histidine kinase for nitrogen regulation. Dependent on the availability of nitrogen, it either autophosphorylates and serves as the phosphodonor for its cognate response regulator, NtrC, or, it promotes the rapid dephosphorylation of NtrC-P. The activity of NtrB depends on the interaction of two subdomains within its transmitter domain, the H-domain and the kinase domain. Both phosphotransfer activity and phosphatase activity reside in the H-domain. When separately expressed, this domain acts as a phosphatase. Interaction with the kinase domain results in the inhibition of the phosphatase activity and the phosphorylation of the conserved histidine of the H-domain.

Introduction

Signal transduction by two-component systems involves three basic steps: the sensing of a stimulus, signal transmission and the response reaction. In classical two-component systems these tasks are assigned to four different domains: the sensor and transmitter domain of the histidine kinase and the receiver and output domain of the response regulator (Chang and Stewart, 1998; Ninfa, 1996; Parkinson and Kofoid, 1992; Perraud *et al.*, 1999). The mechanism of signal transduction by these domains is usually described as a sequence of consecutive phosphotransfer reactions. The sensor domain senses changes in the environment and transmits a conformational signal to the transmitter domain. Depending on the stimulus, the transmitter domain autophosphorylates at a conserved histidine residue. Subsequently, the phosphoryl group is transferred to a conserved aspartate residue of the receiver domain. This induces the conformational change in the output domain, resulting in the activation of the response reaction.

Response Regulators can be Viewed as Phosphatases

The discovery that receiver domains of response regulators can be also phosphorylated by small phosphodonors, such as acetyl phosphate, was an important step for the understanding of the molecular mechanism of signal transduction by two-component systems (Feng *et al.*, 1992; Lukat *et al.*, 1992). It demonstrates that the phosphotransfer from the histidine kinase to the response regulator is actually catalyzed by the receiver domain of the response regulator (Sanders *et al.*, 1992). Therefore, the response regulator can be viewed as a phosphatase, which dephosphorylates its substrate, the autophosphorylated histidine kinase. Furthermore, the phosphorylated response regulator could be viewed as a phospho-enzyme intermediate. When phosphorylated, the response regulator is active and catalyzes the output reaction. Ultimately, however, the phospho-enzyme intermediate will hydrolyze, resulting in the regeneration of the unphosphorylated and inactive response regulator and free phosphate. The latter activity has been termed the autophosphatase activity and its rate varies considerably for different response regulators (Stock *et al.*, 1995). Auxiliary phosphatases, which enhance the rate of hydrolysis, have been identified for some two-component systems (Hess *et al.*, 1988, Perego *et al.*, 1994; Perego, 1998). It has been suggested, that these proteins act as co-phosphatases by stimulating the intrinsic autophosphatase activity of the response regulator (Little, 1993; Parkinson and Kofoid, 1992; Stock *et al.*, 1995). They play an important role in the regulation of two-component systems and contribute to a great extend to a transient signal, which is an important requirement for the regulation in signal transduction.

The final goal in the regulatory cascade of a two-component system is to regulate the concentration of the phosphorylated response regulator. This is achieved by regulating formation and hydrolysis of the phosphorylated response regulator in a reciprocal fashion. In other words, a two-component system is activated, if formation of the phosphorylated response regulator is very efficient due to a high concentration of the autophosphorylated histidine kinase. At the same time hydrolysis is slow, that is, not stimulated by additional effector proteins. Under these conditions, phosphorylated response regulator accumulates and the two-component system is activated. On the other hand, a two-component system is inactivated, if little or no autophosphorylated histidine kinase is present in the cell. Under these conditions some phosphorylated response regulator might form due to crosstalk from a non-cognate histidine kinase or from acetyl phosphate, but at the same time hydrolysis is fast, because it is stimulated by an auxiliary phosphatase.

As a consequence, the concentration of phosphorylated response regulator is very low and the two-component system is shut off.

Bifunctional Histidine Kinases have Both Autokinase and "Phosphatase" Activity

Interestingly, in several systems the histidine kinase itself acts as an auxiliary phosphatase. Such histidine kinases are known to be bifunctional, that is, they are able to switch between two opposing activities. Dependent on the stimulus from the environment they either autophosphorylate and serve as a phosphodonor for the response regulator, or, under different conditions, they change conformation and stimulate the fast dephosphorylation of the response regulator. The mechanism, by which bifunctional histidine kinases promote the fast dephosphorylation of the cognate response regulator, is still disputed. They might either act as a co-phosphatase or, they could promote the dephosphorylation of the cognate response regulator by a reversal of the phosphotransfer reaction. Data obtained for NtrB, the histidine kinase for nitrogen regulation, support a co-phosphatase model (Ninfa and Magasanik, 1986; Atkinson and Ninfa, 1993; Jiang *et al.*, 2000), while recent data obtained for EnvZ, the histidine kinase for osmoregulation, support the reverse phosphotransfer model (Zhu *et al.*, 2000).

NtrB, together with its cognate response regulator NtrC, regulates the expression of the Ntr regulon – genes important for the assimilation of ammonia and the utilization of alternative nitrogen compounds (Reitzer, 1996). NtrC is the transcriptional activator of the Ntr regulon. It is activated by phosphorylation. Both phosphorylation and dephosphorylation are under the control of NtrB, the bifunctional histidine kinase. NtrB senses the nitrogen status of the cell with the help of two additional proteins, the PII protein (product of *glnB*) and the bifunctional UTase/UR (uridylyltransferase/uridylremoving enzyme, product of *glnD*). Dependent on the relative intracellular glutamine and α-ketoglutarate concentrations, PII is reversibly uridylylated by UTase/UR. Under nitrogen excess conditions, PII is deuridylylated. Deuridylylated PII interacts with NtrB and induces the phosphatase activity in NtrB. Uridylylated PII, formed under nitrogen starvation conditions, does not interact with NtrB (Atkinson *et al.*, 1994; Ninfa *et al.*, 2000). Thus, NtrB has autokinase activity in the absence of a signal. To act as a phosphatase, NtrB has to be activated. Accordingly, NtrB behaves as a constitutive positive regulator in a strain lacking PII (Bueno *et al.*, 1985). Similarly, purified NtrB has autokinase activity (Ninfa and Magasanik, 1986; Weiss and Magasanik, 1988). To activate NtrB as a phosphatase *in vitro*, PII, or the paralogue, GlnK, has to be added.

Until recently it was believed, that PII interacts with the sensor domain (Kamberov *et al.*, 1995). Surprisingly however, chemical crosslinking revealed that PII interacts with fragments carrying the kinase domain (Pioszak *et al.*, 2000). This result is unexpected, since mutations conferring a constitutive positive phenotype are found in the sensor domain, the H-domain and the linker domain of NtrB but not in the kinase domain. Interestingly, the smallest fragment, which binds to PII carries aa 190 to 349 of NtrB, that is, it also contains the linker region connecting the H-domain with the kinase domain. An interaction of PII with fragments carrying the sensor domain but lacking the kinase domain could not be detected by chemical crosslinking. Therefore the role of the sensor domain in NtrB has to be newly defined. It has been suggested, that the sensor domain is required to stabilize the "phosphatase" conformation in NtrB (Pioszak *et al.*, 2000). In addition it should be kept in mind that the sensor domain resembles a PAS domain. Therefore, we cannot exclude, that the sensor domain senses an additional unknown stimulus and regulates the H-domain in concert with PII.

The Transmitter Domain Consists of Separate Subdomains

To understand, how NtrB switches between its two opposing activities, we have analyzed the domain organization of NtrB. Our approach was to divide the transmitter domain into separate subdomains, designated the H-, N-, and G-domain (Figure 1). Originally this model was based on sequence alignments with other histidine kinases (Kramer and Weiss, 1999). The model is shown here in a modified version, where the boundaries between the H and the G domain were shifted 30 amino acids towards the N-terminus (Jiang *et al.*, 2000; Dünnebier, Kramer and Weiss, unpublished). It is based on data obtained by limited proteolysis and comparison to the 3D structure of EnvZ and CheA. The H-domain of NtrB carries the active site histidine and is homologous to subdomain A of EnvZ (Park *et al.*, 1998), the structure of which has been solved by NMR (Tomomori *et al.*, 1999). The N- and the G-domains together comprise the kinase domain. Comparison with the 3D-structure of the kinase domains of CheA (Bilwes *et al.*, 1999) and EnvZ (Tanaka *et al.*, 1998) reveals, that the N- and G-domains are subdomains of the kinase domain, with the ATP binding in a cleft between these subdomains.

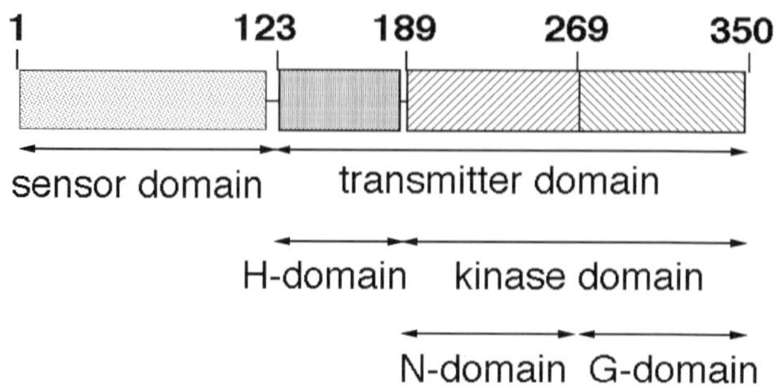

Figure 1. Schematic representation of the domain organization of NtrB.

The H-domain has Both Phospho-Transfer and Phosphatase Activity

To analyze the function of these subdomains, we have performed a deletion analysis of NtrB (Kramer and Weiss, 1999). The analysis revealed that all fragments, which carry the H-domain, but lack all or part of the kinase domain, behave as constitutive negative regulators of glutamine synthetase expression *in vivo*. In cells grown under nitrogen starvation conditions, the level of glutamine synthetase decreased about 20 fold in the presence of these fragments, compared to that of strains carrying wildtype *ntrB*. In accordance with this result, purified fragments carrying the H-domain stimulate the fast dephosphorylation of NtrC-P *in vitro* (Kramer and Weiss, 1999; Jiang *et al.*, 2000). Therefore, the H-domain is the final target in the regulation of NtrB. It carries both the site for the positive regulatory function – the conserved histidine for the phosphotransfer to NtrC – but also the "active site" for the negative regulatory activity of NtrB.

In contrast to the H-domain, the transmitter domain behaves as a positive regulator when separately expressed. Moreover, positive regulation is also achieved by co-expression of a fragment carrying the H-domain together with a fragment carrying the kinase domain, or, by co-expression of a fragment carrying the H-domain and the N-subdomain together with a fragment carrying the G-subdomain (Kramer and Weiss, 1999; Dünnebier, Kramer and Weiss, unpublished). This demonstrates, that the subdomains are able to fold independently and interact correctly to promote the phosphorylation of NtrC. Furthermore, it demonstrates, that the kinase domain inhibits the negative regulatory activity of the H-domain. This requires the intact kinase domain. Neither the N- nor the G-subdomain alone is sufficient for inhibition of the negative regulatory activity of the H-domain, suggesting, that inhibition of the negative regulatory activity of the H-domain requires the phosphorylation of the conserved histidine, His139. In addition, the constitutive positive phenotype is only observed, if the kinase domain is present at a higher copy number than the H-domain. If the H-domain is present at a higher copy number, a constitutive negative phenotype is observed, indicating that excess H-domain counteracts the kinase domain.

The results described above provide the clue to the mechanism of regulation of NtrB. NtrB might be regulated by regulating the interaction of H-domain and kinase domain. Under nitrogen starvation conditions, that is, in the absence of non-uridylylated PII, interaction of kinase and H-domain results in the phosphorylation of the conserved histidine and in the subsequent phosphorylation of NtrC. In this conformation, the "active site" for the negative regulatory activity of the H-domain might be inhibited or the H-domain might adopt a conformation in which it is inactive as a negative regulator. Under nitrogen excess conditions, NtrB is activated as a phosphatase. To this end the H-domain has to be liberated. This is the task of the PII protein in concert with the sensor domain. In other words, NtrB is regulated by an inhibitory cascade: the liberated H-domain acts as a negative regulator, this negative regulatory activity is inhibited by the kinase domain. PII and sensor domain might work together to relieve the inhibition exerted by the kinase domain.

We performed a similar deletion analysis with PhoR, the histidine kinase for phosphate regulation (Kramer, Flotho, and Weiss, unpublished). We found that all fragments, which carried the H-domain of PhoR (codons 199 to 273) but lacked all or part of the kinase domain (codons 274 to 431), behaved as strong negative regulators of *phoA* expression. This indicates, that the proposed mechanism could also apply for other bifunctional histidine kinases.

Another example is EnvZ, the histidine kinase for osmoregulation. In EnvZ two subdomains of the transmitter have been characterized, subdomain A and subdomain B (Park *et al.*, 1998). As mentioned above, subdomain A corresponds to the H-domain of NtrB. When separately expressed, subdomain A behaves as a negative regulator of *ompF* expression (Zhu *et al.*, 2000). In accordance with this, purified subdomain A stimulates the dephosphorylation of OmpR-P *in vitro*. In contrast, in the presence of the kinase domain, subdomain A has the opposing activity: it is phosphorylated in trans by the kinase domain and subsequently serves as the phosphodonor for OmpR. These results are in agreement with the results obtained for NtrB (Kramer and Weiss, 1999). However, the mechanism, by which subdomain A promotes the dephosphorylation of its cognate response regulator, might be different than that of NtrB. While in NtrB the conserved His is not required for the dephosphorylation of NtrC-P, conflicting results have been reported for EnvZ in this regard.

Data obtained with derivatives carrying the intact transmitter domain of EnvZ, in which the conserved His was changed to several other amino acids, suggest, that His243 is not required for the negative regulatory activity of EnvZ. For example, an EnvZ mutant, in which His243 has been changed to Tyr, behaves as a negative regulator of *ompF* transcription *in vivo* (Hsing and Silhavy, 1997). In accordance with this result, soluble derivatives of EnvZ lacking the sensor domain, in which His243 has been changed to Tyr or certain other amino acids, stimulate the dephosphorylation of OmpR-P *in vitro* (Skarphol *et al.*, 1997). These results suggest, that EnvZ, like NtrB, acts as a cophosphatase by stimulating an intrinsic autophosphatase activity in OmpR. However, for subdomain A it has been shown, that the conserved His is required for its negative regulatory activity. Substitution of His243 by other amino acids, including Tyr, results in the inactivation of subdomain A (Zhu *et al.*, 2000). In fact, a phosphorylated intermediate of subdomain A can be observed if this domain is incubated with OmpR-P. This suggests a different mechanism for the dephosphorylation of OmpR-P, which involves the reverse transfer of the phosphoryl group to His243 of EnvZ, followed by the hydrolysis of the phosphohistidine intermediate.

Reverse transfer of the phosphoryl group from the response regulator to the histidine kinase can also occur between NtrC-P and NtrB (Weiss and Magasanik,

1988). However, in marked contrast to EnvZ-P, NtrB-P is much more stable. Hydrolysis of NtrB-P can not be observed at a detectable rate. The phosphoryl group can either be transferred back to NtrC again, or, if present in the reaction mixture, to ADP, resulting in the generation of ATP. Thus, the reverse transfer to NtrC is not the mechanism, by which NtrB catalyses the fast dephosphorylation of NtrC-P. This reaction requires PII, or its paralogue GlnK, and phosphorylation of NtrB by NtrC-P has not been observed in the presence of PII. Moreover, it has been shown, that a mutant NtrB, in which the active site histidine has been changed to an asparagine, behaves as a strong negative regulator of glutamine synthetase expression *in vivo* (Atkinson and Ninfa, 1993). *In vitro*, this mutant stimulates the fast dephosphorylation of NtrC-P in the absence of PII. In the presence of PII the phosphatase activity was even stronger than that of wildtype NtrB and PII (Jiang *et al.*, 2000). In addition, the negative regulatory activity of NtrB fragments, which carry the H-domain but lack all or part of the kinase domain, is not affected, if the conserved His is changed to an Asn (Kramer and Weiss, unpublished).

The notion that the conserved His is not essential for the negative regulatory activity of NtrB is further supported by recent data obtained in a detailed deletion analysis of the H-domain (Dünnebier and Weiss, unpublished). Successive deletions of the C- and N-terminal parts of the H-domain revealed that a small fragment, encoding amino acids 165 to 189 of NtrB, is still active as a negative regulator of glutamine synthetase expression *in vivo*, when fused to the N-terminal domain of λ-repressor or to thioredoxin. In cells grown under nitrogen starvation conditions, expression of this fragment causes a 15 fold reduction of the glutamine synthetase level compared to that in the presence of wildtype NtrB, or to that of the *ntrB* null, in which NtrC is phosphorylated by acetyl phosphate. Interestingly, the homologous peptide in EnvZ (codons 264 to 288) overlaps with helix II of subdomain A, and region X. Region X has been identified in EnvZ (Hsing *et al.*, 1998). It was proposed that this region is important for the phosphatase activity of EnvZ, since several mutations conferring a constitutive positive phenotype localize to this region. However, the fragment lacks the region carrying the H-motif, which belongs to helix I in subdomain A of EnvZ. This suggests, that in NtrB the major determinants for the negative regulatory activity do not overlap with the H-motif.

The Role of the Linker Domain in Signal Transduction

For histidine kinases it is still an open question, how the signal is transduced from the sensor to the transmitter domain. It is believed, that the signal is not transmitted by a direct interaction between sensor and transmitter module, but rather, that the signal is transduced through the linker to the transmitter domain (Cavicchioli *et al.*, 1996; Collins *et al.*, 1992; Kalman and Gunsalus, 1990; Park and Inouye, 1997; Singh *et al.*, 1998; Tokishita *et al.*, 1992; Williams and Stewart, 1999). For NtrB there are several indications, that, as for other histidine kinases, the linker plays an essential role in the signal transduction process. The linker connecting sensor and H-domain is a hotspot for mutations conferring a constitutive positive phenotype (Atkinson and Ninfa, 1993; Kramer, Dünnebier and Weiss, unpublished). Similarly, insertion of an additional flexible linker between sensor and H-domain results in the inactivation of the sensor domain (Kramer and Weiss, unpublished). Such an NtrB derivative behaves as a constitutive positive regulator of glutamine synthetase expression, unable to respond to the PII protein. In addition, mutations conferring a constitutive phenotype can also be found in the linker connecting the H-domain with the kinase domain (Kramer, Dünnebier and Weiss, unpublished). Due to the homology of the H-domain to subdomain A it is likely, that both linkers are close together in the 3D structure and possibly form a linker domain. This linker domain might integrate conformational signals from the sensor and the kinase domain and transduce it to the H-domain. As previously mentioned, such conformational signals could originate from the interaction of PII with the kinase domain or possibly from the interaction of a ligand interacting with the sensor domain.

Interestingly, it is possible, to substitute the sensor domain of NtrB by another, completely unrelated domain, and thereby induce the negative regulatory activity. Using an approach from Cochran and Kim (1996), we have constructed a set of fusion proteins, in which the coiled–coil domain of GCN4 from yeast was fused to the transmitter domain of NtrB (Kramer and Weiss, unpublished). All seven constructs carry the same fragment from GCN4 but differ in the fusion point to the transmitter domain. The smallest fusion protein carries the GCN4-domain fused to codon 127 of NtrB. The other constructs carry each one additional amino acid more of the linker region. For all fusion proteins regulation was lost, six behaved as weak constitutive positive regulators. However, one construct behaved as a strong constitutive negative regulator of glutamine synthetase expression. Thus, in accordance with previous results, the sensor domain is not essential for the phosphatase activity. The data support a model in which the conformation of the H-domain is under the control of the linker domain.

Acknowledgements

This work was supported by a Priority Program "Regulatory Networks in Bacteria" of the Deutsche Forschungsgemeinschaft.

References:

Atkinson, M.R., and Ninfa, A.J. 1993. Mutational analysis of the bacterial signal-transducing protein kinase/phosphatase nitrogen regulator II (NRII or NtrB). J. Bacteriol. 175: 7016–7023.

Atkinson M.R., Kamberov E.S., Weiss R.L., and Ninfa A.J. 1994. Reversible uridylylation of the *Escherichia coli* PII signal transduction protein regulates its ability to stimulate the dephosphorylation of the transcription factor nitrogen regulator I (NRI or NtrC). J. Biol. Chem. 269: 288–293

Bilwes, A.M., Alex, L.A., Crane, B.R., and Simon, M.I. 1999. Structure of CheA, a signal-transducing histidine kinase. Cell 96: 131–41.

Bueno, R., Pahel, G., and Magasanik, B. 1985. Role of *glnB* and *glnD* gene products in regulation of the *glnALG* operon of *Escherichia coli*. J. Bacteriol. 164: 816–822.

Cavicchioli, R., Chiang, R.C., Kalman, L.V., and Gunsalus, R.P. 1996. Role of the dependent periplasmic domain of the *Escherichia coli* NarX sensor-transmitter protein in nitrate-signal transduction and gene regulation. Mol. Microbiol. 21: 901–911.

Chang, C., and Stewart, R.C. 1998. The two-component system. Regulation of diverse signaling pathways in prokaryotes and eukaryotes. Plant Physiol. 117: 723–731.

Cochran, A.G., and Kim, P.S. 1996. Imitation of *Escherichia coli* aspartate receptor signaling in engineered dimers of the cytoplasmic domain. Science 271: 1113–1116.

Collins, L.A., Egan, S.M., and Stewart, V. 1992. Mutational analysis reveals functional similarity between NARX, a nitrate sensor in *Escherichia coli* K-12, and the methyl-accepting chemotaxis proteins. J. Bacteriol. 174: 3667–3675.

Feng, J., Atkinson, M.R., McCleary, W., Stock, J.B., Wanner, B.L., and Ninfa, A.J. 1992. Role of phosphorylated metabolic intermediates in the regulation of glutamine synthetase synthesis in *Escherichia coli*. J. Bacteriol. 174: 6061–6070.

Hess, J.F., Oosawa, K., Kaplan, N., and Simon, M.I. 1988. Phosphorylation of three proteins in the signaling pathway of bacterial chemotaxis. Cell 53: 79–87.

Hsing, W., Russo, F.D., Bernd, K.K., and Silhavy, T.J. 1998. Mutations that alter the kinase and phosphatase activities of the two-component sensor EnvZ. J. Bacteriol. 180: 4538–4546.

Hsing, W., and Silhavy, T.J. 1997. Function of conserved histidine-243 in phosphatase activity of EnvZ, the sensor for porin osmoregulation in *Escherichia coli*. J. Bacteriol. 179: 3729–3735.

Jiang, P., Atkinson, M.R., Srisawat, C., Sun, Q., and Ninfa, A.J. 2000. Functional dissection of the dimerization and enzymatic activities of *Escherichia coli* nitrogen regulator II and their regulation by the PII protein. Biochemistry 39: 13433–13449.

Jiang, P., Peliska, J.A., and Ninfa, A.J. 1998. Enzymological characterization of the signal-transducing uridylyltransferase/uridylyl-removing enzyme (EC 2.7.7.59) of *Escherichia coli* and its interaction with the PII protein. Biochemistry 37: 12782–12794.

Jiang, P., Peliska, J.A., and Ninfa, A.J. 1998. Reconstitution of the signal-transduction bicyclic cascade responsible for the regulation of Ntr gene transcription in *Escherichia coli*. Biochemistry 37: 12795–127801.

Kamberov, E.S., Atkinson, M.R., and Ninfa, A.J. 1995. The *Escherichia coli* PII signal transduction protein is activated upon binding 2-ketoglutarate and ATP. J. Biol. Chem. 270: 17797–17807.

Kalman, L.V., and Gunsalus, R.P. 1990. Nitrate- and molybdenum–independent signal transduction mutations in narX that alter regulation of anaerobic respiratory genes in *Escherichia coli*. J. Bacteriol. 172: 7049–7056.

Kramer, G., and Weiss, V. 1999. Functional dissection of the transmitter module of the histidine kinase NtrB in *Escherichia coli*. Proc. Natl. Acad. Sci. USA 96: 604–609.

Little, J.W. 1993. LexA cleavage and other self-processing reactions. J. Bacteriol. 175: 4943–4950.

Lukat, G.S., McCleary, W.R., Stock, A.M., and Stock, J.B. 1992. Phosphorylation of bacterial response regulator proteins by low molecular weight phospho-donors. Proc. Natl. Acad. Sci. USA. 89: 718–722.

Ninfa, A. 1996. Regulation of gene transcription by extracellular stimuli. In: *Escherichia coli* and *Salmonella*: Cellular and Molecular Biology. F. C. Neidhardt, ed. 2nd ed. American Society for Microbiology, Washington, DC. p. 1246–1262.

Ninfa, A.J., Jiang, P., Atkinson, M.R., and Peliska, J.A. 2000. Integration of antagonistic signals in the regulation of nitrogen assimilation in *Escherichia coli*. Curr Top Cell Regul. 36: 31–75.

Ninfa, A.J., and Magasanik, B. 1986. Covalent modification of the *glnG* product, NRI, by the glnL product, NRII, regulates the transcription of the glnALG operon in *Escherichia coli*. Proc. Natl. Acad. Sci. USA 83: 5909–5913.

Park, H., and Inouye, M. 1997. Mutational analysis of the linker region of EnvZ, an osmosensor in *Escherichia coli*. J. Bacteriol. 179: 4382–4390.

Park, H., Saha, S.K., and Inouye, M. 1998. Two-domain reconstitution of a functional protein histidine kinase. Proc. Natl. Acad. Sci. USA 95: 6728–6732.

Parkinson, J.S., and Kofoid, E.C. 1992. Communication modules in bacterial signaling proteins. Annu. Rev. Genet. 26: 71–112.

Perego, M. 1998. Kinase-phosphatase competition regulates *Bacillus subtilis* development. Trends Microbiol. 6: 366–370.

Perego, M., Hanstein, C., Welsh, K.M., Djavakhishvili, T., Glaser, P., and Hoch, J.A. 1994. Multiple protein-aspartate phosphatases provide a mechanism for the integration of diverse signals in the control of development in *B. subtilis*. Cell 79: 1047–1055.

Pioszak, A.A., Jiang, P., and Ninfa, A.J. 2000. The *Escherichia coli* PII signal transduction protein regulates the activities of the two-component system transmitter protein NRII by direct interaction with the kinase domain of the transmitter module. Biochemistry 39: 13450–13461

Perraud, A.L., Weiss, V., and Gross, R. 1999. Signalling pathways in two-component phosphorelay systems. Trends Microbiol. 7: 115–120.

Reitzer, L.J. 1996. Ammonia assimilation and the biosynthesis of glutamine, glutamate, aspartate, asparagine, L-alanine, and D-alanine. In: *Escherichia coli* and *Salmonella*: Cellular and Molecular Biology. F. C. Neidhardt, ed. 2nd ed. American Society for Microbiology, Washington, DC. p. 391–407.

Sanders, D.A., Gillece-Castro, B.L., Burlingame, A.L., and Koshland, D.E., Jr. 1992. Phosphorylation site of NtrC, a protein phosphatase whose covalent intermediate activates transcription. J. Bacteriol. 174: 5117–5122.

Singh, M., Berger, B., Kim, P.S., Berger, J.M., and Cochran, A.G. 1998. Computational learning reveals coiled coil-like motifs in histidine kinase linker domains. Proc. Natl. Acad. Sci. USA 95: 2738–2743.

Skarphol, K., Waukau, J., and Forst, S.A. 1997. Role of His243 in the phosphatase activity of EnvZ in *Escherichia coli*. J. Bacteriol. 179: 1413–1416.

Stock, J.B., Surette, M.G., Levit, M., and Park, P. 1995. Two-component signal transduction systems: structure-function relationships and mechanisms of catalysis. In: Two-component signal transduction. J. A. Hoch and T. J. Solhavy, (eds.), ASM Press, Washington, D.C. p. 25–51.

Tanaka, T., Saha, S.K., Tomomori, C., Ishima, R., Liu, D., Tong, K.I., Park, H., Dutta, R., Qin, L., Swindells, M.B., Yamazaki, T., Ono, A.M., Kainosho, M., Inouye, M., and Ikura, M. 1998. NMR structure of the histidine kinase domain of the *E. coli* osmosensor EnvZ. Nature 396: 88–92.

Tokishita, S., Kojima, A., and Mizuno, T. 1992. Transmembrane signal transduction and osmoregulation in *Escherichia coli*: functional importance of the transmembrane regions of membrane-located protein kinase, EnvZ. J. Biochem. (Tokyo). 111: 707–713.

Tomomori, C., Tanaka, T., Dutta, R., Park, H., Saha, S.K., Zhu, Y., Ishima, R., Liu, D., Tong, K.I., Kurokawa, H., Qian, H., Inouye, M., and Ikura, M. 1999. Solution structure of the homodimeric core domain of *Escherichia coli* histidine kinase EnvZ. Nat. Struct. Biol. 6: 729–734.

Weiss, V., and Magasanik, B. 1988. Phosphorylation of nitrogen regulator I (NRI) of *Escherichia coli*. Proc. Natl. Acad. Sci. USA 85: 8919–8923.

Williams, S.B., and Stewart, V. 1999. Functional similarities among two-component sensors and methyl-accepting chemotaxis proteins suggest a role for linker region amphipathic helices in transmembrane signal transduction. Mol. Microbiol. 33: 1093–1102.

Zhu, Y., Qin, L., Yoshida, T., and Inouye, M. 2000. Phosphatase activity of histidine kinase EnvZ without kinase catalytic domain. Proc. Natl. Acad. Sci. USA 97: 7808–7813.

Chapter 12

Regulation of Nitrogen Fixation in *Klebsiella pneumoniae* and *Azotobacter vinelandii:* NifL, Transducing Two Environmental Signals to the *nif* Transcriptional Activator NifA

Ruth A. Schmitz, Kai Klopprogge,
Roman Grabbe, and Jessica Stips

Abstract

The enzymatic reduction of molecular nitrogen to ammonia requires high amounts of energy, and the presence of oxygen causes the catalyzing nitrogenase complex to be irreversible inactivated. Thus nitrogen-fixing microorganisms tightly control both the synthesis and activity of nitrogenase to avoid the unnecessary consumption of energy. In the free-living diazotrophs *Klebsiella pneumoniae* and *Azotobacter vinelandii*, products of the nitrogen fixation *nifLA* operon regulate transcription of the other *nif* operons. NifA activates transcription of *nif* genes by the alternative form of RNA-polymerase, σ^{54}-holoenzyme; NifL modulates the activity of the transcriptional activator NifA in response to the presence of combined nitrogen and molecular oxygen. The translationally-coupled synthesis of the two regulatory proteins, in addition to evidence from studies of NifL/NifA complex formation, imply that the inhibition of NifA activity by NifL occurs via direct protein-protein interaction *in vivo*. The inhibitory function of the negative regulator NifL appears to lie in the *C*-terminal domain, whereas the *N*-terminal domain binds FAD as a redox-sensitive cofactor, which is required for signal transduction of the internal oxygen status. Recently it was shown, that NifL acts as a redox-sensitive regulatory protein, which modulates NifA activity in response to the redox-state of its FAD cofactor, and allows NifA activity only in the absence of oxygen. In *K. pneumoniae,* the primary oxygen sensor is Fnr (fumarate nitrate reduction regulator), which is presumed to transduce the signal of anaerobiosis towards NifL by activating the transcription of gene(s) whose product(s) reduce the FAD cofactor. Reduction of the cofactor eventually results in sequestration of NifL to the cytoplasmic membrane allowing NifA to activate *nif* genes in the cytoplasm. In contrast, the reduction of *A. vinelandii*-NifL appears to occur unspecifically in response to the availability of reducing equivalents in the cell. Nitrogen status of the cells is transduced towards the NifL/NifA regulatory system by the GlnK protein, a paralogue PII-protein, which appears to interact with the NifL/NifA regulatory system via direct protein-protein interaction. It is not currently known whether GlnK interacts with NifL alone or affects the NifL/NifA-complex; moreover the effects appear to be the opposite in *K. pneumoniae* and *A. vinelandii.* In addition to these environmental signals, adenine nucleotides also affect the inhibitory function of NifL; in the presence of ATP or ADP the inhibitory effect on NifA activity *in vitro* is increased. The NifL proteins from the two organisms differ, however, in that stimulation of *K. pneumoniae*-NifL occurs only when synthesized under nitrogen excess, and is correlated with the ability to hydrolyze ATP. In general, transduction of environmental signals to the *nif* regulatory system appears to involve a conformational change of NifL or the NifL/NifA complex. However, experimental data suggest that *K. pneumoniae* and *A. vinelandii* employ significantly different species-specific mechanisms of signal transduction.

Introduction

Biological nitrogen fixation, the enzymatic reduction of molecular nitrogen (N_2) to ammonia, is strictly limited to prokaryotes. However, within the prokaryotes nitrogen fixation is found in a large number of species belonging to the bacterial domain, as well as in several methanogenic *Archaea* (Dean and Jacobson, 1992; Young, 1992; Lobo and Zinder, 1992; Fischer, 1994). The reduction of molecular nitrogen is catalyzed by the nitrogenase enzyme complex, and has a high energy demand, two ATP molecules are consumed for each electron transferred to the catalytic site (Burgess and Lowe, 1996; Howard and Rees, 1996; Rees and Howard, 1999). Upon the high energy requirement, in nitrogen fixing cells up to 40% of the ATP is utilized by the nitrogenase resulting in a drop of the energy charge (Daesch and Mortenson, 1972; Upchurch and Mortenson, 1980). In addition the nitrogenase enzyme complex is highly sensitive to molecular oxygen, and becomes irreversibly inactivated by its presence. Thus, to avoid unnecessary consumption of energy nitrogen-fixing microorganisms tightly control both the synthesis and activity of nitrogenase in response to oxygen and nitrogen availability. In all diazotrophic proteobacteria examined, the transcriptional activator NifA is required for expression of the nitrogen fixation (*nif*) genes. NifA expression and activity is regulated in response to the environmental signals of molecular oxygen and combined nitrogen. However, the mechanisms involved in this control vary in different organisms (Fischer, 1994; Fischer, 1996; Dixon, 1998; Halbleib and Ludden, 2000). In free-living and symbiotic diazotrophs belonging to the α and β subgroup of the Proteobacteria (genera *Rhizobium*, *Bradyrhizobium*, *Azospirillum* and *Herbaspirillum*) NifA activity is directly sensitive to molecular oxygen, and, in some cases, is also directly affected by combined nitrogen (Souza *et al.*, 1999; Monteiro *et al.*, 1999a and 1999b; Arsene *et al.*, 1999; Steenhoudt and Vanderleyden, 2000). However, in the free-living diazothrophs *Klebsiella pneumoniae, Azotobacter vinelandii*, and *Enterobacter cloacae*, which belong to the γ-Proteobacteria, NifA activity is not directly sensitve to oxygen. In these organisms NifA activity is regulated in response to molecular oxygen and combined nitrogen through a second regulator NifL; the *nifL* and *nifA* genes together form an operon (Filser *et al.*, 1983; Dixon, 1998, Xiao *et al.,* 1998). In *K. pneumoniae,* the expression of the *nifLA* operon itself is regulated by the nitrogen status of the cell, via the NtrB/NtrC two component regulatory system; in *A. vinelandii,* however, *nifLA* is expressed constitutively (Drummond and Wootton, 1987; Toukdarian and Kennedy, 1986; Blanco *et al.,* 1993). Interestingly, it was recently found that nitrogen fixation in the endophytic diazotroph *Azoarcus sp.* BH72 – belonging to the β-Proteobacteria – is also regulated by the coordinated activities of the homologous *nifL* and *nifA* gene products in response to environmental signals (Reinhold-Hurek *et al.,* 1993; Egener *et al.*, 2002). In this review, we will focus on the transduction of signals for the presence of molecular oxygen and combined nitrogen in *K. pneumoniae* and *A. vinelandii*, and compare the mechanisms by which these signals are mediated to the NifL/NifA system in the two organisms.

NifL Modulates NifA Transcriptional Activity by Direct Protein-Protein Interaction

The transcriptional activator NifA is composed of three domains: an amino (*N*)-terminal domain, apparently involved in regulation, a central catalytic domain, and a carboxy (*C*)-terminal DNA-binding domain (Drummond *et al.*, 1990; Morett and Segovia, 1993). Transcription of *nif* genes by an alternative RNA polymerase (σ^{54}-RNA holoenzyme) is generally activated by NifA, which binds to an upstream activation sequence (UAS) (Morrett and Buck, 1988) and contacts promoter-bound σ^{54}-RNA polymerase by means of a DNA loop (Buck *et al.*,1987). Subsequently, NifA catalyzes the isomerization of closed complexes between σ^{54}-holoenzyme and the *nif* promoter to transcriptionally-productive open complexes (Morett and Buck, 1989; Hoover *et al.,* 1990). This open complex formation requires the hydrolysis of ATP or GTP catalyzed by NifA (Lee *et al.*, 1993a; Austin *et al.,* 1994). In the presence of molecular oxygen or combined nitrogen, NifL inhibits NifA transcriptional activity *in vivo* (Merrick *et al.,* 1982; Hill *et al.,* 1981; Dixon, 1998). Recent findings indicate that *A. vinelandii*-NifL is also competent to inhibit the DNA binding function of NifA (Barett *et al.,* 2001).

The inhibitory protein NifL is composed of two domains, separated by a hydrophilic interdomain linker (Q-linker) (Drummond and Wootton, 1987) (Figure 1). The *C*-terminal domain of NifL shows homology to the histidine protein kinase (Blanco *et al.,* 1993); however no autophosphorylation, or phosphor transfer between the two regulatory proteins NifA and NifL has been detected in *K. pneumoniae* or *A. vinelandii* (Lee et *al.*, 1993b; Austin *et al.*, 1994; Schmitz *et al.*, 1996). The translationally-coupled synthesis of *nifL* and *nifA* in addition to evidence from immunological studies of complex formation, imply that the inhibition of NifA activity by NifL occurs via a direct protein-protein interaction (Govantes *et al.,* 1998; Henderson *et al.*, 1989). Recently, in *A. vinelandii* formation of NifL/NifA complexes has been demonstrated by *in vitro* co-chromatography in the presence of adenine nucleotides and using the "yeast-two-hybrid" system (Money *et al.*, 1999; Lei *et al.*, 1999; Money *et al.*, 2001). Thus, it appears that signal transduction apparently occurs via protein-protein interaction. The *C*-terminal domain of

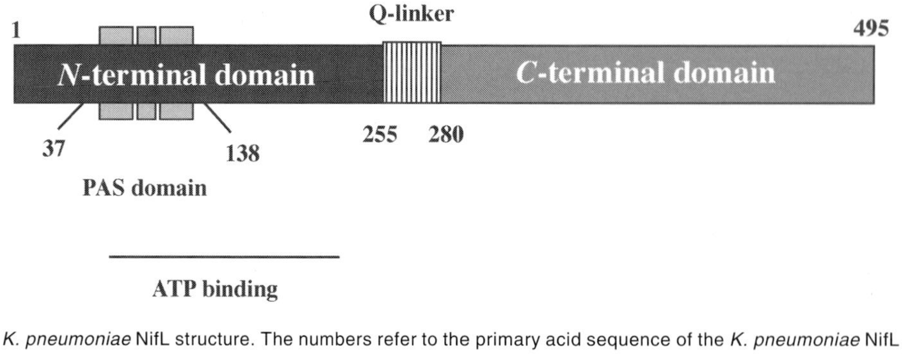

Figure 1. Cartoon of *K. pneumoniae* NifL structure. The numbers refer to the primary acid sequence of the *K. pneumoniae* NifL protein and mark the approximate boundaries of its *N*-terminal and *C*-terminal domains. The locations of the hydrophilic interdomain linker (Q-linker), the PAS domain, and the apparent ATP binding site are indicated.

K. pneumoniae-NifL is sufficient to inhibit transcriptional activation by NifA *in vitro* and *in vivo* (Narberhaus *et al.,* 1995); this indicates that the inhibitory function of the NifL protein appears to lie in its *C*-terminal domain which presumably directly interacts with NifA.

Oxygen Signal Transduction

The *N*-terminal domain of NifL contains conserved S-motifs of PAS-like domains (Figure 1), which are known for a number of regulators that sense oxygen, redox or light (Zhulin *et al.*, 1997; Taylor *et al.,* 1999). This suggests that the *N*-terminal domain is involved in environmental sensing and signal transduction. Biochemical analyses with purified proteins showed that NifL, in both *A. vinelandii* and *K. pneumoniae*, is a flavoprotein with an *N*-terminally bound FAD as cofactor (Hill *et al.*, 1996; Söderbäck *et al.,* 1998; Schmitz, 1997). Analysis of the inhibitory function of NifL-holoenzyme and NifL-apoenzyme on NifA activity in *in vitro* transcription assays showed that the FAD-cofactor is not directly required for the inhibitory function of NifL. This suggests the possibility that FAD may instead function as a redox-sensitive cofactor, involved in the oxygen signal transduction (Schmitz, 1997). The oxidized form of NifL inhibits NifA transcriptional activity *in vitro*, whereas *A. vinelandii*-NifL chemically reduced by sodium dithionite or enzymatically reduced with NADH by the flavoheme protein from *E. coli*, does not inhibit NifA activity *in vitro* (Hill *et al.*, 1996; Machereaux *et al.*, 1998); this indicates that reduction of the flavin moiety of NifL results in a non-inhibitory form of NifL. These findings support the model, that NifL acts as a redox-sensitive regulatory protein, which modulates NifA activity in response to the redox state of its FAD cofactor, and allows NifA activity only in the absence of oxygen. The redox potentials determined for *A. vinelandii*-NifL and *K. pneumoniae*-NifL are -226 mV and -277 mV, respectively at pH 8 (Macheroux *et al.,* 1998; Klopprogge and Schmitz, 1999). However, in neither organism is the physiological electron donor for NifL known.

By reducing the cofactor of NifL, the physiological electron donor is possibly transducing the signal for anaerobiosis to NifL, therefore the physiological electron donor for NifL reduction itself might be a potential component of the oxygen signal transduction. Thus, the key question concerning the oxygen signal transduction is whether NifL senses the oxygen status in the cell directly via a redox-induced conformational change, or through a more general oxygen-sensing system, which then regulates NifL by inducing the oxidation or reduction of the flavin cofactor. In this respect, it is of interest that in *K. pneumoniae* iron is specifically required for relief of NifL inhibition under oxygen and nitrogen limitation. However, analysis of NifL synthesized in the absence or presence of oxygen showed that neither iron nor acid-labile sulfur is present in NifL under either condition (Schmitz *et al.,* 1996; Schmitz, 1997); therefore the iron requirement for *nif* induction is not due to an oxygen-sensitive iron-sulfur cluster of NifL. Early studies showed that redox regulation of *nif* gene expression in a heterologous *Escherichia coli* system is independent of the *E. coli fnr* gene product, suggesting that the global regulator Fnr (fumarate nitrate reduction regulator) is not the primary oxygen sensor for the *nif* gene induction in *K. pneumoniae* (Hill, 1985). In contrast, recent data indicate that in *K. pneumoniae,* Fnr is required for NifL-dependent oxygen regulation of *nif* gene expression; strains carrying *fnr* null mutations fail to release NifL inhibition of NifA transcriptional activity under oxygen limitation. Moreover, Fnr does not affect NifA activity in the absence of NifL, suggesting that the Fnr effect is mediated through NifL towards NifA (Grabbe *et al.*, 2001a, Grabbe *et al.*, 2001b). In the absence of Fnr, NifL apparently does not receive the signal for anaerobiosis, suggesting that Fnr is the primary oxygen sensor in *K. pneumoniae*, and transduces the signal either directly or indirectly to NifL. As Fnr contains oxygen-sensitive [4Fe-4S]-centers responsible for the oxygen sensing mechanism (Khoroshilova *et al.*, 1997; Kiley and Beinert, 1998; Beinert and Kiley, 1999), the requirement for Fnr could account for the iron requirement of *nif*-induction in *K. pneumoniae.* However, from the current experimental data one cannot completely rule out the possibility that the Fnr requirement may be due to some Fnr-dependent metabolic signals not directly related to the lack of oxygen. If Fnr is indeed the primary oxygen sensor and transduces the signal of the oxygen status to the *nif* regulatory system in *K. pneumoniae*, then it remains to be explained how this signal is mediated towards NifL. Fnr can be excluded as the physiological electron donor for NifL reduction under anaerobic conditions; it is more likely that Fnr transduces the signal by transcriptional activation of genes directly involved in the reduction. Candidates for this physiological electron donor for NifL reduction include those components of the anaerobic electron transport system, which are synthesized under Fnr control. This is a particularly attractive model taking into account that the transduction of the oxygen signal towards NifL via the anaerobic electron transport chain would allow the simultaneous signal integration of the cell's energy status, as sensed through the electron flux. Localization experiments showed that *K. pneumoniae* NifL is membrane-associated under oxygen-limiting conditions, but in the cytosolic fraction when in the presence of oxygen (Klopprogge *et al.*, 2002a). Thus, the current working model of oxygen signal transduction proposes that under anaerobic conditions Fnr activates transcription of gene(s) whose product(s) reduce the FAD cofactor of NifL, resulting in a higher membrane affinity of NifL and thus in a sequestration of NifL to the cytoplasmic membrane consequently allowing cytoplasmic NifA to activate *nif* gene promoters (Grabbe *et al.*, 2001b, Klopprogge *et al.*, 2002a, Figure 2).

For *A. vinelandii,* an obligate aerobic microorganism, which apparently protects nitrogenase against oxygen damage by high oxygen consumption (so called 'respiratory protection') (Dixon, 1998; Oelze, 2000), a different mechanism for the transduction of the oxygen signal to NifL is expected. In *in vitro* experiments

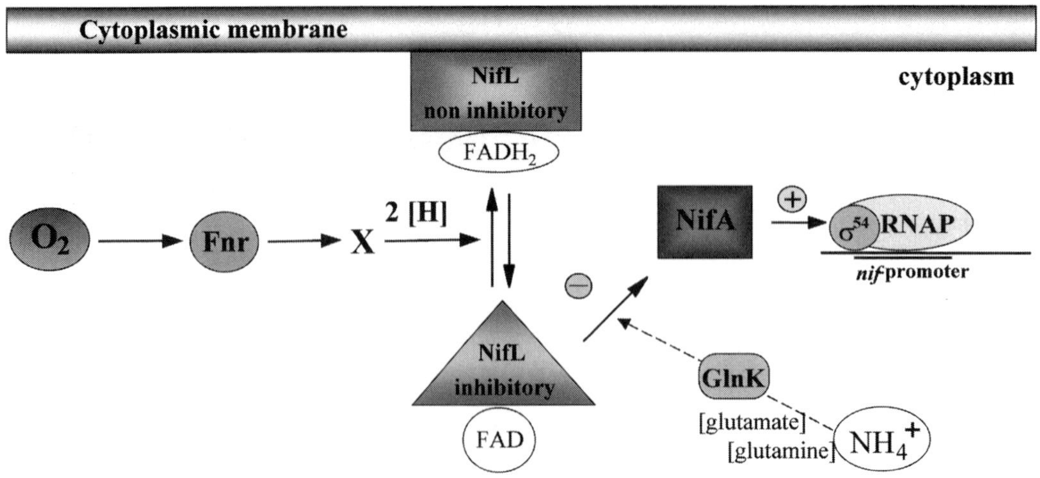

Figure 2. Current working model for oxygen and nitrogen signal transduction in *K. pneumoniae*. NifA activates the transcription of *nif* genes by the alternative RNA-polymerase (σ^{54}-RNAP). NifL modulates the activity of the transcriptional activator NifA in response to molecular oxygen and combined nitrogen. Molecular oxygen is sensed by Fnr, which transduces the signal of anaerobiosis to NifL by activating the transcription of a gene, the product of which (X) reduces the FAD cofactor of NifL resulting in membrane sequestration of NifL. The nitrogen status reflected by the internal glutamine and glutamate pools is transduced towards the NifL/NifA system by GlnK, which antagonizes the inhibitory effects of NifL under nitrogen limitation resulting in active NifA.

A. vinelandii-NifL is reduced by NADH and the reaction is catalyzed by the *E. coli* flavoheme protein (HMP), which is proposed to be a global oxygen sensor (Poole, 1994; Macheroux *et al.*, 1998); however, the functional and physiological relevance of the reduction of NifL by HMP has not been demonstrated *in vivo* to date. It is currently hypothesized, that the reduction of *A. vinelandii*-NifL occurs non-specifically and dependant on the availability of reducing equivalents in the cell; because of the relatively high redox potential of NifL, they are a number of electron donors and NAD(P)H-dependent enzymes that could be potentially be involved in reduction of NifL (Dixon, 1998; Machereux *et al.*, 1998).

Nitrogen Signal Transduction

Analyses of internal pool sizes of glutamine and glutamate – the two central nitrogen intermediates – showed that under nitrogen-limiting growth conditions the glutamine pool in *K. pneumoniae* is 6 to 9-fold lower than that under nitrogen sufficiency; under these same conditions the glutamate pool remains stable (Schmitz, 2000). This indicates that *K. pneumoniae* apparently senses external nitrogen limitation as a drop in the internal glutamine pool; thus the size of the intracellular glutamine pool appears to reflect the cellular nitrogen status, as has been shown for *Salmonella typhimurium* (Ikeda *et al.,* 1996). Changes in the internal glutamine pool, reflecting the internal nitrogen status, are mediated to the NtrB/NtrC two-component regulatory system via uridylylation and deuridylylation of the regulatory protein PII (encoded by *glnB*) by GlnD (uridylyl-transferase and uridylyl-removing-enzyme) (Merrick and Edwards, 1995; Ninfa and Atkinson, 2000).

A shift from nitrogen limitation to nitrogen sufficiency results in repression of *nif* gene induction, through inhibition of NifA transcriptional activity by NifL (Arnott *et al.*, 1989; Blanco *et al.*, 1993; Narberhaus *et al.,* 1995). This indicates that NifL either senses nitrogen availability directly, or the nitrogen status is sensed in a NifL-independent manner, and the signal is subsequently transduced to NifL or the NifL/NifA complex. Interestingly, as in *E. coli* a second PII-like protein encoded by *glnK* was recently discovered in *K. pneumoniae*, which is expressed in a NtrC dependent manner, and cotranscribed with *amtB* (encoding for an ammonium transporter). Because of high similarities to PII, the GlnK-protein is a candidate for sensing changes in the glutamine pool size – as it reflects the internal nitrogen status – and mediating the signal of nitrogen status to the *nif* regulatory system. Strong evidence was obtained from studying *nif* regulation in *glnK* mutant strains, that GlnK is indeed required to release NifL inhibition under nitrogen-limiting growth conditions in *K. pneumoniae* (He *et al.*, 1997; He *et al.*, 1998; Jack *et al.*, 1999). This indicates that changes of the internal nitrogen status are not sensed by NifL directly, but are apparently mediated through GlnK to the NifA/NifL regulatory system. Whereas NifL is a negative regulator, GlnK apparently acts positively to antagonize inhibitory effects of NifL under nitrogen-limiting conditions. Surprisingly, and in contrast with PII, uridylylation of GlnK appears not to be essential for its nitrogen signaling function in nitrogen fixation (He *et al.*, 1998, Arcondeguy *et al.*, 1999). The T-loop of PII-like proteins, which includes the site of uridylylation, is thought to be the site of interaction with other proteins involved in signal transduction. In *K. pneumoniae,* the T-loop of GlnK and PII differ in only three amino acid residues.

It has been shown that the conserved asparagine 54 in the T-loop of GlnK, which discriminates GlnK from GlnB in *K. pneumoniae,* is a crucial amino acid residue for nitrogen signal transduction to the *nif* regulatory system, possibly directly involved in the interaction with NifL/NifA (Arcondeguy *et al.*, 2000). NifL inhibitory function is apparently rapidly restored, when ammonium is added back to the growth medium by the formation of non-functional heterotrimers between GlnK and PII (Arcondeguy *et al.*, 1999). Although GlnK function has been clearly demonstrated the question arises, as to how GlnK mediates the signal towards the NifL/NifA regulatory system. The nitrogen signal is apparently mediated by direct protein-protein interaction but it has yet to be determined, whether GlnK interacts directly with NifL or affects the NifL/NifA complex formation. For diazothrophs not belonging to the γ-proteobacteria and not containing a NifL protein (e.g. *Herbaspirillum seropedicae* and *Azospirillum brasilense*), experimental data indicate that the PII proteins participate in signaling the nitrogen status to the *N*-terminal domain of NifA (Steenhoudt and Vanderleyden, 2000; Souza *et al.,* 1999; Monteiro *et al.,* 1999a; Monteiro *et al.,* 1999b; Arsene *et al.*, 1999).

A. vinelandii, has only one PII-like protein, which is encoded in a *glnK/amtB*-operon and is constitutively expressed (Meletzus *et al.,* 1998). Interestingly, *A. vinelandii*-GlnK has a T-loop structure, which more closely resembles the 'GlnB-like' T-loop rather than the 'GlnK-like' T-loop. Recent studies concerning the role of *A. vinelandii*-GlnK in nitrogen sensing and transducing the nitrogen status to the *nif* regulatory system showed that GlnK is not required for derepression in *A. vinelandii.* In contrast to *K. pneumoniae*, where GlnK has a positive role in relieving NifL inhibition under nitrogen-limiting conditions, yeast-two-hybride studies and *in vitro* experiments suggest that the inhibitory function of *A. vinelandii*-NifL is activated under nitrogen excess through direct protein-protein interaction with the non-uridylylated form of GlnK. Under nitrogen limitation, the inhibitory activity of *A. vinelandii*-NifL is apparently relieved by elevated levels of 2-oxoglutarate when PII-like regulatory proteins are uridylylated (Reyes-Ramirez *et al.,* 2000, Reyes-Ramirez *et al.*, 2001, Little *et al.*, 2000, Little *et al.*, 2002, Rudnick, *et al.*, 2002). These findings for GlnK function in *K. pneumoniae* and *A. vinelandii* indicate that interactions between GlnK and the NifL/NifA complex are likely to be specific to each organism.

Biochemical analysis of *K. pneumoniae*-NifL synthesized under different nitrogen availabilities showed that the nitrogen status of the cells does not affect the redox potential of NifL. However, NifL appears to undergo an as yet unidentified covalent modification when synthesized in the presence of ammonium, as determined by absorbance at 420 nm (Klopprogge and Schmitz, 1999). A similar phenomenon was observed for *A. vinelandii*-NifL synthesized under different nitrogen availabilities (Klopprogge and Schmitz, unpublished results). Thus the question arises, whether this covalent modification in the presence of ammonium is of relevance for nitrogen signal transduction; if this is so, it remains to be determined, whether the modification occurs in a GlnK-dependent manner.

Effects of Adenine Nucleotides on NifL

The *C*-terminal domain of *A. vinelandii*-NifL shows sequence similarity to members of the histidine kinase family and binds adenine nucleotides, indicating that the *C*-terminal domain may have evolved from a classical histidine protein kinase to a domain responsive to nucleotides (Söderbäck *et al.,* 1998). Interestingly, ADP-binding to the *C*-terminal domain specifically increases the inhibitory activity of *A. vinelandii*-NifL on open complex formation by NifA *in vitro,* independent of the redox response (Eydmann *et al.,* 1995; Söderbäck *et al.,* 1998). Proteolysis protection analyses further indicated that ADP-binding to the *C*-terminal domain induces a change in NifL conformation (Söderbäck *et al.,* 1998; Dixon, 1998), which apparently promotes complex formation between purified NifL and NifA *in vitro,* as shown by co-chromatography experiments (Money *et al.,* 1999). These findings and the higher affinity of NifL for ADP versus ATP, suggest that the formation of the inhibitory complex between NifL and NifA might be regulated by the ADP/ATP ratio in response to the energy charge *in vivo* (Söderbäck *et al.,* 1998). The *C*-terminal domain of *K. pneumoniae*-NifL, however, is distinct from that of *A. vinelandii*-NifL and lacks three out of the five sequence motifs found in the histidine protein kinase family (Woodley and Drummond, 1994); this raises the question, of whether adenine nucleotides also affect *K. pneumoniae*-NifL. Recent studies have shown, that the inhibitory function of *K. pneumoniae*-NifL – synthesized in the presence of sufficient nitrogen – is also stimulated in the presence of ATP or ADP prior to open complex formation *in vitro.* Unexpectedly, and in contrast to *A. vinelandii* NifL, *K. pneumoniae*-NifL synthesized in the presence of sufficient nitrogen exhibits significant ATP-hydrolyzing activity. However, when synthesized under nitrogen starvation, NifL showed neither hydrolysis of ATP, nor an increase in the inhibitory function in the presence of adenine nucleotides (Klopprogge *et al.*, 2002b). These findings suggest the possibility that stimulation of the inhibitory function and ATP-hydrolysis in *K. pneumoniae*-NifL may depend on a specific NifL conformation induced by ammonium. Furthermore it has been shown that changing from nitrogen sufficiency to nitrogen limitation results in an increase of the ADP/ATP ratio in *K. pneumoniae* (Upchurch and Mortenson, 1980). Thus, under nitrogen fixing conditions – when NifL is in its non-inhibitory form – the ADP concentration is high as compared with levels seen for growth under nitrogen sufficiency. These findings strongly indicate that the signal transduction of energy charge in *K. pneumoniae* is not mediated by NifL to NifA in response to the ADP/ATP ratio as proposed for *A. vinelandii.*

The finding that *K. pneumoniae*-NifL catalyzes ATP-hydrolysis, when synthesized under nitrogen sufficiency was unexpected, and raises questions about the physiological function of this activity. Interestingly, stimulation of the inhibitory function of NifL by adenine nucleotides only occurs under nitrogen sufficiency, when NifL exhibits both an ATP- and an ADP-hydrolysing activity, suggesting that hydrolysis of adenine nucleotides may play a key role in the stimulation of NifL inhibitory function observed in the presence of adenine nucleotides. If this is indeed the case, one could speculate that hydrolysis of adenine nucleotides induces a NifL conformational change promoting tighter or more stable complex formation between NifA and NifL.

Conclusions

In *Klebsiella pneumoniae* and *Azotobacter vinelandii,* NifL inhibits NifA transcriptional activity in response to environmental changes, so as to tightly control nitrogen fixation and avoid the unnecessary consumption of energy. The inhibition of NifA activity by NifL occurs via direct protein-protein interaction as shown by immunological studies and the demonstration of complex formation between NifL and NifA by co-chromatography and the "yeast-two-hybrid"-system (Henderson *et al.,* 1989; Money *et al.,* 1999; Lei *et al.,* 1999). Thus, one can expect that changes in the oxygen, nitrogen or energy level result in either conformational changes of the NifL protein, promoting complex formation with NifA, or affecting the NifL/NifA-complex formation directly. This hypothesis is supported by the findings that (i) the presence of molecular oxygen apparently results in a redox induced conformational change of NifL, which strongly interacts with NifA (Hill *et al.,* 1996; Machereux *et al.,* 1998; Schmitz, 1997), (ii) the nitrogen signal is transduced by GlnK to the *nif* system by affecting either NifL conformation or the NifL/NifA complex, and (iii) adenine nucleotides stimulate inhibitory function of NifL apparently through a conformational change of NifL (Eydmann *et al.*, 1995; Söderbäck *et al.*, 1998; Klopprogge *et al.*, 2002b). Although, signal transduction to the *nif* regulatory system appears to result in conformational changes of NifL or the NifL/NifA complex, experimental data suggest that the mechanisms of signal transduction pathways in *A. vinelandii* and *K. pneumoniae* differ significantly. In *K. pneumoniae,* the primary oxygen sensor is the global regulator Fnr, which transduces the signal to NifL resulting in membrane sequestration of NifL. *In A. vinelandii* the primary oxygen sensor appears to be NifL, which is reduced dependant on availability of reducing equivalents in the cell. The nitrogen signal is mediated in both organisms via the GlnK protein; however, in *A. vinelandii,* GlnK apparently promotes the formation of the inhibitory NifL/NifA complex under nitrogen excess, whereas in *K. pneumoniae,* GlnK functions to relieve NifL inhibition under nitrogen-limiting conditions. Binding of adenine nucleotides to the *C*-terminal domain induces a conformational change of *A. vinelandii*-NifL, which promotes complex formation between NifA and NifL. In contrast to *A. vinelandii*, the stimulatory effects of adenine nucleotides on the inhibitory function of *K. pneumoniae*-NifL only occur when NifL is synthesized under nitrogen sufficiency, and correlate with the ability of NifL to hydrolyze ATP; hydrolysis of adenine nucleotides apparently induces a conformational change of NifL resulting in increased formation of inhibitory NifL/NifA complexes. Thus, *K. pneumoniae* and *A. vinelandii* employ species-specific mechanisms for the transduction of environmental signals towards NifL.

Acknowledgments

We thank Gerhard Gottschalk for generous support and helpful discussions and Andrea Shauger for critical reading the manuscript. This work was supported by the Deutsche Forschungsgemeinschaft (SCHM1052/4-3) and the Fonds der Chemischen Industrie.

References

Arcondeguy, T., Lawson, D., and Merrick, M. 2000. Two residues in the T-loop *of Klebsiella pneumoniae* GlnK determine NifL-dependent nitrogen control of *nif* gene expression. J. Biol. Chem. 275: 38452–38456.

Arcondeguy, T., van Heeswijk, W.C., and Merrick, M. 1999. Studies on the roles of GlnK and GlnB in regulating *Klebsiella pneumoniae* NifL-dependent nitrogen control. FEMS Microbiol. Lett. 179: 263–270.

Arnott, M., Sidoti, C., Hill, S., and Merrick, M. 1989. Deletion analysis of the nitrogen fixation regulatory gene *nifL* of *Klebsiella pneumoniae.* Arch. Microbiol. 151: 180–182.

Arsene, F., Kaminski, P.A., and Elmerich, C. 1999. Control of *Azospirillum brasilense* NifA activity by P(II): effect of replacing Tyr residues of the NifA N-terminal domain on NifA activity. FEMS Microbio. Lett. 179: 339–343.

Austin, S., Buck, M., Cannon, W., Eydmann,T., and Dixon, R. 1994. Purification and *in vitro* activities of the native nitrogen fixation control proteins NifA and NifL. J. Bacteriol. 176: 3460–3465.

Barrett, J., Ray, P., Sobczyk, A., Little, R., and Dixon, R. 2001. Concerted inhibition of the transcriptional activation functions of the enhancer-binding protein NIFA by the anti-activator NIFL. Mol. Microbiol. 39: 480–494.

Beinert, H., and Kiley, P.J. 1999. Fe-S proteins in sensing and regulatory functions. Opin. Chem. Biol. 3: 152–157.

Blanco, G., Drummond. M., Woodley, P., and Kennedy, C. 1993. Sequence and molecular analysis of the *nifL* gene of *Azotobacter vinelandii.* Mol. Microbiol. 9: 869–880.

Buck, M., Cannon, W., and Woodcock, J. 1987. Transcriptional activation of the *Klebsiella pneumoniae* nitrogenase promoter may involve DNA loop formation. Mol. Microbiol. 1: 243–249.

Burgess, B.K., and Lowe, D.J. 1996. Mechanism of molybdenum nitrogenase. Chem. Rev. 96: 2983–3011.

Daesch, G., and Mortenson, L.E. 1972. Effect of ammonia on the synthesis and function of the N_2-fixing enzyme systeme in *Clostridium pasteurianum.* J. Bacteriol. 110: 103–109.

Dean, D.R., and Jacobson, M.R. 1992. Biochemical genetics of nitrogenase In: Biological nitrogen fixation. G. Stacey, R.H. Burris and H.J. Evans, (eds.), Chapman & Hall, New York. p. 763–834.

Dixon, R. 1998. The oxygen-responsive NifL-NifA complex: a novel two-component regulatory system controlling nitrogenase synthesis in gamma-proteobacteria. Arch. Microbiol. 169: 371–380.

Drummond, M.H., Contreras, A., and Mitchenall, L.A. 1990. The function of isolated domains and chimaeric proteins constructed from the transcriptional activators NifA and NtrC of *Klebsiella pneumoniae.* Mol. Microbiol. 4: 29–37.

Drummond, M.H., and Wootton, J.C. 1987. Sequence of *nifL* from *Klebsiella pneumoniae*: mode of action and relationship to two families of regulatory proteins. Mol. Microbiol. 1: 37–44.

Egener, T., Sarkar, A., Martin, D.E., and Reinhold-Hurek, B. 2002. Identification of a NifL-like protein in a diazotroph of the beta-subgroup of the Proteobacteria, *Azoarcus sp.* strain BH72. Microbiology 148: 3203–3212.

Eydmann, T., Söderbäck, E., Jones, T., Hill, S., Austin, S., and Dixon, R. 1995. Transcriptional activation of the nitrogenase promoter *in vitro*: adnenosine nucleotides are required for inhibition of NifA activity by NifL. J. Bacteriol. 177: 1186–1195.

Filser, M., Merrick, M., and Cannon, F. 1983. Cloning and characterization of *nifLA* regulatory mutations from *Klebsiella pneumoniae*. Mol. Gen. Genet. 191: 485–491.

Fischer, H.M. 1994. Genetic regulation of nitrogen fixation in rhizobia. Microbiol. Reviews 58: 352–386.

Fischer, H.M. 1996. Environmental regulation of rhizobial symbiotic nitrogen fixation genes. Trends Microbiol. 4: 317–320.

Govantes, F., Andújar, E., and Santero, E. 1998. Mechanism of translational coupling in the *nifLA* operon of *Klebsiella pneumoniae*. EMBO J. 17: 2368–2377.

Grabbe, R., Kuhn, A., and Schmitz, R.A. 2001a. Cloning, sequencing and characterization of Fnr from *Klebsiella pneumoniae*. Antonie Van Leeuwenhoek, 79: 319–326.

Grabbe, R., Klopprogge, K., and Schmitz, R.A. 2001b. Fnr is Required for NifL-dependent oxygen control of *nif* gene expression in *Klebsiella pneumoniae*. J. Bacteriol. 183: 1385–1393.

Halbleib, C.M., and Ludden, P.W. 2000. Regulation of biological nitrogen fixation. J. Nutr. 130: 1081–1084.

He, L., Soupene, E., and Kustu, S. 1997. NtrC is required for control of *Klebsiella pneumoniae* NifL activity. J. Bacteriol. 179: 7446–7455.

He, L., Soupene, E., Ninfa, A., and Kustu, S. 1998. Physiological role for the GlnK protein of enteric bacteria: relief of NifL inhibition under nitrogen-limiting conditions. J. Bacteriol. 180: 6661–6667.

Henderson, N., Austin, S., and Dixon, R. A. 1989. Role of metal ions in negative regulation of nitrogen fixation by the *nifL* gene produkt from *Klebsiella pneumoniae*. Mol. Gen. Genet. 216: 484–491.

Hill, S. 1985. Redox regulation of enteric *nif* expression is independent of the *fnr* gene product. FEMS Microbiol. Letters 29: 5–9.

Hill, S., Austin, S., Eydmann, T., Jones, T., and Dixon, R. 1996. *Azotobacter vinelandii* NifL is a flavoprotein that modulates transcriptional activation of nitrogen-fixation genes via a redox-sensetive switch. Proc. Natl. Acad. Sci. USA 93: 2143–2148.

Hill, S., Kennedy, C., Kavanagh, E., Goldberg, R., and Hanau, R. 1981. Nitrogen fixation gene (*nifL*) involved in oxygen regulation of nitrogenase synthesis in *Klebsiella pneumoniae*. Nature 290: 424–426.

Hoover, T.R., Santero, E., Porter, S., and Kustu, S. 1990. The integration host factor stimulates interaction of RNA polymerase with NIFA, the transcriptional activator for nitrogen fixation operons. Cell 63: 11–22.

Howard, J.B., and Rees, D.C. 1996. Structural basis of biological nitrogen fixation. Chem. Rev. 96: 2965–2982.

Ikeda, T.P., Shauger, A.E., and Kustu, S. 1996. *Salmonella typhimurium* apparently perceives external nitrogen limitation as internal glutamine limitation. J. Mol. Biol. 259: 589–607.

Jack, R., DeZamaroczy, M., and Merrick, M. 1999. The signal transduction protein GlnK is required for NifL-dependent nitrogen control of nif gene expression in *Klebsiella pneumoniae*. J. Bacteriol. 181: 1156–1162.

Khoroshilova, N., Popescu, C., Munck, E., Beinert, H., and Kiley, P.J. 1997. Iron-sulfur cluster disassembly in the FNR protein of *Escherichia coli* by O_2: [4Fe-4S] to [2Fe-2S] conversion with loss of biological activity. Proc. Natl. Acad. Sci. USA. 94: 6087–6092.

Kiley, P.J., and Beinert, H. 1998. Oxygen sensing by the global regulator, FNR: the role of the iron-sulfur cluster. FEMS Microbiol. Rev. 5: 341–352.

Klopprogge, K., and Schmitz, R.A. 1999. NifL of *Klebsiella pneumoniae*: redox characterization in relation to the nitrogen source. Biochim. Biophys. Acta 1431: 462–470.

Klopprogge, K., Grabbe, R., Hoppert, M., and Schmitz, R.A. 2002a. Membrane association of *Klebsiella pneumoniae* NifL is affected by molecular oxygen and combined nitrogen. Archives of Microbiology, 177: 223–234.

Klopprogge, K., Stips, J., and Schmitz, R.A. 2002b. The inhibitory form of NifL from *Klebsiella pneumoniae* exhibits ATPase activity only when synthesized under nitrogen sufficiency. Biochem. Biophys. Acta, 1594: 243–254.

Lee, H.S., Berger, D.K., and Kustu, S. 1993a. Activity of purified NIFA, a transcriptional activator of nitrogen fixation genes. Proc. Natl. Acad. Sci. USA 90: 2266–2270.

Lee, H.S., Narberhaus, F., and Kustu, S. 1993b. *In vitro* activity of NifL, a signal transduction protein for biological nitrogen fixation. J. Bacteriol. 175: 7683–7688.

Lei, S., Pulakat, L., and Gvani, N. 1999. Genetiv analysis of *nif* regulatory genes by utilizing the yeast two-hybrid system detected formation of a NifL-NifA complex that is implicated in regulated expression of *nif* genes. J. Bacteriol. 181: 6535–6539.

Little, R., Reyes-Ramirez, F., Zhang, Y., van Heeswijk, W.C., and Dixon, R. 2000. Signal transduction to the *Azotobacter vinelandii* NIFL-NIFA regulatory system is influenced directly by interaction with 2-oxoglutarate and the PII regulatory protein. EMBO J. 19: 6041–6050.

Little, R., Colombo, V., Leech, A. and Dixon, R. 2002. Direct interaction of the NifL regulatory protein with the GlnK signal transducer enables the *Azotobacter vinelandii* NifL-NifA regulatory system to respond to conditions replete for nitrogen. J. Biol. Chem. 277: 15472–15481.

Lobo, A.L., and Zinder, S.H. 1992. Nitrogen fixation by methanogenic bacteria. In: Biological nitrogen fixation. G. Stacey, R.H. Burris and H.J. Evans, (eds.), Chapman & Hall, New York. p. 736–762.

Macheroux, P., Hill, S., Austin, S., Eydmann, T., Jones, T., Kim, S.O., Poole, R., and Dixon, R. 1998. Electron donation to the flavoprotein NifL, a redox-rensing transcriptional regulator. Biochem. J. 332: 413–419.

Meletzus, D., Rudnick, P., Doetsch, N., Green, A., and Kennedy, C. 1998. Characterization of the *glnK-amtB* operon of *Azotobacter vinelandii*. J. Bacteriol. 180: 3260–3264.

Merrick, M.J., and Edwards, R.A. 1995. Nitrogen control in bacteria. Microbiol. Rev. 59: 604–622.

Merrick, M., Hill, S., Hennecke, H., Hahn, M., Dixon, R., and Kennedy, C. 1982. Repressor properties of the *nifL* gene product in *Klebsiella pneumoniae*. Mol. Gen. Genet. 185: 75–81.

Money, T., Jones, T., Dixon, R., and Austin, S. 1999. Isolation and properties of the complex between the enhancer binding protein NifA and the sensor NifL. J. Bacteriol. 181: 4461–4468.

Money, T., Barrett, J., Dixon, R., and Austin, S. 2001. Protein-Protein Interactions in the Complex between the enhancer binding protein NifA and the sensor NifL from *Azotobacter vinelandii*. J. Bacteriol. 183: 1359–1368.

Monteiro, R.A., Souza, E.M., Yates, M.G., Pedrosa, F.O., and Chubatsu, L.S. 1999a. In-trans regulation of the N-truncated-NIFA protein of *Herbaspirillum seropedicae* by the N-terminal domain. FEMS Microbio. Lett. 180: 157–161.

Monteiro, R.A., Souza, E.M., Funayama, S., Yates, M.G., Pedrosa, F.O., and Chubatsu, L.S. 1999b. Expression and functional analysis of an N-truncated NifA protein of *Herbaspirillum seropedica*. FEBS Lett. 447: 283–286.

Morett, E., and Segovia, L. 1993. The sigma 54 bacterial enhancer-binding protein family: mechanism of action and phylogenetic relationship of their functional domains. J. Bacteriol. 175: 6067–6074.

Morett, E., and Buck, M. 1988. NifA-dependent *in vivo* protection demonstrates that the upstream activator sequence of *nif* promoters is a protein binding site. Proc. Natl. Acad. Sci. USA 85: 9401–9405.

Morett, E., and Buck, M. 1989. *In vivo* studies on the interaction of RNA polymerase-σ^{54} with the *Klebsiella pneumoniae* and *Rhizobium meliloti nifH* promoters. J. Mol. Biol. 210: 65–77.

Narberhaus, F., Lee, H.-S., Schmitz, R.A., He, L., and Kustu, S. 1995. The C-terminal domain of NifL is sufficient to inhibit NifA activity. J. Bacteriol. 177: 5078–5087.

Ninfa, A.J., and Atkinson, M.R. 2000. PII signal transduction proteins. Trends Microbiol. 8: 172–179.

Oelze, J. 2000. Respiratory protection of nitrogenase in *Azotobacter* species: is a widely held hypothesis unequivocally supported by experimental evidence? FEMS Microbiol. Rev. 24: 321–333.

Poole, R.K. 1994. Oxygen reactions with bacterial oxidases and globins: binding, reduction and regulation. Antonie van Leuwenhoek 65: 289–310.

Rees, D.C., and Howard, J.B. 1999. Structural bioenergetics and energy transduction mechanisms. J. Mol. Biol. 293: 343–350.

Reinhold-Hurek, B., Hurek, T., Gillis, M., Hoste, B., Vancanneyt, M., Kersters, K., and Deley, J. 1993. *Azoarcus gen. nov.*, nitrogen-fixing proteobacteria associated withroots of kallar grass (*Leptochloa fusca* (L.) Kunth), and description of two species, *Azoarcus indigens sp. nov.* and *Azoarcus communis sp. nov.* Int. J. Syst. Bacteriol. 43: 574–584.

Reyes-Ramirez, F., Little, R., and Dixon, R. 2001. Role of *Escherichia coli* nitrogen regulatory genes in the nitrogen response of the *Azotobacter vinelandii* NifL-NifA complex. J. Bacteriol. 183: 3076–3082.

Reyes-Ramirez, F., Little, R., Hill, S., van Heeswijk, W., and Dixon, R. 2000. Regulation of *Azotobacter vinelandii* NifA activity by NifL: role of PII-like proteins in nitrogen sensing. In: Nitrogen fixation; from molecules to crop productivity. F.O. Pedrosa, M. Hungria, M.G. Yates and W.E. Newton, (eds.), Kluwer Academic, Dordrecht. p. 97–98.

Rudnick, P., Kunz, C., Gunatilaka, M.K., Hines, E.R. & Kennedy, C. 2002. Role of GlnK in NifL-mediated regulation of NifA activity in *Azotobacter vinelandii*. J. Bacteriol. 184: 812–820.

Schmitz, R.A, He, L., and Kustu, S. 1996. Iron is required to relieve inhibitory effects of NifL on transcriptional activation by NifA in *Klebsiella pneumoniae*. J. Bacteriol. 178: 4679–4687.

Schmitz, R.A. 1997. NifL of *Klebsiella pneumoniae* carries an N-terminally bound FAD cofactor, which is not directly required for the inhibitiory function of NifL. FEMS Microbiol. Lett. 157: 313–318.

Schmitz, R.A. 2000. Internal glutamine and glutamate pools in *Klebsiella pneumoniae* grown under different conditions of nitrogen availability. Curr. Microbiol. 41: 357–362.

Söderbäck, E., Reyes-Ramirez, F., Eydmann, T., Austin, S., Hill, S., and Dixon R. 1998. The redox and fixed nitrogen-responsive regulatory protein NifL from *Azotobacter vinelandii* comprises discrete flavin and nucleotide-binding domains. Mol. Microbiol. 28: 179–192.

Souza, E.M., Pedrosa, F.O., Drummond, M., Rigo, L.U., and Yates, M.G. 1999. Control of *Herbaspirillum seropedicae* NifA activity by ammonium ions and oxygen. J. Bacteriol. 181: 681–684.

Steenhoudt, O., and Vanderleyden, J. 2000. *Azospirillum*, a free-living nitrogen-fixing bacterium closely associated with grasses: genetic, biochemical and ecological aspects. FEMS Microbiol. Rev. 24: 487–506.

Taylor, B.L., Zhulin, I.B., and Johnson, M.S. 1999. Aerotaxis and other energy-sensing behavior in bacteria. Annu. Rev. Microbiol. 53: 103–128.

Toukdarian, A., and Kennedy, C. 1986. Regulation of nitrogen metabolism in *Azotobacter vinelandii*: isolation of *ntr* and *glnA* genes and construction of *ntr* mutants. EMBO J. 5: 399–407.

Upchurch, R.G., and Mortenson, L.E. 1980. *In vivo* energetics and control of nitrogen fixation: changes in the adenylate energy charge and adenosine 5′-diphosphate/adenosine 5′-triphosphate ratio of cells during growth on dinitrogen versus growth on ammonia. J. Bacteriol. 143: 274–284.

Woodley, P., and Drummond, M. 1994. Redundancy of the conserved His residue in *Azotobacter vinelandii* NifL, a histidine protein kinase homologue which regulates transcription of nitrogen fixation genes. Mol. Microbiol. 13: 619–626.

Xiao, H., Shen, S., and Zhu, J. 1998. NifL, an antagonistic regulator of NifA interacting with NifA. Science in China (Series C) 41: 303–308.

Young, J.P.W. 1992. Phylogenetic classification of nitrogen fixing organisms. In: Biological nitrogen fixation. G. Stacey, R.H. Burris and H.J. Evans, (eds.), Chapman & Hall, New York. p. 43–86.

Zhulin, I.B., Tayler, B.L., and Dixon, R. 1997. PAS domain S-boxes in Archaea, Bacteria and sensors for oxygen and redox. Trends Biochem. Sci. 22: 331–333.

Chapter 13

Regulation of Synthesis and Activity of Molybdenum Nitrogenase and the Alternative Nitrogenase in *Rhodobacter capsulatus*

Thomas Drepper, Bernd Masepohl, Annette Paschen, Silke Groß, Alice Pawlowski, Karsten Raabe, Kai-Uwe Riedel, and Werner Klipp

Abstract

In *R. capsulatus* synthesis and activity of both molybdenum nitrogenase and the alternative nitrogenase is controlled at three levels by the environmental factors ammonium, molybdenum, light, and oxygen. At the first level, transcription of the *nifA1*, *nifA2*, and *anfA* genes – which encode the transcriptional activators of all other *nif* and *anf* genes, respectively – is controlled by the Ntr system in dependence on ammonium availability. Mutations in *glnB* (coding for the signal transduction protein PII) result in significant expression of *nifA* and *anfA* in the presence of ammonium. In contrast to GlnB, the PII-paralogue GlnK is not involved in the Ntr signal transduction mechanism. In addition to ammonium control, transcription of *anfA* is inhibited by traces of molybdenum via the molybdate-dependent repressor proteins MopA and MopB. At the second level of regulation, activity of NifA1, NifA2, and AnfA is inhibited by ammonium in an NtrC-independent manner. The post-translational ammonium control of NifA activity is partially released in the absence of GlnK, and completely abolished in a *glnB*/*glnK* double mutant. In contrast, AnfA activity is still inhibited by ammonium in the *glnB*/*glnK* mutant background. At the third level of regulation, both GlnB and GlnK as well as the (methyl)-ammonium transporter AmtB are involved in ammonium control of the DraT/DraG system, which mediates reversible ADP-ribosylation of both nitrogenase reductases (NifH and AnfH) in response to changes in ammonium availability or light intensity. Most remarkably, in a *glnB*/*glnK* double mutant ammonium control of the molybdenum (but not of the alternative) nitrogenase is completely relieved, leading to synthesis of active nitrogenase in the presence of high concentrations of ammonium.

Introduction

Rhodobacter capsulatus is a non-sulfur phototrophic purple bacterium which can grow with a variety of different nitrogen sources, such as ammonium, almost all amino acids, purines (e.g. xanthine, hypoxanthine), urea, polyamines (e.g. putrescine, spermidine) and with molecular nitrogen (Masepohl and Klipp, 1996; Moreno-Vivian *et al.*, 1992; Leimkühler *et al.*, 1998). *R. capsulatus* is able to fix atmospheric dinitrogen either by a molybdenum-containing nitrogenase (*nif*-encoded) or by an alternative iron-only nitrogenase (*anf*-encoded; Schneider *et al.*, 1991; Schüddekopf *et al.*, 1993). The organization of *nif* and *anf* genes has been reviewed recently (Masepohl and Klipp, 1996).

Like in many other bacteria, ammonium is the preferred nitrogen source in *R. capsulatus*, and consequently, the highly energy-demanding nitrogen fixation process is regulated by ammonium availability. In addition, N_2 fixation is controlled by the environmental factors molybdenum, light, and oxygen. Although the regulatory influence of these factors is taken into consideration, this review will mainly focus on the regulation of nitrogen fixation by ammonium.

The cellular nitrogen status controls expression of nitrogen fixation genes, as well as synthesis and activity of both nitrogenases at three levels (Figure 1). *R. capsulatus* measures the cellular nitrogen status by a nitrogen regulation (Ntr) system similar to that of enteric bacteria (see below). Under conditions of nitrogen depletion, NtrC activates transcription of the regulatory *nifA1*, *nifA2*, and *anfA* genes (level 1). In turn, in the absence of ammonium NifA and AnfA activate expression of all the other *nif* and *anf* genes resulting in synthesis of nitrogenase. In addition to ammonium control of NtrC-mediated transcriptional regulation, ammonium leads to inhibition of NifA and AnfA activity (level 2) and to DraT/G-mediated reversible inactivation of both nitrogenases (level 3).

Review

The Nitrogen Regulation (Ntr) System of *R. capsulatus*

In enteric bacteria a general nitrogen regulation system that senses the cellular nitrogen status is responsible for controlling the expression of many genes concerned with nitrogen metabolism (for a review, see Ninfa and Atkinson, 2000). This Ntr system comprises five gene products: a bifunctional uridylyltransferase/uridylyl-removing enzyme encoded by *glnD*, a trimeric signal transduction protein (PII) encoded by *glnB*, a PII-paralogue encoded by *glnK*, and a two-component

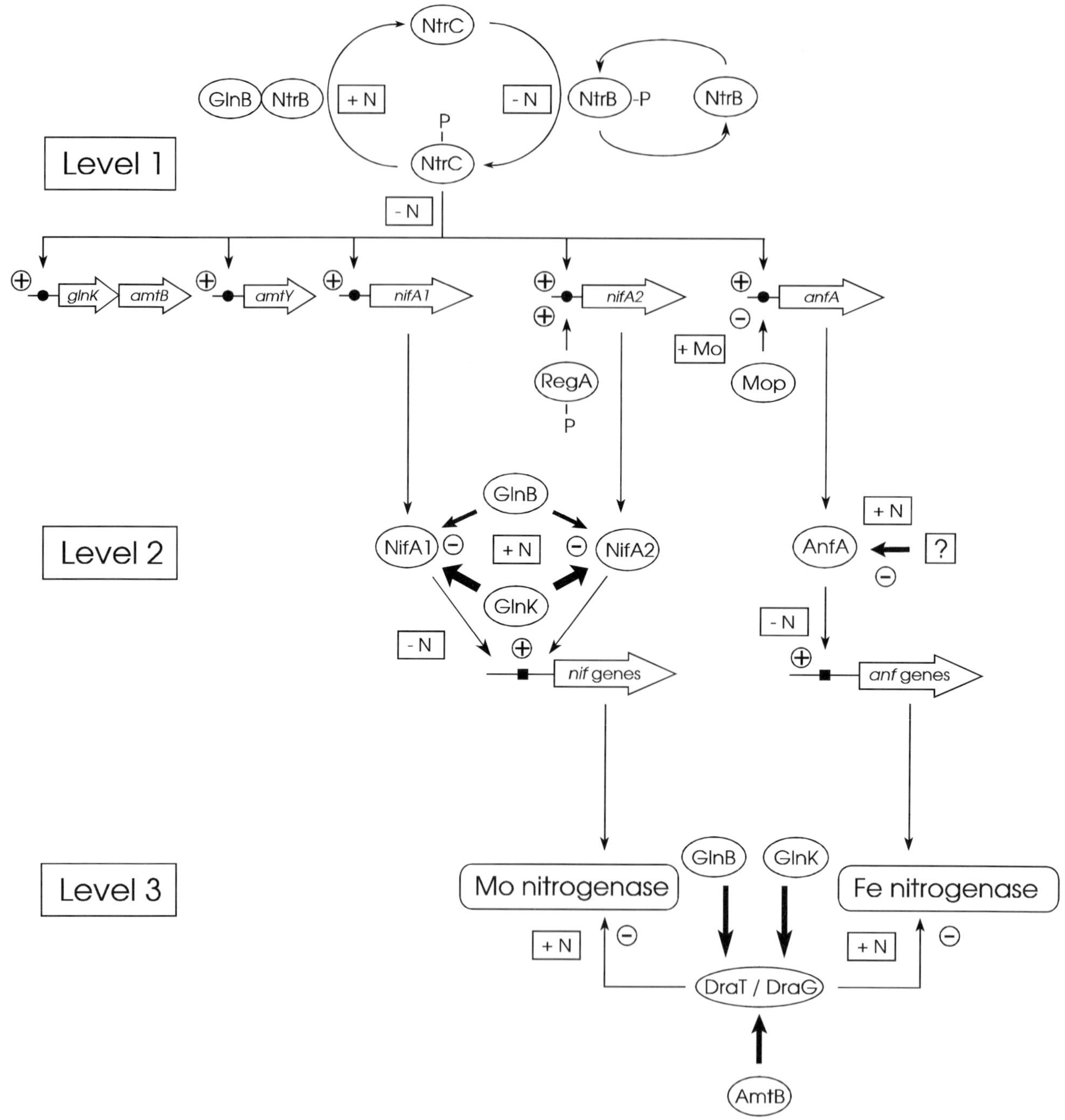

Figure 1. Model of the nitrogen fixation regulon in *R. capsulatus*. The model decribes the regulatory cascade controlling expression of nitrogen fixation genes and activity of the molybdenum nitrogenase (Mo nitrogenase, *nif*-encoded) and the alternative nitrogenase (Fe nitrogenase, *anf*-encoded) in response to ammonium availability. Nitrogenase-repressing conditions (presence of ammonium) or -derepressing conditions are symbolized by [+N] or [– N]. Presence of molybdenum is indicated by [+ Mo]. For description of the regulatory effects of other environmental factors, see text. Filled circles symbolize σ^{70}-dependent promoters activated by NtrC~P, and filled squares mark σ^{54}-dependent promoters activated by NifA or AnfA. Yeast two-hybrid studies suggest direct GlnB-NtrB, GlnB-NifA1, GlnB-NifA2, GlnB-DraT, GlnK-NifA1, and GlnK-NifA2 interactions.

regulatory system encoded by *ntrB-ntrC*. In response to the cellular glutamine/2-ketoglutarate ratio, PII is regulated by reversible GlnD-mediated uridylylation at tyrosine residue 51. When cells are N-limited, PII is uridylylated, and thus is unable to interact with the sensor kinase NtrB. Under these conditions NtrB promotes the phosphorylation of the response regulator NtrC, and NtrC~P can function as a transcriptional activator of its target genes.

R. capsulatus contains genes homologous to *glnD*, *glnB*, *ntrB*, and *ntrC*, and therefore, regulatory mechanisms similar to those of the Ntr system of enteric bacteria have been proposed (Hübner *et al.*, 1991; Kranz and Foster-Hartnett, 1990; Masepohl and Klipp, 1996). This assumption was verified by *in vitro* reconstitution of the NtrB/NtrC system (Cullen *et al.*, 1996). Like many other bacteria, *R. capsulatus* contains a second *glnB*-like gene (*glnK*), which is closely linked to the *amtB* gene coding for a putative (methyl)-ammonium transporter (Drepper *et al.*, 2000; Klipp *et al.*, 2000; Yakunin and Hallenbeck, 2002). Both GlnB and GlnK contain the conserved tyrosine residue at position 51, suggesting that the activity of the two signal transduction proteins is regulated by GlnD-mediated uridylylation.

Under conditions of nitrogen starvation *R. capsulatus* NtrC~P activates transcription of a number of genes including *glnB-glnA*, *glnK-amtB*, *amtY*, *nifA1*, *nifA2*, *anfA*, *mopA-modABCD*, and the *ure* operon (Figure 1; Bowman and Kranz, 1998; S. Groß and W. Klipp, unpublished; Klipp *et al.*, 2000; Masepohl *et al.*, 2001). These genes code for PII, glutamine synthetase (GlnA), the PII-paralogue GlnK, two putative ammonium transporters (AmtB, AmtY), the transcriptional activators of the two nitrogenase systems (NifA1, NifA2, AnfA), a molybdenum repressor of the Anf system (MopA), a high affinity molybdate uptake system (ModABCD), and urease, respectively. However, mutations in *R. capsulatus ntrC* do not result in a "classical" Ntr phenotype as found in enterobacteria, since *R. capsulatus* NtrC is not required for the utilization of amino acids as an N source (Keuntje *et al.*, 1995; Moreno-Vivian *et al.*, 1992). In contrast to all other members of the NtrC family, *R. capsulatus* NtrC does not require σ^{54} (NtrA, RpoN) but instead activates transcription of the above mentioned genes together with RNA polymerase (RNA-P) containing the σ^{70}-like house-keeping sigma factor (Bowman and Kranz, 1998; Foster-Hartnett *et al.*, 1994).

R. capsulatus contains two functional copies of *nifA* (Figure 1. *nifA1* and *nifA2*), which can substitute for each other (Masepohl *et al.*, 1988). Both NifA proteins are almost identical to each other, and differ only in their 19 N-terminal amino acid residues (Hübner *et al.*, 1993; Paschen *et al.*, 2001). In the absence of ammonium both NifA1 and NifA2 activate expression of all the other *nif* genes in concert with RNA-P containing σ^{54}, leading to synthesis of molybdenum nitrogenase.

NtrC-mediated expression of *R. capsulatus anfA* under conditions of nitrogen limitation is inhibited by traces of molybdate (Kutsche *et al.*, 1996; Wang *et al.*, 1993). This molybdate-dependent inhibition is released in mutants carrying either lesions in the high-affinity molybdenum uptake system (*modABCD*) or a double deletion of *mopA* and *mopB*, which encode ModE-like transcriptional repressor proteins binding to dyad symmetric promoter elements that overlap the transcription start (Pau *et al.*, 1997). Although MopA and MopB can substitute for each other in repression of the *anfA* and the *mopA-modABCD* operon, only MopB (but not MopA) is essential for the molybdate-dependent expression of the *dorCDA* operon encoding dimethylsulfoxide reductase in *R. capsulatus* (Solomon *et al.*, 2000). In the absence of ammonium AnfA activates expression of the *anfHDGK* operon and of several *nif* genes required for the alternative nitrogenase in concert with RNA-P containing σ^{54}, resulting in synthesis of the iron-only nitrogenase (Schüddekopf *et al.*, 1993).

In enteric bacteria NtrB (in concert with GlnB or GlnK) promotes the dephosphorylation of NtrC under conditions of N-excess (Ninfa and Atkinson, 2000), and as a result, the activator properties of NtrC are diminished and expression of NtrC-dependent promoters is switched off. In *R. capsulatus*, a *glnB* mutant shows significant expression of NtrC-dependent promoters in the presence of ammonium indicating that GlnK is unable to fully substitute for GlnB in regulating NtrB activity. This result was corroborated by analyzing protein-protein interactions using the yeast two-hybrid system. A strong interaction between GlnB and NtrB could be shown, whereas GlnK was unable to interact with NtrB (A. Pawlowski, S. Groß, and W. Klipp, unpublished).

In addition to NtrC, the Hfq-like protein NrfA is involved in control of expression of *nifA1*, *nifA2* and *anfA* (Drepper *et al.*, 2002). Disruption of the *nrfA* gene leads to clear decrease in diazotrophic growth rates compared to the wild-type situation.

Regulation of NifA and AnfA Activity

A *glnB* mutation results in significant expression of *nifA1* and *nifA2* in the presence of ammonium, but NifA-mediated *nifH* transcription is still inhibited under these conditions (Table 1; Drepper *et al.*, 2000). This inhibition is independent of NtrC as demonstrated by analysis of *R. capsulatus ntrC* mutant strains over-expressing *nifA1* or *nifA2* from a constitutive promoter (Drepper *et al.*, 2000; Hübner *et al.*, 1993; Kern *et al.*, 1998). Two lines of evidence suggest that ammonium regulation of *nifH* gene expression occurs at the post-translational level controlling NifA activity (Figure 1; level 2). (i) *R. capsulatus* strains overexpressing *nifA1* accumulate high levels of NifA1 both in the absence or presence of ammonium as shown by Western analysis using NifA1-specific antibodies (Drepper *et al.*, 2000). As shown by time course experiments, NifA1 is highly stable in the presence of ammonium suggesting that inactivation of NifA in response to ammonium is not due to proteolytic degradation of NifA. (ii) Single mutations in the N-terminal domain of NifA1 resulted in ammonium-tolerant NifA mutant proteins mediating *nifH* gene expression under N-excess (Paschen *et al.*, 2001).

In contrast to lesions in *glnB*, a *glnK* mutation leads to low level expression of *nifH* in the presence of ammonium (in *R. capsulatus* strains constitutively expressing *nifA*), indicating that NifA activity partially escapes ammonium inhibition in this mutant background (Table 1). Furthermore, in a *glnB*/*glnK* double mutant ammonium control of NifA activity is completely abolished. Therefore, GlnB can only partially substitute for GlnK in mediating ammonium inhibition of NifA activity. As shown by yeast two-hybrid studies, both GlnB and GlnK interact with NifA1 and NifA2 (S. Groß, and W. Klipp, unpublished), suggesting that ammonium

Table 1. Effect of GlnB and GlnK on the expression and activity of both nitrogenase systems.

	glnB$^-$	*glnK*$^-$	*glnB*$^-$/*glnK*$^-$
Level 1:			
transcription			
– of *nifA1* and *nifA2*	constitutive	N-regulated	constitutive
– of *anfA*	constitutive	N-regulated	constitutive
Level 2:			
post-translational control[(a)]			
– of NifA activity	+	±	–
– of AnfA activity	+	+	+
Level 3:			
ADP-ribosylation[(b)]			
– of NifH	±	+	–
– of AnfH	±	+	–
Activity[(c)]			
– of the Mo nitrogenase	–	–	+
– of the Fe nitrogenase	–	–	–

[(a)] Analysis of post-translational control of NifA and AnfA activity requires NtrC-independent expression of *nifA* and *anfA* from constitutive promoters. Activity of the activator proteins is defined by the ability to mediate transcription of *nifH-lacZ* or *anfH-lacZ* fusions, or can be followed by Western analysis using nitrogenase reductase-specific antibodies. Post-translational control of the activity of the transcriptional activators is present (+), partially abolished (±), or completely abolished (–).
[(b)] DraT-mediated modification of nitrogenase reductase is not affected (+), reduced (±), or completely abolished (–) as measured by the appearance of the ADP-ribosylated form of NifH and AnfH.
[(c)] Nitrogenase activity was measured by the acetylene reduction assay in the presence of ammonium.

inhibition of NifA activity is directly mediated by these signal transduction proteins.

A *glnB* mutation results in constitutive expression of *anfA* (Drepper *et al.*, 2000). In contrast to the *nif* regulon, *anfH* expression is still inhibited by ammonium in a *glnB*/*glnK* double mutant. This inhibition is also independent of NtrC, and therefore, ammonium control of AnfA activity involves a GlnB/GlnK-independent regulatory mechanism (T. Drepper, and W. Klipp, unpublished).

DraT/G-Mediated Post-Translational Regulation of Nitrogenase Activity

The activity of nitrogenase is controlled by reversible ADP-ribosylation of NifH and AnfH mediated by DraT and DraG in response to ammonium or darkness (Figure 1; level 3; Masepohl *et al.*, 1993). Inactivation of dinitrogenase reductase by ADP-ribosylation is catalyzed by the *draT* gene product (dinitrogenase reductase ADP-ribosyltransferase), whereas restoration of nitrogenase activity is performed by the DraG protein (dinitrogenase reductase-activating glycohydrolase). In addition, nitrogenase activity is regulated by a second DraT/G-independent mechanism (Förster *et al.*, 1999; Pierrard *et al.*, 1993; Yakunin *et al.*, 1999). In a *glnB*/*glnK* double mutant both post-translational ammonium control mechanisms of nitrogenase activity are relieved (Table 1; Klipp *et al.*, 2000). In contrast, ADP-ribosylation of NifH and AnfH in response to darkness is not affected in a *glnB*/*glnK* double mutant.

As mentioned above, *R. capsulatus* contains two genes (*amtB*, *amtY*) coding for putative (methyl)-ammonium transporters. It has been suggested that AmtB could act as a signal transducer for ammonium availability (Thomas *et al.*, 2000). Both *R. capsulatus* *glnK-amtB* and *amtY* are part of the NtrC regulon (Figure 1; Klipp *et al.*, 2000; S. Groß, and W. Klipp, unpublished). An *amtB* mutant (but not an *amtY* mutant), in addition to being ineffective in methyl-ammonium uptake, is also completely defective in carrying out ADP-ribosylation of NifH in response to ammonium addition (Yakunin and Hallenbeck, 2002). As shown for the *glnB*/*glnK* double mutant, ADP-ribosylation in response to darkness is also unaltered in the *amtB* mutant.

Linkage of Nitrogen Fixation and Photosynthesis

The RegB/RegA signal transduction system is a global two-component regulatory system that activates photosynthesis, carbon assimilation, and nitrogen fixation genes, and represses hydrogenase structural genes and its own expression (Elsen *et al.*, 2000). RegA indirectly activates nitrogenase synthesis by binding to and activating the expression of *nifA2* but not of *nifA1*. Thus, although NtrC is absolutely required for *nifA2* expression, RegA acts as a coactivator of *nifA2*.

The *regA* gene is cotranscribed with *hvrA*, which encodes a histone-like protein homologous to *E. coli* H-NS. *R. capsulatus* HvrA is involved in low-light activation of the photosynthetic apparatus (Buggy *et al.*, 1994). A mutation in *hvrA* releases both ammonium and oxygen control of *nifH* expression in an *R. capsulatus* strain constitutively expressing *nifA1* (Kern *et al.*, 1998). This regulatory effect of the HvrA protein is mediated by binding of HvrA to the *nifH* promoter as demonstrated by gel shift experiments (Raabe *et al.*, 2002).

Conclusions

This review describes mechanisms regulating synthesis and activity of the molybdenum-containing and the alternative nitrogenase in *R. capsulatus* in response to ammonium availability and other environmental factors. Regulation occurs on at least three levels including (i) transcriptional activation of the regulatory genes *nifA1*, *nifA2* and *anfA*, (ii) post-translational regulation of NifA and AnfA activity, and (iii) post-translational control of nitrogenase activity by reversible modification of NifH and AnfH. The complexity of these regulatory mechanisms is reflected by the number and diversity of regulatory proteins involved in control of nitrogen fixation. Among these are two two-component regulatory systems (NtrB/NtrC, RegB/RegA), two signal transduction proteins (GlnB, GlnK), three specific transcriptional activator proteins (NifA1, NifA2, AnfA), two molybdate-dependent repressor proteins (MopA, MopB), an ADP-ribosyl-transferase/glycohydrolase system (DraT, DraG), two (methyl)-ammonium transporter proteins (AmtB, AmtY), and two nucleoid-associated proteins (HvrA, NrfA).

Acknowledgements

This work was supported by financial grants from the Deutsche Forschungsgemeinschaft, Studienstiftung des Deutschen Volkes, and Fonds der Chemischen Industrie.

References

Bowman, W.C., and Kranz, R.G. 1998. A bacterial ATP-dependent, enhancer binding protein that activates the housekeeping RNA polymerase. Genes and Development 12: 1884–1893.

Buggy, J.J., Sganga, M.W., and Bauer, C.E. 1994. Characterization of a light-responding *trans*-activator responsible for differentially controlling reaction center and light-harvesting-I gene expression in *Rhodobacter capsulatus*. J. Bacteriol. 176: 6936–6943.

Cullen, P.J., Bowman, W.C., and Kranz, R.G. 1996. *In vitro* reconstitution and characterization of the *Rhodobacter capsulatus* NtrB and NtrC two-component system. J. Biol. Chem. 271: 6530–6536.

Drepper, T., Groß, S., Masepohl, B., and Klipp, W. 2000. Ammonium and molybdenum regulation of the molybdenum nitrogenase and the alternative nitrogenase in *Rhodobacter capsulatus*. In: Nitrogen Fixation: From Molecules to Crop Productivity. F.O. Pedrosa, M. Hungria, G. Yates, and W.E. Newton (eds.), Kluwer Academic Publishers, Dordrecht, Boston, London. p. 129.

Drepper, T., Raabe, K., Giaourakis, D., Gendrullis, M., Masepohl, B., and Klipp, W. 2002. The Hfq-like protein NrfA of the phototrophic purple bacterium *Rhodobacter capsulatus* controls nitrogen fixation via regulation of *nifA* and *anfA* expression. FEMS Microbiol. Lett. 215: 221–227.

Elsen, S., Dischert, W., Colbeau, A., and Bauer, C.E. 2000. Expression of uptake hydrogenase and molybdenum nitrogenase in *Rhodobacter capsulatus* is coregulated by the RegB-RegA two-component regulatory system. J. Bacteriol. 182: 2831–2837.

Förster, B., Maner, K., Fassbinder, F., and Oelze, J. 1999. Reversible inactivation of nitrogenase in *Rhodobacter capsulatus* strain W107I deleted in the *draTG* gene region. FEMS Microbiol. Lett. 170: 167–171.

Foster-Hartnett, D., Cullen, P.J., Monika, E.M., and Kranz, R.G. 1994. A new type of NtrC transcriptional activator. J. Bacteriol. 176: 6175–6187.

Hübner, P., Masepohl, B., Klipp, W., and Bickle, T.A. 1993. *nif* gene expression studies in *Rhodobacter capsulatus*: *ntrC*-independent repression by high ammonium concentrations. Mol. Microbiol. 10: 123–132.

Hübner, P., Willison, J.C., Vignais, P.M., and Bickle, T.A. 1991. Expression of regulatory *nif* genes in *Rhodobacter capsulatus*. J. Bacteriol. 173: 2993–2999.

Kern, M., Kamp, P.-B., Paschen, A., Masepohl, B., and Klipp, W. 1998. Evidence for a regulatory link of nitrogen fixation and photosynthesis in *Rhodobacter capsulatus* via HvrA. J. Bacteriol. 180: 1965–1969.

Keuntje, B., Masepohl, B., and Klipp, W. 1995. Expression of the *putA* gene encoding proline dehydrogenase from *Rhodobacter capsulatus* is independent of NtrC regulation but requires an Lrp-like activator protein. J. Bacteriol. 177: 6432–6439.

Klipp, W., Drepper, T., Groß, S., Masepohl, B., Raabe, K., Riedel, K.-U., Yakunin, A.F., and Hallenbeck, P.C. 2000. Genetics of nitrogen fixation in *Rhodobacter capsulatus*: ammonium and molybdenum control of both nitrogenase systems. In: Nitrogen Fixation: From Molecules to Crop Productivity. F.O. Pedrosa, M. Hungria, G. Yates, and W.E. Newton (eds.), Kluwer Academic Publishers, Dordrecht, Boston, London. p. 141–142.

Kranz, R.G., and Foster-Hartnett, D. 1990. Transcriptional regulatory cascade of nitrogen-fixation genes in anoxygenic photosynthetic bacteria: oxygen- and nitrogen-responsive factors. Mol. Microbiol. 4: 1793–1800.

Kutsche, M., Leimkühler, S., Angermüller, S., and Klipp, W. 1996. Promoters controlling expression of the alternative nitrogenase and the molybdenum uptake system in *Rhodobacter capsulatus* are activated by NtrC, independent of σ^{54}, and repressed by molybdenum. J. Bacteriol. 178: 2010–2017.

Leimkühler, S., Kern, M., Solomon, P.S., McEwan, A.G., Schwarz, G., Mendel, R.R., and Klipp, W. 1998. Xanthine dehydrogenase from the phototrophic purple bacterium *Rhodobacter capsulatus* is more similar to its eukaryotic counterparts than to prokaryotic molybdenum enzymes. Mol. Microbiol. 27: 853–869.

Masepohl, B., Kaiser, B., Isakovic, N., Richard, C.L., Kranz, R.G., and Klipp, W. 2001. Urea utilization in the phototrophic bacterium *Rhodobacter capsulatus* is regulated by the transcriptional activator NtrC. J. Bacteriol. 183: 637–643.

Masepohl, B., and Klipp, W. 1996. Organization and regulation of genes encoding the molybdenum nitrogenase and the alternative nitrogenase in *Rhodobacter capsulatus*. Arch. Microbiol. 165: 80–90.

Masepohl, B., Klipp, W., and Pühler, A. 1988. Genetic characterization and sequence analysis of the duplicated *nifA*/*nifB* gene region of *Rhodobacter capsulatus*. Mol. Gen. Genet. 212: 27–37.

Masepohl, B., Krey, R., and Klipp, W. 1993. The *draTG* gene region of *Rhodobacter capsulatus* is required for post-translational regulation of both the molybdenum and the alternative nitrogenase. J. Gen. Microbiol. 139: 2667–2675.

Moreno-Vivian, C., Soler, G., and Castillo, F. 1992. Arginine catabolism in the phototrophic bacterium *Rhodobacter capsulatus* E1F1. Eur. J. Biochem. 204: 531–537.

Ninfa, A.J., and Atkinson, M.R. 2000. PII signal transduction proteins. Trends Microbiol. 8: 172–179.

Paschen, A., Drepper, T., Masepohl, B., and Klipp, W. 2001. *Rhodobacter capsulatus nifA* mutants mediating *nif* gene expression in the presence of ammonium. FEMS Microbiol. Lett. 200: 207–213.

Pau, R.N., Klipp, W., and Leimkühler, S. 1997. Molybdenum transport, processing and gene regulation. In: Transition Metals in Microbial Metabolism. G. Winkelmann, and C.J. Carrano (eds.), Harwood Academic Publishers, Amsterdam. p. 217–234.

Pierrard, J., Ludden, P.W., and Roberts, G.P. 1993. Posttranslational regulation of nitrogenase in *Rhodobacter capsulatus*: existence of two independent regulatory effects of ammonium. J. Bacteriol. 175: 1358–1366.

Raabe, K., Drepper, T., Riedel, K.-U., Masepohl, B., and Klipp, W. 2002. The H-NS-like protein HvrA modulates expression of nitrogen fixation genes in the phototrophic purple bacterium *Rhodobacter capsulatus* by binding to selected *nif* promoters. FEMS Microbiol. Lett. 216: 151–158.

Schneider, K., Müller, A., Schramm, U., and Klipp, W. 1991. Demonstration of a molybdenum- and vanadium-independent nitrogenase in a *nifHDK*-deletion mutant of *Rhodobacter capsulatus*. Eur. J. Biochem. 195: 653–661.

Schüddekopf, K., Hennecke, S., Liese, U., Kutsche, M., and Klipp, W. 1993. Characterization of *anf* genes specific for the alternative nitrogenase and identification of *nif* genes required for both nitrogenases in *Rhodobacter capsulatus*. Mol. Microbiol. 8: 673–684.

Solomon, P.S., Shaw, A.L., Young, M.D., Leimkühler, S., Hanson, G.R., Klipp, W., and McEwan, A.G. 2000. Molybdate-dependent expression of dimethylsulfoxide reductase in *Rhodobacter capsulatus*. FEMS Microbiol. Lett. 190: 203–208.

Thomas, G., Coutts, G., and Merrick, M. 2000. The *glnKamtB* operon. Trends Genet. 16: 11–14.

Wang, G., Angermüller, S., and Klipp, W. 1993. Characterization of *Rhodobacter capsulatus* genes encoding a molybdenum transport system and putative molybdenum-pterin-binding proteins. J. Bacteriol. 175: 3031–3042.

Yakunin, A.F., Laurinavichene, T.V., Tsygankov, A.A., and Hallenbeck, P.C. 1999. The presence of ADP-ribosylated Fe protein of nitrogenase in *Rhodobacter capsulatus* is correlated with cellular nitrogen status. J. Bacteriol. 181: 1994–2000.

Yakunin, A.F., and Hallenbeck, P.C. 2002. AmtB is necessary for NH_4^+ induced nitrogenase switch-off and ADP-ribosylation in *Rhodobacter capsulatus*. J. Bacteriol. 184: 4081–4088.

Chapter 14

Oxygen-Regulated Expression of Genes for Pigment Binding Proteins in *Rhodobacter capsulatus*

Jutta Gregor and Gabriele Klug

Abstract

Oxygen is the major external factor affecting the expression of photosynthesis genes in facultatively photosynthetic bacteria. Many investigations over the last years mainly carried out on the closely related species *Rhodobacter capsulatus* and *Rhodobacter sphaeroides* have identified a number of proteins involved in the oxygen-regulated signal pathway, in which the RegB/RegA two component system plays a central role. While the RegB/RegA system activates photosynthesis genes under low oxygen tension other proteins like CrtJ and PPBP have a repressing function under high oxygen tension. Additional DNA binding proteins like the integration host factor can modulate the expression of photosynthesis genes. The role of alternative sigma factors in this signal pathway is still unclear.

Introduction

Facultatively phototrophic bacteria like *Rhodobacter capsulatus* can adapt quickly to changing environmental conditions by using different pathways for energy conversion. If high levels of oxygen are present, *Rhodobacter* gains energy by aerobic respiration and synthesizes only low amounts of photosynthetic complexes. If, however, the oxygen tension drops below a threshold value, the synthesis of pigments and pigment binding proteins is strongly increased and photosynthetic complexes are assembled into a newly formed intracytoplasmic membrane system. This process of bacterial differentiation is independent of light. As long as some oxygen is available the cells still perform aerobic respiration. This allows the investigation of mutant strains unable to perform photosynthesis under conditions inducing the formation of photosynthetic components. Under anaerobic conditions *Rhodobacter* can use light energy for anoxygenic photosynthesis.

The influence of oxygen on the formation of the photosynthetic apparatus in *R. capsulatus* has been known for many years. However, the molecular mechanisms for sensing the oxygen signal, for signal transmission within the cell and for regulation of gene expression could not be studied until genetic tools became available for this organism and the genes encoding the components of the photosynthetic apparatus were identified and sequenced. The genes for the pigment binding proteins of the reaction center and the light harvesting I complex are organized in the polycistronic *puf* operon which is part of a 44 kb photosynthesis gene cluster that also harbours genes for the non pigment binding protein of the reaction center, for pigment synthesis and for regulatory proteins. The *puc* operon which is localized elsewhere on the chromosome encodes the proteins of the light harvesting II complex. Over the last years many factors involved in the oxygen-regulated expression of photosynthesis genes in *Rhodobacter* have been identified and characterized (Bauer, 1995; Gregor and Klug, 1999).

The RegB/RegA Two Component System

A central regulator in the oxygen-dependent signal transduction in *R. capsulatus* is the RegB/RegA two component system. RegB is a membrane bound histidine kinase which can transfer a phosphatidyl residue *in vitro* to the response regulator RegA in a redox-dependent manner (Mosley *et al.*, 1994; Inoue *et al.*, 1995). It is, however, not clear what signal is sensed by RegB *in vivo* and which domains of RegB contribute to signal sensing. For PrrB, the homologue of RegB in *R. sphaeroides*, evidence was provided that it receives a signal from PrrC which depends on the electron flow through the *cbb3* cytochrome c oxidase (Oh and Kaplan, 2000).

In order to define the RegB domains which are involved in redox sensing, it is important to know the topology of this membrane protein. Mosley *et al.* (1994) suggested a model with five membrane spanning regions of RegB on the basis of computer prediction, with the N-terminus extending into the periplasm. More recent analysis of the *regB* sequence revealed that a C nucleotide was missing in the original sequence. Correction of the sequence resulted in an extension of the deduced amino acid sequence at the N-terminus and added an extra membrane spanning domain (Chen *et al.*, 2000). A model with six trans membrane domains was confirmed for RegB by analyzing RegB-LacZ and RegB-PhoA fusion proteins (Chen *et al.*, 2000). This model is in agreement with a topology model published on the basis of similar experiments for PrrB of *R. sphaeroides* (Ouchane and Kaplan, 1999).

When the response regulator RegA was identified (Sganga and Bauer, 1992) it was suggested that it indirectly acts on the transcription of photosynthesis genes since the authors did not detect a known DNA binding motif. It was shown later that RegA harbours a helix-turn-helix motif at its C-terminus and that it binds to the *puf* and *puc* promoter regions (Kirndörfer *et al.*, 1997; Du *et al.*, 1998; Kirndörfer *et al.*, 1998; Hemschemeier *et al.*, 1999; Hemschemeier *et al.*, 2000a). Although it is now generally accepted that RegA is a DNA binding protein, there is still a controversy about the role of its phosphorylation. Bird *et al.* (1999) observed that phosphorylation enhances binding of RegA to its target DNA. On the other hand it was reported that unphosphorylated RegA and phosphorylated RegA show similar DNA binding affinity *in vitro* (Hemschemeier *et al.*, 2000a). There is evidence from several investigations that a conformational change of RegA after phosphorylation leads to activation of photosynthesis genes. Du *et al.* (1998) isolated a constitutively active RegA protein with an amino acid exchange in near proximity to the region linking the N-terminal receiver domain and the C-terminal DNA binding domain. A truncated RegA protein which contains only the C-terminal activation domain still leads to oxygen-regulated activation of *puf* and *puc* transcription, although to a lesser extent than wild type RegA (Hemschemeier *et al.*, 2000b). Mutant RegA protein which can not be phosphorylated due to an exchange of aspartate 63 can still bind to the *puf* and *puc* promoter regions *in vitro* but is unable to activate transcription of these genes *in vivo* (Hemschemeier *et al.*, 2000b).

As mentioned above, an *R. capsulatus* mutant strain harbouring the truncated RegA protein consisting of the activaton domain only, still showed oxygen-regulated expression of the *puf* and *puc* genes (Hemschemeier *et al.*, 2000b). This result clearly shows that mechanisms independent of RegA contribute to this oxygen-regulated process. The CrtJ protein (PpsR in *R. sphaeroides*) has been shown to repress *puc* expression at high oxygen tension (Ponnampalam *et al.*, 1995) by redox-dependent, co-operative binding to distant sites in the *puc* promoter region (Ponnampalam *et al.*, 1997; Elsen *et al.*, 1998). Binding of CrtJ to the *puf* promoter was not reported, but a mutant strain lacking this protein shows altered oxygen-regulated *puf* expression (unpublished results). The PPBP (*puf* promoter binding protein) has been shown to bind to the *puf* promoter region and was suggested to repress *puf* expression under high oxygen tension (Klug, 1991). Isolation and sequence analysis of this protein or the identification of the corresponding gene, however, were unsuccessful. Analysis of different mutant strains revealed that PPBP is not identical to RegA or CrtJ (unpublished results). Since the binding sites of CrtJ and PPBP in part overlap the binding sites of RegA, it is conceivable that RegA can only bind to the *puf* and *puc* promoter regions, when CrtJ and PPBP no longer bind to these targets at low oxygen tension. This model is in agreement with the observation of an oxygen-regulated activation of photosynthesis genes by the isolated RegA activator domain which can not be phosphorylated in a redox-dependent manner by RegB.

Mutant strains lacking RegB or RegA show almost no increase of *puc* expression after reduction of oxygen tension, whereas *puf* operon expression shows a significant increase after change of growth conditions (Hemschemeier *et al.*, 2000a). Indeed the *puf* operon is part of a large cluster of genes which are partially cotranscribed and have been named "superoperon" (Young *et al.*, 1989; Wellington and Beatty, 1991). The *puf* operon is localized within the coding region of the *bchZ* gene. A low level of *puf* transcription occurs from promoters of upstream bacteriochlorophyll and carotenoid genes (Wellington *et al.*, 1991). It has been proposed that this type of gene organization guarantees low level of *puf* transcription in *R. capsulatus* at high oxygen tension and allows a faster adaptation to changes in growth conditions.

The fact that RegB mutants still show some oxygen-regulated expression of photosynthesis genes also proves that other components beside the histidine kinase RegB are involved in oxygen or redox sensing. One of these additional sensing molecules is thioredoxin, an ubiquitous protein which undergoes a redox-dependent disulfid / dithiol exchange and affects the function of many cellular proteins. An *R. sphaeroides* strain with altered levels of reduced thioredoxin also shows altered expression of the *puf* and *puc* genes (Pasternak *et al.*, 1999). Results from a yeast two hybrid screen suggest the interaction of thioredoxin A from *R. sphaeroides*, thioredoxin A and thioredoxin C from *R. capsulatus* with FnrL and gyrase (Härtig *et al.*, 2000). An involvement of gyrase in oxygen-regulated expression of photosynthesis genes has been described before in *R. capsulatus* (Zhu and Hearst, 1988). FnrL was shown to participate in photosynthesis gene regulation in *R. sphaeroides* (Zeilstra-Ryalls and Kaplan, 1995). FnrL mutants of *R. capsulatus* however show similar expression of photosynthesis genes as wild type cells (Zeilstra-Ryalls *et al.*, 1997). The exact signaling pathway from thioredoxin to photosynthesis gene expression is currently under investigation.

It is now well established that the RegB/RegA two component system does not only control the expression of photosynthesis genes but is also involved in oxygen-regulated expression of genes for nitrogen fixation, carbon dioxide assimilation and hydrogenase synthesis (Joshi and Tabita, 1996; Elsen *et al.*, 2000). The constitutively active RegA protein RegA* has been shown to bind to the promoters of the *cbbI* and *cbbII* operons encoding components for the Calvin cycle (Dubbs *et al.*, 2000; Vivhivanives *et al.*, 2000), to the promoter of the *nifA2* gene encoding a *nif*-specific transcriptional activator, and to the promoter of the hydrogenase gene operon (Elsen *et al.*, 2000).

Since binding of wild type RegA protein to DNA *in vitro* is weak it was postulated that additional factors are involved in facilitating DNA binding and activation of transcription by RegA. Results from a yeast two hybrid screen for proteins interacting with RegA suggest an interaction of RegA with proteins involved in nitrogen fixation and an enzyme of the bacteriochlorophyll biosynthesis pathway (unpublished). It has been known for many years that mutants with defects in bacteriochlorophyll synthesis are affected in oxygen-regulated transcription of *puf* and *puc* operon (i.e. Klug *et al.*, 1986; Rödig *et al.*, 1999). Interaction of RegA with enzymes of the bacteriochlorophyll synthesis is an attractive model to explain this coregulation which needs further studies for confirmation.

Involvement of Integration Host Factor (IHF) in the Expression of Photosynthesis Genes

It has been shown for many bacterial response regulators that other proteins bind in close proximity to their DNA targets (i.e. Zhang and DeMoss, 1996; Charlton *et al.*, 1993). One of these proteins is the integration host factor (IHF) which induces a bending of the DNA after binding (Rice *et al.*, 1996). An *R. capsulatus* strain which carries a mutation in the *himA* gene encoding one of the subunits of IHF shows altered oxygen-regulated expression of some photosynthesis genes. After lowering of oxygen tension *pufQ-lacZ* and *pucB-lacZ* gene fusions show lower expression in the *himA* mutant compared to the wild type. The expression of a *bchD-lacZ* gene fusion, however, is not affected by the mutation (Kirndörfer *et al.*, 1998). When *puf* and *puc* expression was monitored by Northern blot analysis the maximal mRNA levels were similar in mutant and wild type strains. The increase of *puf* and *puc* mRNA levels after a shift of cultures from high to low oxygen tension was clearly delayed in the mutant strain. Purified IHF protein from *R. capsulatus* was shown to bind to the *puf* and *puc* promoter regions close to the RegA binding sites (Kirndörfer *et al.*, 1998). A typical consensus sequence for IHF binding as reported for *E. coli* and also described for the *puc* promoter region of *R. sphaeroides*, however, is not present in this region (Lee *et al.*, 1993).

Are Alternative Sigma Factors Involved in Oxygen-Regulated Expression of Photosynthesis Genes ?

As soon as the DNA sequences for photosynthesis genes in *R. capsulatus* were identified they were analyzed for homology to typical bacterial promoter regions. While some photosynthesis genes are preceeded by promoters showing a typical sigma-70 consensus sequence, other promoters lack such homologies. Thus, the involvement of alternative sigma factors in the transcription of photosynthesis genes had to be considered. The vegetative sigma factor gene of *R. capsulatus rpoD* was not analyzed until 1996 (Pasternak *et al.*). For many years the only alternative sigma factor known in *R. capsulatus* was RpoN (sigma 54) which is required for transcription of certain genes involved in nitrogen fixation (Cannon *et al.*, 1996). More recently the *rpoH* gene of *R. capsulatus* was cloned and characterized (Emetz and Klug, 1998). The amount of the 5′ ends of the *rpoH* transcript increased in *R. capsulatus* after a shift to high temperature but was independent of oxygen tension (Emetz and Klug, 1998).

Decreasing oxygen tension in cultures of *R. capsulatus* leads to increased amounts of a 34 kD protein which cross-reacts with antibodies raised against a bacterial sigma factor (Emetz and Klug, 1998). However, final proof is lacking that this protein is indeed sigma-32 and that the *rpoH* gene product is involved in transcription of photosynthesis genes. Using an *in vitro* transcription system with purified RNA polymerase from *R. capsulatus* it was shown that the *puf* and *puc* promoters can be activated by RegA in the presence of the housekeeping sigma factor (Bowman *et al.*, 1999). These results suggest that no alternative sigma factor may be required for the oxygen-regulated transcription of photosynthesis genes, but no *in vivo* data confirming this assumption are available.

Conclusion

Much effort has been taken over the last years to elucidate the mechanisms of oxygen-regulated gene expression in *Rhodobacter*. One of the remarkable discoveries was the fact that the RegB/RegA two component system is a global regulator for many metabolic pathways which are directly or indirectly affected by oxygen. As for most bacterial two component systems the mechanism of signal sensing is not clear. Evidence was presented that the RegB homologue PrrB in *R. sphaeroides* is indirectly affected by the electron flow through the Cbb3 terminal oxidase (Oh and Kaplan, 2000). Although *R. capsulatus* and *R. sphaeroides* are closely related, they differ significantly in regard to the organization of their electron transport chains. While *R. sphaeroides* contains a branched aerobic electron transport chain that is terminated with two functional cytochrome c oxidases (*cbb3* and *aa3*), *R. capsulatus* does not contain the *aa3* cytochrome c oxidase. Experimental proof for a similar mechanism of redox sensing in *R. capsulatus* is lacking to date.

Although the central role of RegB/RegA in the oxygen-dependent signal pathway has been proven, the exact molecular mechanisms of signaling need further investigation. The role of phosphorylation of RegA is discussed controversially, and the exact mechanism of activation of RNA polymerase by RegA is not known. The domains of RegB and RegA which are involved in signal recognition and signal transmission need to be determined. Despite the central function of RegB/RegA many other factors have been shown to be involved in the redox-dependent signal pathways, and the picture of a complex regulatory network has emerged. The knowledge of the complete genome sequences of *R. capsulatus* and *R. sphaeroides* makes it possible to apply DNA chip technologies and proteomics which may be the tools to better understand this regulatory network and to asign the functions for individual components in the future.

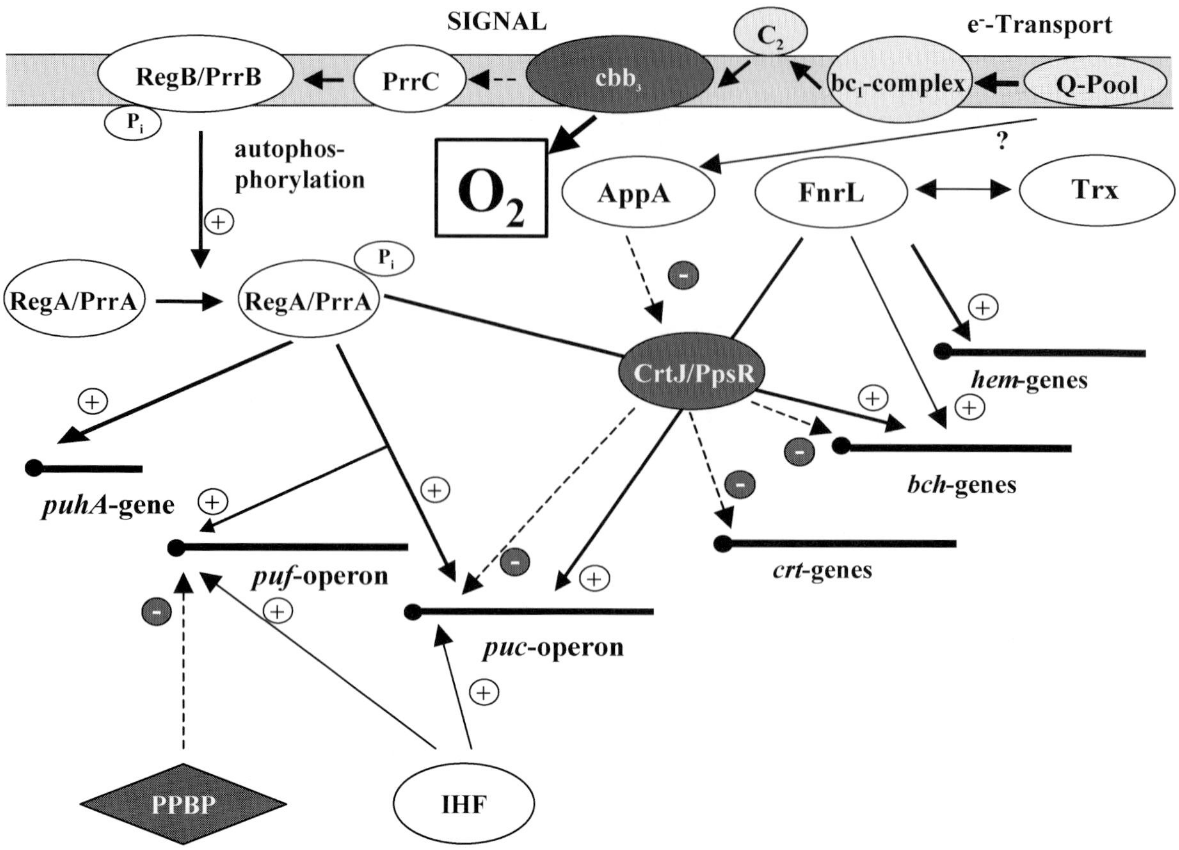

Figure 1. Schematic overview of proteins which are involved in the transcriptional regulation of photosynthesis genes in *Rhodobacter*. Proteins which activate transcription are shown in white, proteins which negatively affect tramscription in black. Proteins which are found only in *Rhodobacter capsulatus* are shown in rectangles, proteins which are known for both organisms, *R. capsulatus* and *R. sphaeroides,* are surrounded with circles. Negative action of proteins on other proteins are indicated by broken arrows, stimulating effects by solid arrows. Thioredoxin (*trxA*) has been shown to affect the *puf*-operon transcription, but the mechanism of signal transmission has not been elucidated so far. Not all proteins which act on transcription of photosynthesis genes are included for reason of clarity; *hem*: genes for heme synthesis; *bch*: genes for bacteriochlorophyll synthesis; *crt*: genes for carotenoid synthesis; *puc*-operon: encodes structural and regulatory proteins for the formation of the LHII antenna complex; *puf*-operon: encodes structural proteins of the LHI antenna complex, the reaction center proteins and proteins involved in the assembly of these complexes; *puhA* gene: encodes the non-pigment-binding subunit of the reaction center.

References

Bauer, C.E. 1995. Regulation of photosynthesis gene expression. In: R.E. Blankenship, M.T. Madigan, and C.E. Bauer (eds.). Anoxygenic photosynthetic bacteria. Kluwer Academic Publishers, Dordrecht, The Netherlands, 1221–1234.

Bird, T.H., Du, S., and Bauer, C.E. 1999. Autophosphorylation, phosphotransfer, and DNA- binding properties of the RegB/RegA two-component regulatory system in *Rhodobacter capsulatus*. J. Biol. Chem. 274: 16343–16384.

Bowman, W.C., Bauer, C.E., and Kranz, R.G. 1999. In vitro activation and repression of photosynthesis gene transcription in *Rhodobacter capsulatus*. Mol. Microbiol. 33: 429–437.

Cannon, W., Missailidis, S., Austin, S., Moore, M., Drake, A., and Buck, M. 1996. Purification and activities of the *Rhodobacter capsulatus* RpoN (sigma N) protein. Mol. Microbiol. 21: 233–245.

Charlton, W., Cannon, W., and Buck, M. 1993. The *Klebsiella pneumoniae* nifJ promoter: analysis of promoter elements regulating activation by the NifA protein. Mol. Microbiol. 7: 1007–1021.

Chen, W., Jäger, A., and Klug, G. 2000. Correction of the DNA sequence of the regB gene of *Rhodobacter capsulatus* with implications for the membrane topology of the sensor kinase RegB. J. Bacteriol. 182: 818–820.

Du, S. Bird, T.H., and Bauer, C.E. 1998. DNA binding characteristics of RegA. A constitutively active anaerobic activator of photosynthesis gene expression in *Rhodobacter capsulatus*. J. Biol. Chem. 273: 18509–18513.

Dubbs, J.M., Bird, T.H., Bauer, C.E., and Tabita, F.R. 2000. Interaction of CbbR and RegA* transcription regulators with the *Rhodobacter sphaeroides* cbbI promoter-operator region. J Biol Chem. 2000 23: 19224–19230.

Elsen, S., Ponnampalam, S.N., and Bauer, C.E. 1998. CrtJ bound to distinct binding sites interacts cooperatively to aerobically repress photopigment biosynthesis and light harvesting II gene expression in *Rhodobacter capsulatus*. J. Biol. Chem. 273: 30762–30769.

Elsen, S., Dischert, W., Colbeau, A., and Bauer, C.E. 2000. Expression of uptake hydrogenase and molybdenum nitrogenase in *Rhodobacter capsulatus* is coregulated by the RegB-RegA two-component regulatory system. J Bacteriol. 182: 2831–2837.

Emetz, D., and Klug, G. 1998. Cloning and characterization of the rpoH gene of *Rhodobacter capsulatus*. Mol. Gen. Genet. 260: 212–217.

Gregor, J., and Klug, G. 1999. Regulation of bacterial photosynthesis genes by oxygen and light. FEMS Microbiol. Lett. 179: 1–9.

Haertig, E., Pasternak, C., Haberzettl, K., Li, K., and Klug, G. 2000. Involvement of thioredoxin in oxygen-dependent regulation of bacterial photosynthesis genes. abstr. S3-2/2. In: 10th International Symposium on Photosynthetic Prokaryotes. Book of Abstracts, Barcelona, Spain.

Hemschemeier, S.K., Kirndörfer, M., Ebel, U., and Klug, G. 1999. Transcriptional regulation of *puf* and *puc* expression in *Rhodobacter capsulatus* by the DNA binding protein RegA. In: The Phototrophic Procaryotes. Peschek *et al.* (eds.), Kluwer Academic, Plenum Publishers, New York, 127–130.

Hemschemeier, S.K., Kirndörfer, M., Hebermehl, M., and Klug, G. 2000. DNA binding of wild type RegA protein and its differential effect on the expression of pigment binding proteins in *Rhodobacter capsulatus*. J. Mol. Microbiol. Biotechnol. 2: 235–243.

Hemschemeier, S.K., Ebel, U., Jäger, A., Balzer, A., Kirndörfer, M., and Klug, G. 2000. In vivo and in vitro analysis of RegA response regulator mutants of *Rhodobacter capsulatus*. J. Mol. Microbiol. Biotechnol. 2: 291–300.

Inoue, K., Kouadio, J.K., Mosley, C.S., and Bauer, C.E. 1995. Isolation and in vitro phosphorylation of sensory transduction components controlling anaerobic induction of light harvesting and reaction center gene expression in *Rhodobacter capsulatus*. Biochemistry 34: 391–396.

Joshi, H.M., and Tabita, F.R. 1996. A global two component signal transduction system that integrates the control of photosynthesis, carbon dioxide assimilation, and nitrogen fixation. Proc. Natl. Acad. Sci. USA 93: 14515–14520.

Kirndörfer, M., Hebermehl, M., Hemschemeir, K., Jäger, A., Ebel, U., and Klug, G. 1997. Transcriptional regulation of *puf* and *puc* operon expression in *Rhodobacter capsulatus* by the DNA binding protein RegA and IHF, abstr. 69B. In G. A. Peschek, W. Löffelhardt, and G. Schmetterer (eds.), IX International Symposium on Photosynthetic Prokaryotes. Book of Abstracts, Vienna, Austria.

Kirndörfer, M., Jäger, A., and Klug, G. 1998. Integration host factor affects the oxygen-regulated expression of photosynthesis genes in *Rhodobacter capsulatus*. Mol. Gen. Genet. 258: 297–305.

Klug, G. 1991. A DNA sequence upstream of the *puf* operon of *Rhodobacter capsulatus* is involved in its oxygen-dependent regulation and functions as a protein binding site. Mol. Gen. Genet. 226: 167–176.

Klug, G., Liebetanz, R., and Drews, G. 1986. The influence of bacteriochlorophyll biosynthesis on formation of pigment-binding proteins and assembly of pigment-protein complexes in *Rhodopseudomonas capsulata*. Arch. Microbiol. 146: 284–291.

Lee, J.K., Wang, S., Eraso, J.M., Gardner, J., and Kaplan, S. 1993. Transcriptional regulation of *puc* operon expression in *Rhodobacter sphaeroides*. Involvement of an integration host factor-binding sequence. J. Biol. Chem. 268: 24491–24497.

Mosley, C.S., Suzuki, J.Y., and Bauer, C.E. 1994. Identification and molecular genetic characterization of a sensor kinase responsible for coordinately regulating light harvesting and reaction center gene expression in response to anaerobiosis. J. Bacteriol. 176: 7566–7573.

Oh, J., and Kaplan, S. 2000. Redox sigaling: globilization of gene expression. EMBO J. 19: 4237–4247.

Ouchane, S., and Kaplan, S. 1999. Topology analysis of the membrane-localized redox-responsive sensor kinase PrrB from *Rhodobacter sphaeroides* 2.4.1. J. Biol. Chem: 274: 17290–17296.

Pasternak, C., Chen, W., Heck, C., and Klug, G. 1996. Cloning, nucleotide sequence and characterization of the *rpoD* gene encoding the primary sigma factor of *Rhodobacter capsulatus*. Gene 176: 177–184.

Pasternak, C., Haberzettl K., and Klug, G. 1999 Thioredoxin is involved in formation of the photosynthetic apparatus of *Rhodobacter sphaeroides*. J. Bacteriol. 181: 100–106.

Ponnanpalam, S.N., Buggy, J.J., and Bauer, C.E. 1995. Characterization of an aerobic repressor that coordinately regulates bacteriochlorophyll, carotenoid, and light – harvesting-II expression in *Rhodobacter capsulatus*. J. Bacteriol. 177: 2990–2997.

Ponnampalam, S.N., and Bauer, C.E. 1997. DNA Binding Characteristics of CrtJ. A redox-responding repressor of bacteriochlorophyll, carotenoid, and light harvesting-II gene expression in *Rhodobacter capsulatus*. J. Biol. Chem. 272: 18391–18396.

Rice, P.A., Yang, S.-W., Mizzuchi, A., and Nash, H.A. 1996. Crystal structure of an IHF-DNA complex: a protein-induced U-turn. Cell 87: 1295–1306.

Rödig, J., Jock, S., and Klug, G. 1999. Coregulation of the syntheses of bacteriochlorophyll and pigment binding proteins in *Rhodobacter capsulatus*. Arch. Microbiol. 171: 198–204.

Sganga, M.W., and Bauer, C.E. 1992. Regulatory factors controlling photosynthetic reaction center and light – harvesting gene expression in *Rhodobacter capsulatus*. Cell 68: 945–954.

Vichivanives, P., Bird, T.H., Bauer, C.E., and Tabita, F.R. 2000. Multiple regulators and their interactions *in vivo* and *in vitro* with the *cbb* regulons of *Rhodobacter capsulatus*. J. Mol. Biol. 300: 1079–1099.

Wellington, C.L., and Beatty, J.T., 1991. Overlapping mRNA transcripts of photosynthesis gene operons in *Rhodobacter capsulatus*. J Bacteriol. 173: 1432–1443.

Wellington, C.L., Taggart, A.K., and Beatty, J.T. 1991. Functional significance of overlapping transcripts of *crtEF*, *bchCA and puf* photosynthesis gene operons in *Rhodobacter capsulatus*. J. Bacteriol. 173: 2954–2961.

Young, D.A., Bauer, C.E., Williams, J.C., and Marrs, B.L. 1989. Genetic evidence for superoperonal organization of genes for photosynthetic pigments and pigment-binding proteins in *Rhodobacter capsulatus*. Mol. Gen. Genet. 218: 1–12.

Zeilstra-Ryalls, J.H., and Kaplan, S. 1995. Aerobic and anaerobic regulation in *Rhodobacter sphaeroides* 2.4.1: the role of the *fnrL* gene. J. Bacteriol. 177: 6422–6431.

Zeilstra-Ryalls, J.H., Gabbert, K., Mouncey, N.J., Kaplan, S., and Kranz, R.G. 1997. Analysis of the *fnrL* gene and its function in *Rhodobacter capsulatus*. J. Bacteriol. 179: 7264–7273.

Zhang, X., and DeMoss, J. 1996. Structure modification induced in the *narG* promoter by binding of intergration host factor and NarL-P. J. Bacteriol. 178: 3971–3973.

Zhu, Y.S., and Hearst, J.E. 1988. Transcription of oxygen-regulated photosynthetic genes requires DNA gyrase in *Rhodobacter capsulatus*. Proc. Natl. Acad. Sci. USA. 85: 4209–4213.

Chapter 15

The Hydrogen-Sensing Apparatus in *Ralstonia eutropha*

Oliver Lenz, Michael Bernhard, Thorsten Buhrke, Edward Schwartz, and Bärbel Friedrich

Abstract

Molecular hydrogen is widely used by microorganisms as a source of energy. One of the best studied aerobic hydrogen oxidizers, the β-proteobacterium *Ralstonia eutropha* (formerly *Alcaligenes eutrophus*), harbors two distinct [NiFe]-hydrogenases which catalyze the heterolytic cleavage of H_2 into $2H^+$ and $2e^-$. The genes encoding the hydrogenase subunits are arranged in two large operons together with accessory and regulatory genes involved in hydrogenase biosynthesis. Both operons are transcribed from strong σ^{54}-dependent promoters. Transcription requires the activation by the HoxA protein, a member of the NtrC family of response regulators. HoxA is only active when H_2 is present in the environment. H_2 recognition is mediated by a signal transduction complex consisting of the soluble histidine protein kinase HoxJ and a regulatory [NiFe]-hydrogenase which acts as an H_2 receptor. Biochemical and genetic data suggest that signal transduction between the RH and HoxJ involves an electron transport process. According to our current model the histidine protein kinase HoxJ inactivates HoxA by phosphorylation in the absence H_2. This property of the HoxJ-HoxA regulator pair is quite different from the behaviour of common two-component regulatory systems. Phosphorylation of HoxA is blocked in the presence of H_2 provided the RH can contact HoxJ and transmit the signal to the kinase. Furthermore, hydrogenase gene expression is subject to a global regulatory network in response to the carbon and energy source. HoxA is a major component of this epistatic control the molecular mechanism of which is not yet understood.

Metabolism of Hydrogen is a Common Microbial Trait

Dihydrogen, the smallest molecule in nature, may well be a major fuel in the world's future energy supply. Biological H_2 conversion represents an ancient process which developed in the early stages of evolution in microorganisms including archaea, bacteria, green algae and some protozoa. During fermentation microbes dispose of excess reducing equivalents by the release of H_2. Dihydrogen is also evolved as a costly side-product in the course of nitrogen fixation. The biologically produced H_2 serves as an attractive source of energy and reducing power for other groups of prokaryotes e.g. methanogens, anoxic phototrophs and "Knallgas" bacteria (Adams, 1994; Friedrich and Schwartz, 1993, Wu and Mandrand, 1993; Vignais *et al.*, 2002). The physiological role of H_2 metabolism in a number of organisms, including cyanobacteria and human pathogens, is less well defined and awaits further exploration.

The nature of the catalyst hydrogenase which mediates the reversible cleavage of H_2 into $2H^+$ and $2e^-$ is an important factor for an efficient H_2 metabolism. Two major classes of hydrogenases are distinguished on the basis of their metal content. i) The O_2-sensitive, Fe-only hydrogenases found in anaerobes catalyze, with few exceptions, H_2 evolution (Adams, 1990; Peters, 1999). ii) The [NiFe]-hydrogenases, most widely distributed in nature, are predominantly involved in H_2-consuming processes (Vignais *et al.*, 2002). [NiFe]-hydrogenases are the subject of this article and will be described in more detail.

The first crystal structure of a [NiFe]-hydrogenase was obtained from the sulfate-reducer *Desulfovibrio gigas* (Volbeda *et al.*, 1995). This periplasmic, heterodimeric protein is considered as a prototypical [NiFe]-hydrogenase. The H_2 active site for H_2 oxidation, which is deeply embedded in the protein, is a [NiFe] dinuclear center coordinated by thiolates of four highly conserved cysteine residues in the large subunit (Figure 1). The electron-transferring small subunit contains three Fe-S clusters which guide the electrons released from H_2 oxidation to the protein surface (Volbeda *et al.*, 1995). Structural and chemical analysis and Fourier transform infrared spectroscopy (FTIR), uncovered an unusual architecture of the active site. Three diatomic molecules, two CN^- groups and one CO molecule, were identified as intrinsic ligands to the Fe (Figure 1). The diatomic ligands are thought to anchor the active site cluster to the protein matrix and to help maintain a low redox state at the metal ions (Volbeda *et al.*, 1996; Happe *et al.*, 1997).

Biosynthesis of the Two Energy-Conserving Hydrogenases in *Ralstonia eutropha*

The strictly respiratory proteobacterium *Ralstonia eutropha* (formerly *Alcaligenes eutrophus*) served as a model organism to study the molecular basis of H_2 metabolism in more depth (Friedrich and Schwartz, 1993). *R. eutropha* is capable of growing autotrophically

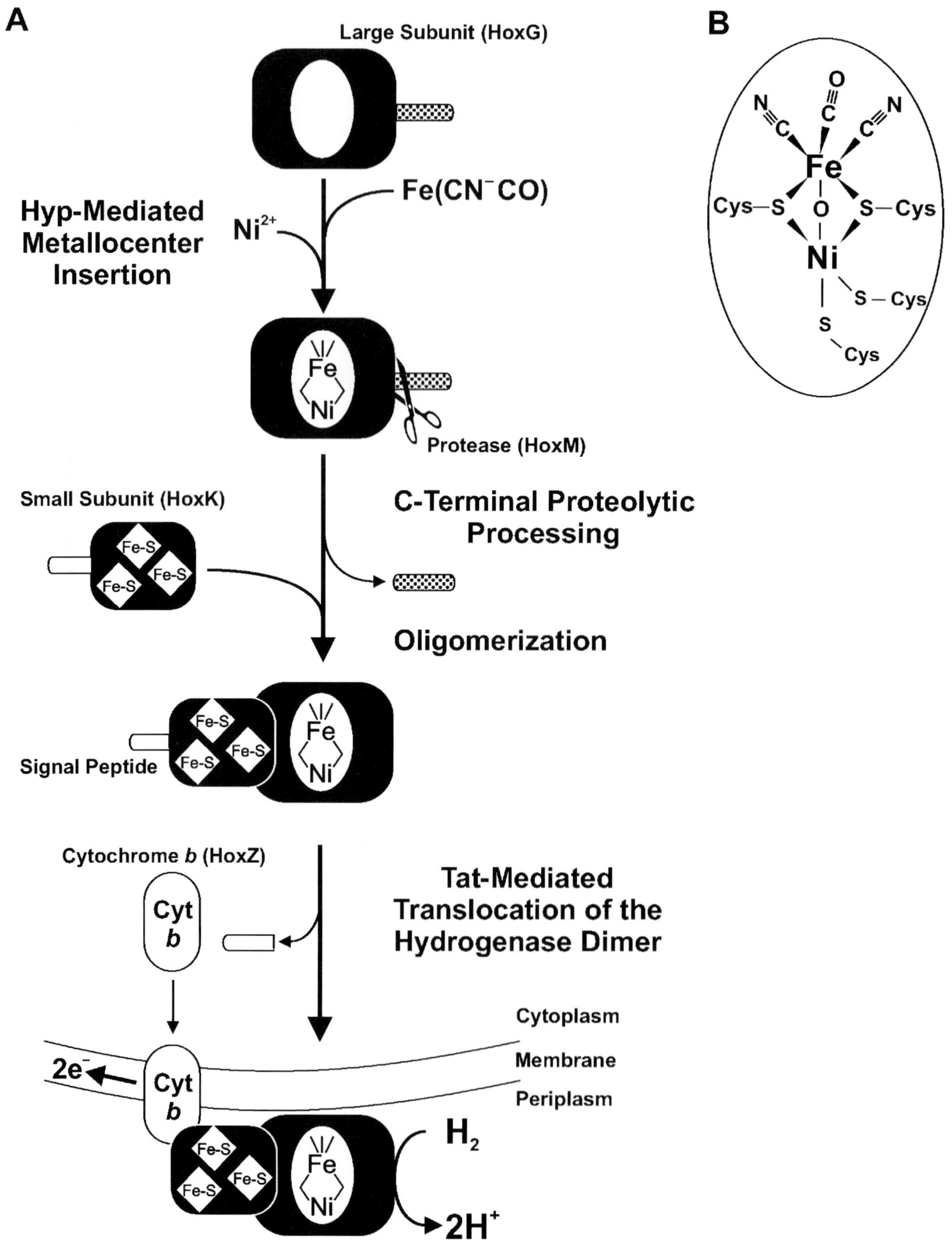

Figure 1. **A**. Model of the maturation of the membrane-bound hydrogenase (MBH) from *Ralstonia eutropha*. **B**. Schematic representation of the catalytic site of prototypical [NiFe]-hydrogenases.

under aerobic conditions with H_2 as a sole source of energy. Alternatively the organism grows heterotrophically on a wide range of organic compounds.

In *R. eutropha* energy conservation from H_2 is mediated by two different [NiFe]-hydrogenases which are synthesized coordinately. One of the two enzymes, the MBH, is bound to the cytoplasmic membrane facing the periplasm. It resembles the prototypical hydrogenase of *Desulfovibrio gigas* and is anchored via the C-terminal end of the small subunit to a membrane-integral *b*-type cytochrome. The cytochrome *b* serves as the primary electron acceptor and connects the MBH with the respiratory chain (Figure 1). The MBH is primarily involved in electron transport coupled phosphorylation (Schink and Schlegel, 1979; Bernhard *et al.*, 1996). The second hydrogenase in *R. eutropha* is a soluble heterotetramer (SH) that resides in the cytoplasm. The protein consists of a large subunit containing the [NiFe] active site and a truncated form of the electron-transferring small subunit, harboring a single Fe-S cluster. This minimal version of an active [NiFe]-hydrogenase is attached to an FMN-containing NADH oxidoreductase dimer which provides the ability to reduce NAD at the expense of H_2. Thus the SH provides reducing equivalents to the cell (Schneider and Schlegel, 1976; Massanz *et al.*, 1998).

In *Escherichia coli* and in several other proteobacteria the synthesis of active [NiFe]-hydrogenase is a protein-assisted process involving sets of accessory proteins (Figure 1). Hydrogenase biosynthesis in *R. eutropha* involves: i) A high affinity nickel permease which specifically provides Ni ions to the cell (Eitinger and Friedrich, 1997); ii) seven *hyp* (*hy*drogenase *p*leiotropic) gene products which are involved in the assembly of the [NiFe] site in both the MBH and the SH (Dernedde *et al.*, 1996), iii) a hydrogenase-specific endopeptidase which removes approximately 20 amino acids from the C-terminus of the large subunit after incorporation of the [NiFe] site, and iv) the oligomerization of the respective hydrogenase subunits (Thiemermann *et al.*, 1996; Massanz *et al.*, 1997). In the case of the MBH, the protein is exported as a cofactor-containing dimer across the membrane by the Tat (twin-arginine translocation) system which is independent of the general secretion (Sec) pathway (Figure 1) (Sargent *et al.*, 1998; Bernhard *et al.*, 2000).

The Hydrogenase Regulon of *R. eutropha*

The hydrogenase (*hox, hyp*) gene cluster spans a region of approximately 80 kbp on the indigenous megaplasmid pHG1 of *R. eutropha*. The hydrogenase genes together with the corresponding accessory genes are arranged in two major transcriptional units, the MBH and the SH operon, respectively (Figure 2). Primer extension analysis and transcriptional fusion studies using *lacZ* as a reporter gene identified two strong promoters, P_{MBH} and P_{SH}, upstream of *hoxK* and *hoxF*, respectively (Schwartz *et al.*, 1998). Since the two promoters are regulated coordinately the MBH and the SH operon form a regulon.

Sets of hydrogenase-specific accessory gene products are encoded adjacent to the hydrogenase structural genes. Closely linked to the MBH-specific genes are seven pleiotropic *hyp* genes of which three, *hypA*, *hypB* and *hypF*, are duplicated and are located in the SH operon (Figure 2) (Wolf *et al.*, 1998). Hydrogenase-specific regulatory genes, *hoxA*, *hoxB*, *hoxC*, and *hoxJ*, were identified in the 3′ region of the MBH operon. *hoxA* and *hoxJ* encode the response regulator and the cognate histidine protein kinase, respectively, of a two-component regulatory system whereas *hoxB* and *hoxC* code for a [NiFe]-hydrogenase-like protein (Lenz and Friedrich, 1998).

Both the P_{MBH} and the P_{SH} promoter are recognized by the alternative sigma factor σ^{54} (RpoN) of the RNA polymerase. Mutants deficient in *rpoN* fail to express the SH and MBH genes under all growth conditions (Schwartz *et al.*, 1998). Screening for additional transcriptional start sites in the *hox/hyp* gene cluster uncovered rather weak but constitutive promoters with typical σ^{70} consensus elements (Figure 2). These promoters direct low expression of the accessory and regulatory genes independent of the adjustable MBH promoter (Schwartz *et al.*, 1999). However, deletion of the MBH promoter and interposon mutagenesis resulted in a dramatic decrease of the HoxA protein under derepressing conditions. This indicates that the strong MBH promoter contributes to the transcription of the downstream located *hyp* and regulatory genes suggesting the formation of an unusually large transcript of at least 17 kbp (Schwartz *et al.*, 1999).

Two physiological conditions must be met to permit MBH and SH gene expression: i) H_2 must be present in the environment and ii) preferentially utilized substrates such as organic acids have to be absent; otherwise hydrogenase gene transcription is subject to a superimposed catabolite control. The degree of repression directly correlates with the quality of the substrate (Figure 2). On poor growth-supporting compounds like glycerol, carbon catabolite repression is released allowing H_2 to be used as an additional energy source, whereas on succinate or pyruvate, hydrogenase gene expression is completely blocked indicating that under these conditions H_2 is dispensable as a substrate (Schwartz *et al.*, 1998; Lenz and Friedrich, 1998).

The Hydrogenase Regulon is Controlled by Three Proteins

Structure and function of the regulatory *hox* gene locus were analyzed with the aid of mutants with in-frame deletions and point mutations in *hoxA, hoxB, hoxC* and *hoxJ*. The product of *hoxA* shows the typical tripartite modular composition of an NtrC-like response regulator. The N-terminal receiver module contains the conserved signatures which are indicative of phosphorylation (Figure 3) (Eberz and Friedrich, 1991). A nucleotide binding site in the input domain of

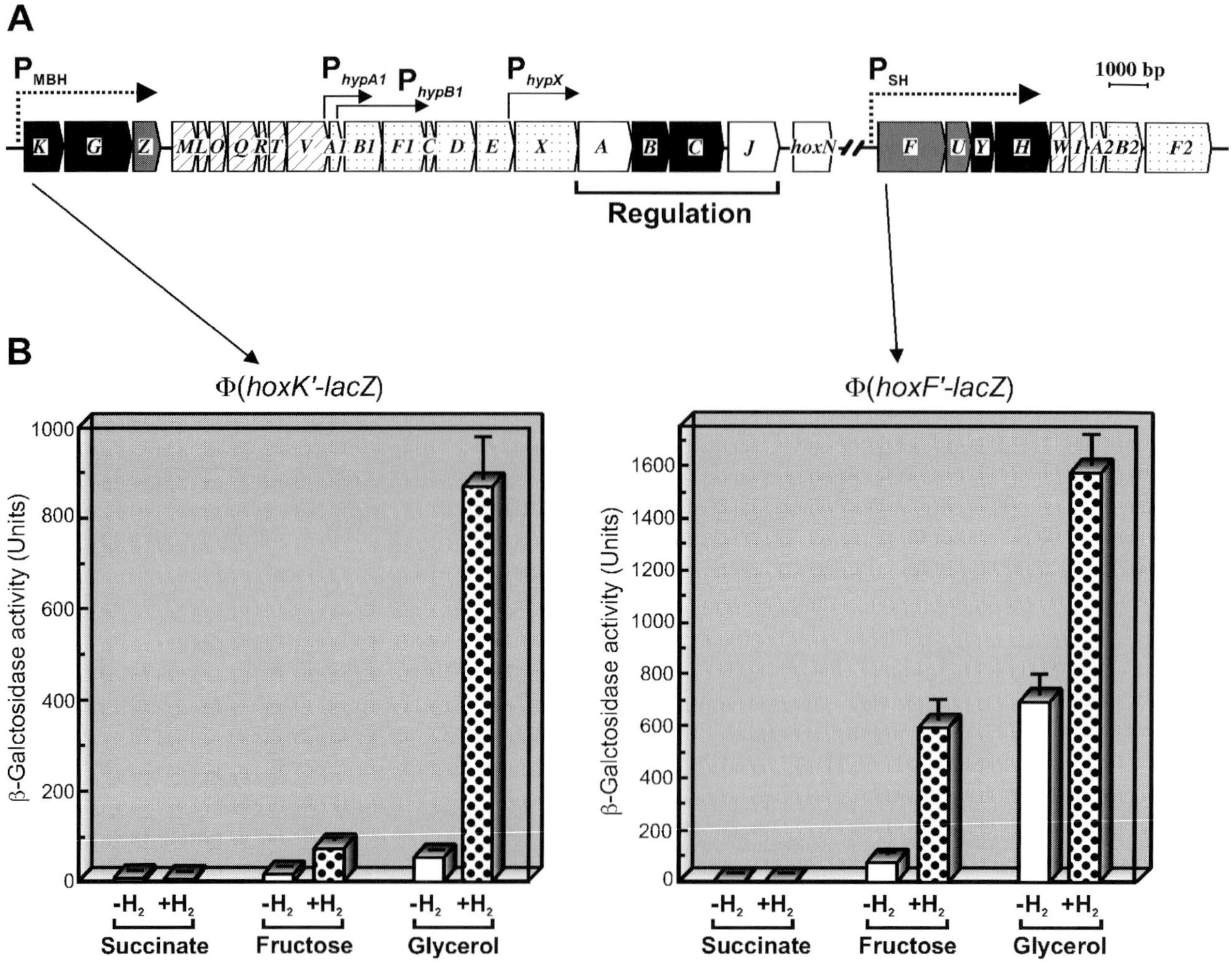

Figure 2. H_2-dependent expression of the hydrogenase regulon of *R. eutropha*. **A.** Hydrogenase gene arrangement on megaplasmid pHG1. Hydrogenase subunit genes (*hox*) are shown in black, MBH- and SH-specific accessory genes (*hox*) are illustrated in hatched boxes, genes involved in metallocenter assembly (*hyp*) are shown in dotted boxes. The regulatory *hox* genes are located at the 3′ end of the MBH gene cluster. *hoxN* encodes a nickel permease. Transcriptional start sites of regulated and constitutive promoters, respectively, are indicated by dotted and solid arrows above the gene cluster. **B.** Activity of the MBH and SH promoters in response to various growth conditions. *Ralstonia* cells were grown in minimal medium containing various carbon sources either in the absence (white bars) or in the presence of molecular hydrogen (dotted bars). Hydrogenase gene expression was determined in the logarithmic growth phase by monitoring β-galactosidase activity derived from vector-based ϕ(*hoxK'-lacZ*) and ϕ(*hoxF'-lacZ*) translational fusions.

HoxA is consistent with the assumption that HoxA initiates transcription by interacting with a σ^{54}–containing RNA polymerase in an ATP-requiring process (Buck *et al.*, 2000). A helix-turn-helix motif is located at the C-terminal end of HoxA. Gel retardation assays in combination with a deletion analysis of the P_{SH} promoter showed that HoxA binds specifically to a region of 170 nucleotides upstream of the *hoxF* transcriptional start site (Zimmer *et al.*, 1995). HoxA is absolutely necessary for the expression of the MBH and SH operons, irrespective of the growth conditions. A large in-frame deletion in *hoxA* abolishes transcription of the MBH and SH genes (Figure 3) (Lenz *et al.*, 1997). The phenotype of a second class of *hoxA* mutants points to an uncommon feature of the regulator. Replacement of the highly conserved aspartate-55 either by glutamate or asparagine led to an unexpected phenotype: constitutive, H_2-independent transcription of the MBH and SH genes (Figure 3). Comparable mutations in other response regulators prevented phosphorylation and eliminated, in most cases, transcription of the corresponding genes (Hoch and Silhavy, 1995).

The product of *hoxJ* has the typical characteristics of a histidine protein kinase (Figure 3). The isolated protein showed autokinase activity in the presence of [γ-^{32}P] ATP as phosphoryl donor (Lenz and Friedrich, 1998). Moreover phosphoryl transfer from HoxJ to HoxA was recently demonstrated *in vitro*, supporting the hypothesis that the proteins are constituents of a hydrogenase-specific two-component regulatory system (M. Forgber, O. Lenz, and B. Friedrich, unpublished results). A deletion of the C-terminal kinase module or an exchange of the highly conserved glycine-422 by serine impaired the autophosphorylation capacity of

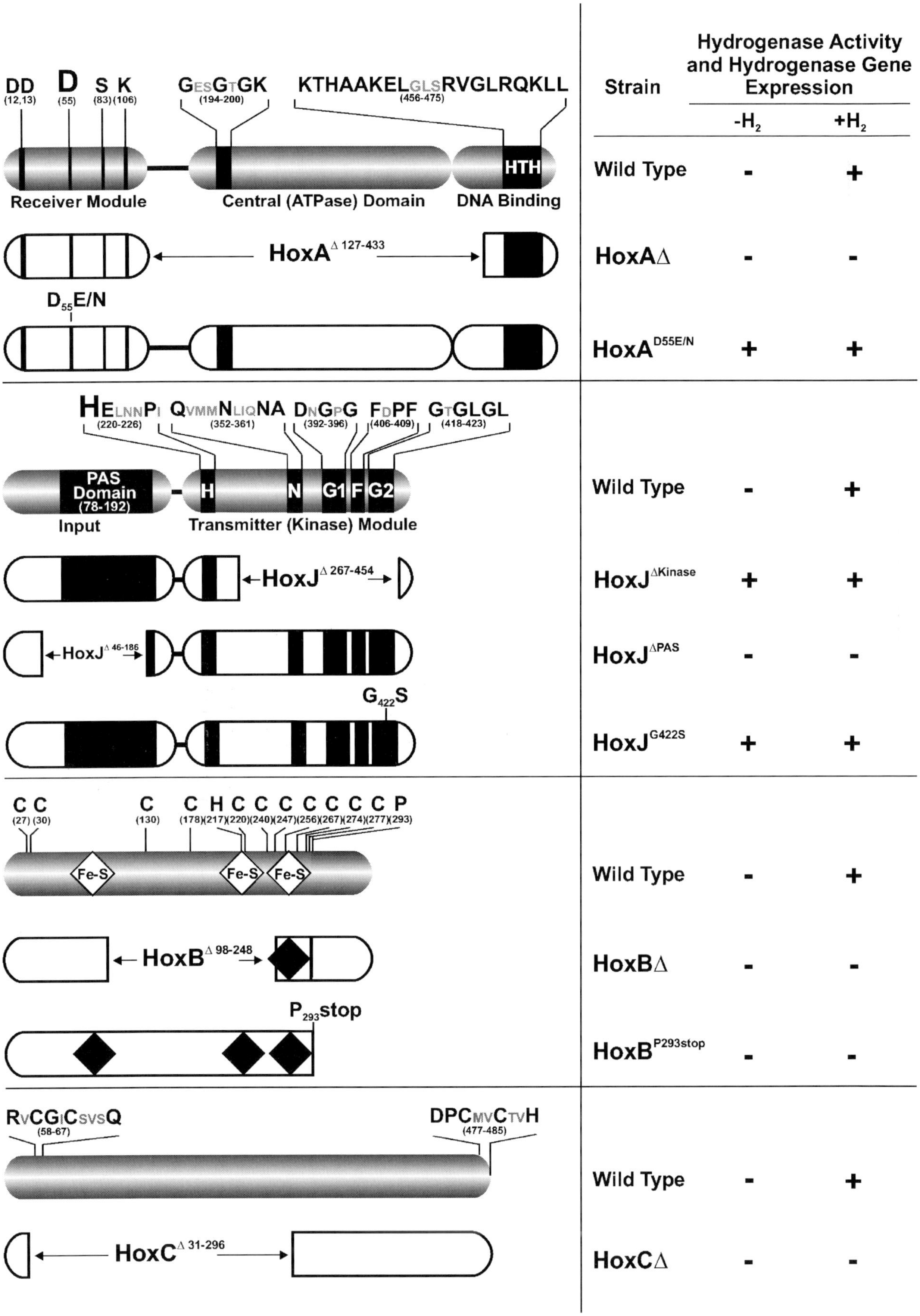

Figure 3. Primary structure of the hydrogenase-specific regulatory proteins and phenotypes of mutant strains carrying in-frame deletions or site-directed mutations in the corresponding genes. Activity measurements were performed with cells grown in fructose-glycerol minimal medium either in the absence or in the presence of molecular hydrogen. Activities of the MBH and the SH were determined by monitoring H_2-dependent methylene blue reduction in the membrane fraction and H_2-dependent NAD reduction in the soluble fraction, respectively. Hydrogenase gene expression was determined as described in the legend of Figure 2. Symbols: (–) denote activities which did not exceed background levels; (+) activity present.

HoxJ. As a consequence MBH and SH gene transcription proceeded constitutively (Figure 3). This result is in good agreement with the phenotype of the HoxA aspartate-55 exchange mutants and supports the conclusion that unlike other systems phophorylation of HoxA has a negative effect on the initiation of gene transcription (Lenz and Friedrich, 1998).

HoxJ is a member of the superfamily of PAS domain proteins, which includes sensors of light, oxygen, and redox state (Taylor and Zhulin, 1999). The removal of the PAS motif, which is located in the N-terminal input domain, by deletion abolished hydrogenase gene transcription completely (Figure 3) (O. Lenz and B. Friedrich, unpublished results) suggesting that this part of HoxJ communicates either directly or via a receptor protein with the stimulus, H_2. A [NiFe]-hydrogenase encoded by the genes *hoxB* and *hoxC* could serve as receptor for H_2. Indeed, deletion mutants lacking either the [NiFe] site-containing large subunit HoxC or the electron-transferring small subunit HoxB were impaired in MBH and SH gene transcription and failed to grow on H_2 (Figure 3). This observation clearly pointed to a regulatory function for the HoxBC protein (Lenz and Friedrich, 1998, Kleihues *et al.*, 2000).

The H_2 Sensing Complex

A double mutant of *R. eutropha* harboring deletions in the large subunit genes of the MBH and SH failed to grow autotrophically with H_2 as the sole source of energy, although the *hoxBC* genes were left intact (Kleihues *et al.*, 2000). This result initially raised the question whether the *hoxBC* genes were really expressed, and, if so, whether their product had any catalytic activity coupled to an energy-conserving process.

SH-/MBH-negative cells of *R. eutropha* contained only trace amounts of H_2-oxidizing activity based on redox dye reduction. Immunochemical analysis with fractionated cell extracts revealed that HoxBC is a soluble protein which resides in the cytoplasm. This observation is compatible with the absence of an N-terminal translocation signal sequence in the HoxB polypeptide (Kleihues *et al.*, 2000). Since H_2 is accessible by diffusion, an H_2-sensing protein need not be attached to the cytoplasmic membrane.

To characterize the HoxBC protein, henceforth designated as the regulatory hydrogenase (RH), in more detail, an overexpression system was developed for *R. eutropha*. This system allowed the isolation of the RH on a preparative scale (Kleihues *et al.*, 2000). The RH showed several biochemical characteristics which are clearly distinct from those of prototypical [NiFe]-hydrogenases. The catalytic activity of the RH was two orders of magnitude lower than that of standard hydrogenases. The H_2-oxidizing activity as well as the regulatory function of the RH were strictly dependent on nickel. Unlike most hydrogenases, the RH is insensitive to O_2 and CO and does not require reductive activation prior to catalysis. The latter observation is in line with EPR spectroscopic data showing that the RH can occur only in two active states, the so-called Ni_a-S and Ni_a-C* states (Pierik *et al.*, 1998; Kleihues *et al.*, 2000). The RH, as isolated, forms an $\alpha_2\beta_2$ tetramer (Bernhard *et al.*, 2001). Metal analysis showed that the RH contains a [NiFe] cofactor similar to that of prototypical [NiFe]-hydrogenases. FTIR spectroscopy identified two CN^- and one CO molecule in the RH active site. A puzzling result, which is not compatible with a standard cofactor composition, was obtained by EPR spectroscopy. An electromagnetic coupling between the active site and the proximal [4Fe-4S] cluster usually observed in standard hydrogenases was not detected in the RH. Preliminary UV-VIS spectroscopy revealed the presence of an additional non-metal cofactor in the RH which may interfere with the cofactor coupling normally observed in [NiFe]-hydrogenases. We postulate that the extra cofactor in the RH is crucial for H_2 signal transduction and the communication with the sensor kinase HoxJ (Bernhard *et al.*, 2001).

A direct interaction between the RH and HoxJ was demonstrated by native polyacrylamide electrophoresis (Figure 4). Upon mixing the two purified components, a high molecular weight complex was formed which remained stable during electrophoresis and retained its hydrogenase activity (Bernhard *et al.*, 2001). The removal of 55 amino acids from the C-terminus of HoxB blocked both complex formation with HoxJ and oligomerization to the $\alpha_2\beta_2$ tetramer. The mutant RH protein retained hydrogenase activity but lost its H_2-sensing regulatory function. The mutant failed to express the MBH and the SH genes even in the presence of H_2 (T. Buhrke and B. Friedrich, unpublished results). It is interesting to note that the sequence of the C-terminal tail of the small subunits of H_2-sensing

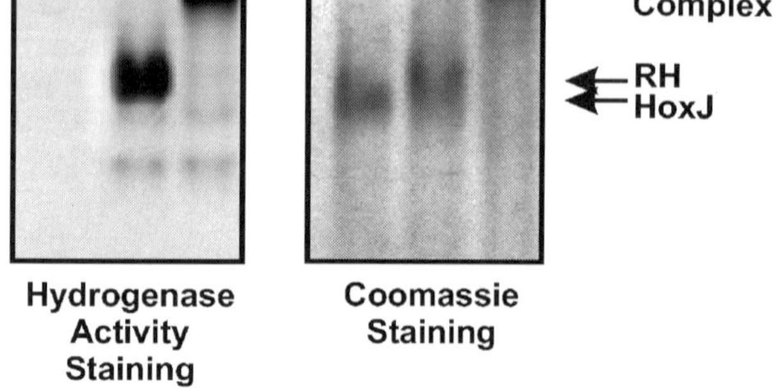

Figure 4. Complex formation of the RH and the histidine protein kinase HoxJ. Purified RH, HoxJ and a mixture of equal amounts of RH and HoxJ, preincubated for 10 min, were applied to native PAGE gel. Panel **A** depicts an in-gel hydrogenase activity staining (H_2-dependent reduction of the artificial electron acceptor phenazine methosulfate). Panel **B** shows the corresponding Coomassie staining.

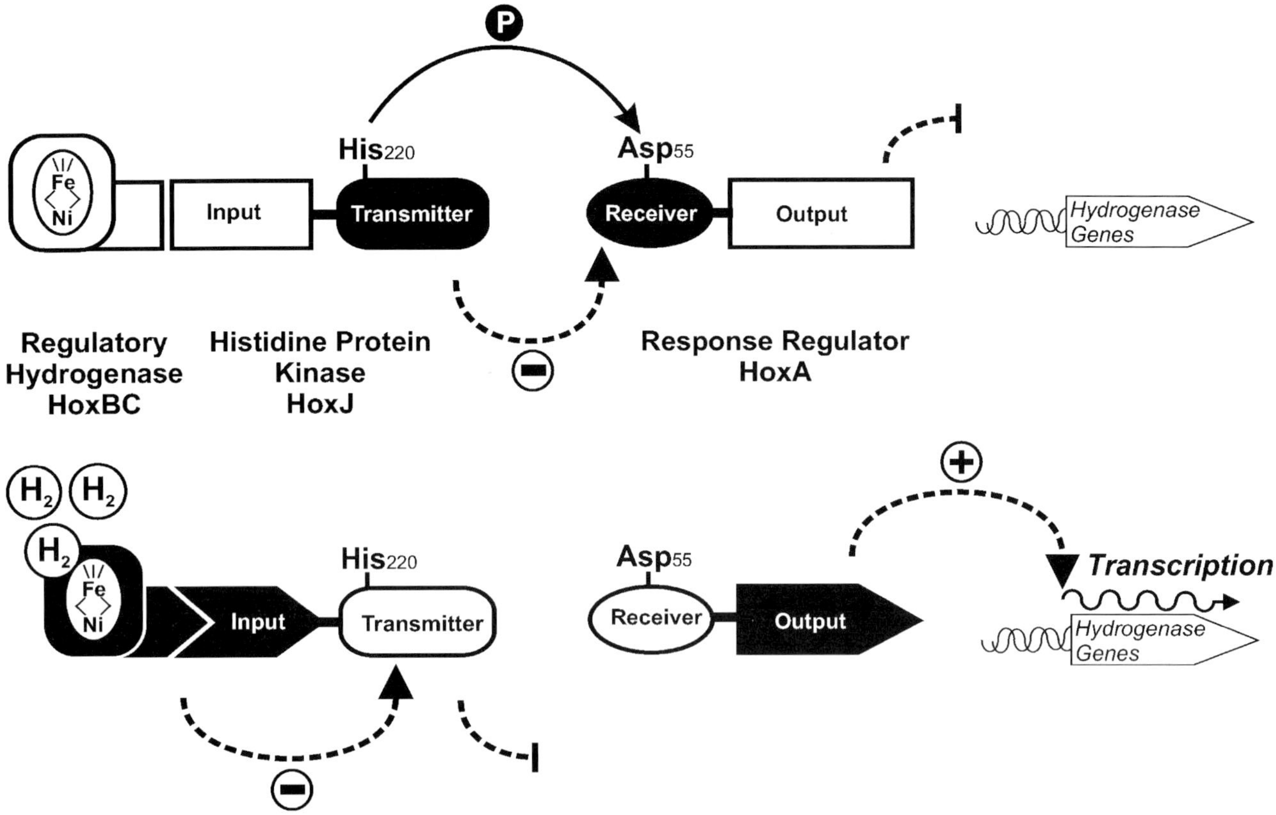

Figure 5. Model of the H_2-dependent signal transduction pathway which controls hydrogenase gene transcription in *R. eutropha*. The proteins and/or domains which are active in regulation in the absence (upper part) or in the presence of molecular hydrogen (lower part) are highlighted in black.

hydrogenases is conserved in other H_2-sensing proteins e.g from *Bradyrhizobium japonicum* and *Rhodobacter capsulatus* (Black *et al.*, 1994, Elsen *et al.*, 1996). The sequence is quite distinct from the C-terminus of the small subunits of MBH-like hydrogenases which interact with a *b*-type cytochrome. Thus the small subunit seems to play not only a role in electron transfer but also in contacting the appropriate reaction partner (Kleihues *et al.*, 2000).

Conclusion and Future Perspectives

Hydrogenase maturation is a complex process which requires a cascade of posttranslational reactions to obtain a catalytically active enzyme. To guarantee a concerted and balanced function of the individually and pleiotropically acting proteins, the whole apparatus requires a finely tuned regulation. The hydrogenase regulon of *R. eutropha*, which codes for the synthesis of two distinct energy-conserving [NiFe]-hydrogenases, is coordinately controlled by two different mechanisms: i) a substrate-specific H_2-sensing signal transduction system, and ii) a global epistatic control which responds to the quality of the energy and carbon source available to the cells. So far our knowledge of the molecular mechanism of the global control is very limited, whereas accumulating data on the H_2 sensing system provide the basis for a regulatory model which is depicted in Figure 5 (Lenz and Friedrich, 1998).

Three proteins are the major players in the H_2-sensing control circuit: HoxA, HoxJ, and the RH. Initiation of transcription is completely dependent on activation by the response regulator HoxA. The activity of HoxA is modified via phosphorylation by the cognate histidine protein kinase HoxJ in conjunction with the regulatory hydrogenase RH (Lenz and Friedrich, 1998).

In the absence of H_2-transcription of the hydrogenase genes is blocked. Experimental evidence suggests that under these conditions HoxA is phosphorylated and incapable of activating transcription. This situation is the opposite of that observed for other two-component systems, in which the phosphorylated form of the protein is active in regulation (Hoch and Silhavy, 1995).

The moment H_2 is available to the cells, transcription of hydrogenase genes is initiated. H_2 signal transmission requires the regulatory hydrogenase HoxBC which forms a tight complex with the sensor kinase HoxJ. It is not clear at present if H_2 signal transduction involves an electron transfer process between the RH and HoxJ or a simple conformational change caused by H_2 binding. The fact that the RH contains an additional non-metal cofactor in the small subunit, HoxB, and the occurrence of the putative redox-signaling PAS domain in HoxJ points to an electron transfer reaction (Bernhard *et al.*, 2001). We postulate that H_2 signal transduction triggers the kinase activity of HoxJ, i.e., HoxA is not phosphorylated or becomes dephosphorylated resulting in a switch to an active HoxA conformer (Figure 5).

It is interesting to note that mutants of *R. eutropha* lacking H_2-sensing control arise rather frequently. The most prominant example is *R. eutropha* H16, the best studied strain, which was considered as the wild-type and kept as such for decades in the culture collection. The H_2-insensitive derivatives form the MBH and SH constitutively but still respond to the global carbon and energy control (Lenz and Friedrich, 1998). Genetic analysis revealed that the H_2-constitutive phenotype is imparted by insertions or point mutations in *hoxJ* which knock out its kinase activity (A. Gleiche, O. Lenz, and B. Friedrich, unpublished results).

References

Adams, M.W. 1990. The structure and mechanism of iron-hydrogenases. Biochim. Biophys. Acta 1020: 115–145.

Adams, M.W. 1994. Biochemical diversity among sulfur-dependent, hyperthermophilic microorganisms. FEMS Microbiol. Rev. 15: 261–277.

Bernhard, M., Buhrke, T., Bleijlevens, B., De Lacey, A.L., Fernandez, V.M., Albracht, S.P.J., and Friedrich, B. 2001. The H_2 sensor of *Ralstonia eutropha*. Biochemical characteristics, spectroscopic properties, and its interaction with a histidine protein kinase. J. Biol. Chem. 276: 15592–15597.

Bernhard, M., Schwartz, E., Rietdorf, J., and Friedrich, B. 1996. The *Alcaligenes eutrophus* membrane-bound hydrogenase gene locus encodes functions involved in maturation and electron transport coupling. J. Bacteriol. 178: 4522–4529.

Bernhard, M., Friedrich, B., and Siddiqui, R.A. 2000. *Ralstonia eutropha* TF93 is blocked in *tat*-mediated protein export. J. Bacteriol. 182:581–588.

Black, L.K., Fu, C., and Maier, R.J. 1994. Sequence and characterization of *hupU* and *hupV* genes of *Bradyrhizobium japonicum* encoding a possible nickel-sensing complex involved in hydrogenase expression. J. Bacteriol. 176: 7102–7106.

Buck, M., Gallegos, M.T., Studholme, D.J., Guo, Y., and Gralla, J.D. 2000. The bacterial enhancer-dependent sigma(54) (sigma(N)) transcription factor. J. Bacteriol. 182: 4129–4136.

Dernedde, J., Eitinger, T., Patenge, N., and Friedrich, B. 1996. *hyp* gene products in *Alcaligenes eutrophus* are part of a hydrogenase maturation system. Eur. J. Biochem. 235: 351–358.

Eberz, G., and Friedrich, B. 1991. Three *trans*-acting regulatory functions control hydrogenase synthesis in *Alcaligenes eutrophus*. J. Bacteriol. 173: 1845–1854.

Eitinger, T., and Friedrich, B. 1997. Microbial nickel transport and incorporation into hydrogenases. In: Transition Metals in Microbial Metabolism. G. Winkelmann, and C.J. Carrano (eds.), Harwood Academic Publishers, Amsterdam, The Netherlands. p. 235–256.

Elsen, S., Colbeau, A., Chabert, J., and Vignais, P.M. 1996. The *hupTUV* operon is involved in negative control of hydrogenase synthesis in *Rhodobacter capsulatus*. J. Bacteriol. 178: 5174–5181.

Friedrich, B., and Schwartz, E. 1993. Molecular biology of hydrogen utilization in aerobic chemolithotrophs. Annu. Rev. Microbiol. 47: 351–383.

Happe, R.P., Roseboom, W., Pierik, A.J., and Albracht S.P.J. 1997. Biological activation of hydrogen. Nature 385: 126.

Hoch, J.A., and Silhavy, T.J. 1995. Two-Component Signal Transduction. American Society for Microbiology, Washington D.C.

Kleihues, L., Lenz, O., Bernhard, M., Buhrke, T., and Friedrich, B. 2000. The H_2 sensor of *Ralstonia eutropha* is a member of the subclass of regulatory [NiFe]-hydrogenases. J. Bacteriol. 182: 2716–2724.

Lenz, O., Strack, A., Tran-Betcke, A., and Friedrich, B. 1997. A hydrogen-sensing system in transcriptional regulation of hydrogenase gene expression in *Alcaligenes* species. J. Bacteriol. 179: 1655–1663.

Lenz, O., and Friedrich, B. 1998. A novel multicomponent regulatory system mediates H_2 sensing in *Alcaligenes eutrophus*. Proc. Natl. Acad. Sci. USA. 95: 12474–12479.

Massanz, C., Fernandez, V.M., and Friedrich, B. 1997. C-terminal extension of the H_2-activating subunit, HoxH, directs maturation of the NAD-reducing hydrogenase in *Alcaligenes eutrophus*. Eur. J. Biochem. 245: 441–448.

Massanz, C., Schmidt, S., and Friedrich, B. 1998. Subforms and *in vitro* recontitution of the NAD-reducing hydrogenase of *Alcaligenes eutrophus*. J. Bacteriol. 180: 1023–1029.

Peters, J.W. 1999. Structure and mechanism of iron-only hydrogenases. Curr. Opin. Struct. Biol. 9: 670–6.

Pierik, A.J., Schmelz, M., Lenz, O., Friedrich, B., and Albracht, S.P.J. 1998. Characterization of the active site of a hydrogen sensor from *Alcaligenes eutrophus*. FEBS Lett. 438: 231–235.

Sargent, F., Bogsch, E.G., Stanley, N.R., Wexler, M., Robinson, C., Berks, B.C., and Palmer, T. 1998. Overlapping functions of components of a bacterial Sec-independent protein export pathway. EMBO J. 17: 3640–50.

Schink, B., and Schlegel, H.G. 1979. The membrane-bound hydrogenase of *Alcaligenes eutrophus*. I. Solubilization, purification and biochemical properties. Biochem. Biophys. Acta 567: 315–324.

Schneider, K., and Schlegel, H.G. 1976. Purification and properties of the soluble hydrogenase from *Alcaligenes eutrophus* H16. Biochim. Biophys. Acta 452: 66–80.

Schwartz, E., Buhrke, T., Gerischer, U., and Friedrich, B. 1999. Positive Transcriptional feedback controls hydrogenase expression in *Alcaligenes eutrophus* H16. J. Bacteriol. 181: 5684–5692.

Schwartz, E., Gerischer, U., and Friedrich, B. 1998. Transcriptional regulation of *Alcaligenes eutrophus* hydrogenase genes. J. Bacteriol. 180: 3197–3204.

Taylor, B.L., and Zhulin, I.B. PAS domains: internal sensors of oxygen, redox potential, and light. Microbiol. Mol. Biol. Rev. 63: 479–506.

Thiemermann, S., Dernedde, J., Bernhard, M., Schröder, W., Massanz, C., and Friedrich, B. 1996. Carboxy-terminal processing of the soluble, NAD-reducing hydrogenase of *Alcaligenes eutrophus* requires the *hoxW* gene product. J. Bacteriol. 178: 2368–2374.

Vignais, P.M., Billoud, B., and Meyer, J. 2002. Classification and Phylogeny of hydrogenases. FEMS Microbiol. Rev. 25: 455–501.

Volbeda, A., Charon, M.H., Piras, C., Hatchikan, E.C., Frey, M., and Fontecilla-Camps, J.C. 1995. Crystal structure of the nickel-iron hydrogenase from *Desulfovibrio gigas*. Nature 373: 580–587.

Volbeda, A., Garcin, E., Piras, C., De Lacey, A.L., Fernandez, V.M., Hatchikian, E.C., Frey, M., and Fontecilla-Camps, J.C. 1996. Structure of the [NiFe] hydrogenase active site: evidence for biologically uncommon Fe ligands. J. Am. Chem. Soc. 118: 12989–12996.

Wolf, I., Buhrke, T., Dernedde, J., Pohlmann, A., and Friedrich, B. 1998. Duplication of *hyp* genes involved in maturation of [NiFe] hydrogenases in *Alcaligenes eutrophus* H16. Arch. Microbiol. 170: 451–459.

Wu, L.F., and Mandrand, M.A. 1993. Microbial hydrogenases: primary structure, classification, signatures and phylogeny. FEMS Microbiol. Rev. 104: 243–270.

Zimmer, D., Schwartz, E., Tran-Betcke, A., Gewinner, P., and Friedrich, B. 1995. Temperature tolerance of hydrogenase expression in *Alcaligenes eutrophus* is conferred by a single amino acid exchange in the transcriptional activator HoxA. J. Bacteriol. 177: 2373–2380.

Chapter 16

Control of the O_2 Sensor/regulator FNR of *Escherichia coli* by O_2 and Reducing agents *in vivo* and *in vitro*

S. Achebach, Y. Zeuner, and G. Unden

Abstract

The synthesis of the enzymes constituting the electron transport chain of *Escherichia coli* is controlled by electron acceptors in order to achieve high ATP yields and high metabolic rates as well. High ATP yields (or efficiency) are obtained by the use of electron acceptors for respiration which allow high ATP yields, preferentially O_2, and nitrate in the absence of O_2. The rate of metabolism is adjusted by use of respiratory isoenzymes which differ in the rate and the efficiency of energy conservation, such as the non-coupling NADH dehydrogenase II (*ndh* gene) and the coupling NADH dehydrogenase I (*nuo* genes). By combination of the contrary principles (rate versus efficiency), growth is optimized for growth yields and rates.

One of the major transcriptional regulators controlling the switch from aerobic to anaerobic respiration is FNR (*f*umarate *n*itrate reductase *r*egulator). FNR is located in the cytoplasm and contains a [4Fe–4S] cluster in the active (anaerobic) state. By reaction with O_2 the cluster is converted to a [2Fe–2S] cluster and finally to apoFNR. O_2 diffuses into the cytoplasm even at very low O_2-tensions (1 μM) where it inactivates [4Fe–4S]·FNR. *In vitro*, glutathione or other thiol compounds can be used as reducing agents for the reconstitution of [4Fe-4S] FNR. A model for the control of the functional state of FNR by O_2 and glutathione or other reducing agents is discussed. According to this model the functional state of FNR is determined by a (rapid) inactivation of FNR by O_2, and a slow (constant) reactivation or reconstitution of [4Fe-4S] FNR by reducing agents like glutathione.

Introduction

In facultatively anaerobic bacteria, O_2 is an important environmental signal for the control of gene expression. For O_2-sensing many different types of sensors have been identified in bacteria and archaea, most of which react directly with O_2 by O_2-reactive groups, like heme, FeS clusters, cysteine pairs, and FAD (for recent reviews see Bauer *et al.*, 1999; Patschkowski *et al.*, 2000). Other sensors like ArcB or PrrB sense the presence of O_2 indirectly, either by the function of respiratory chains or from characteristic metabolites of aerobic or anaerobic energy metabolism (Oh and Kaplan, 2000; Iuchi and Lin, 1993, Georgellis *et al.*, 2001). The proton potential Δp, or the membrane potential $\Delta\psi$ over the cytoplasmic membrane, as well as the energy charge of the bacteria decrease only slightly during transition from aerobic to anaerobic respiration (from −170 to −150 mV) (Tran and Unden, 1998). These parameters therefore most likely do not provide a signal for the switch from aerobic to anaerobic respiration.

Many of the O_2-sensors are located in the cytoplasmic membrane and directly accessible to signal molecules from the environment. Other O_2-sensors, like FNR (*f*umarate and *n*itrate reductase *r*egulator) from *E. coli* or NifL from *Klebsiella* are located in the cytoplasm of the bacteria (Dixon,1998; Klopprogge and Schmitz, 1999; Unden *et al.*, 1995; Becker *et al.*, 1996). This location requires diffusion of O_2 into the cells for reaction with the sensor (Unden *et al.*, 1995).

In *E. coli*, FNR is one of the major regulators controlling the switch from aerobic to anaerobic metabolism at the transcriptional level (Spiro and Guest, 1990; Gunsalus, 1992; Unden *et al.*, 1995; Kiley and Beinert, 1999). Three functionally different forms of FNR have been described (Figure 1). Under anoxic conditions, FNR is a dimer and carries a $[4Fe–4S]^{2+}$ cluster per monomer ([4Fe–4S]·FNR) which is bound by 4 cysteine residues. This form binds to target DNA-sites, and controls the expression of the corresponding genes. The $[4Fe–4S]^{2+}$ cluster is essential for O_2^- sensing, and is converted to a $[2Fe–2S]^{2+}$ cluster ([2Fe–2S]·FNR) in the presence of O_2 (Kiley and Beinert, 1999; Khoroshilova *et al.*, 1995, 1997; Green *et al.*, 1996). [2Fe–2S]·FNR is no longer active in transcriptional activation. After prolonged exposure to O_2, the [2Fe–2S] cluster disintegrates, and FNR devoid of Fe and sulfide is formed (apoFNR) (Lazazzera *et al.*, 1996).

The review will deal with the switch from aerobic to anaerobic metabolism which takes place in a way such as to optimize for adverse principles, energetic efficiency (ATP yield) and high growth rates. The second part will discuss low molecular weight factors like cytoplasmic O_2 and reducing agents controlling

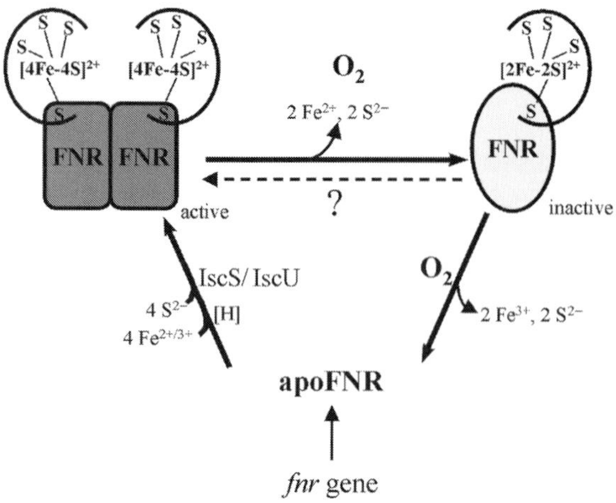

Figure 1. Biochemically and functionally different forms of FNR and their interconversion induced by O_2 as oxidizing agent and by reducing agents. The scheme gives the different forms of FNR which have been identified, and the supposed roles of O_2 and reducing agents (GSH, glutathione) in their conversion. [4Fe-4S]·FNR and [2Fe-2S]·FNR have been demonstrated *in vivo* and *in vitro*, apoFNR produced by O_2-inactivation only *in vitro*. The cysteine desulfurase (IscS in *E. coli*, or NifS from *Azotobacter vinelandii*) provides HS^- from cysteine for the formation of the $[4Fe-4S]^{2+}$ cluster. The reaction is part of *de novo* formation of [4Fe-4S]·FNR, but also of the reactivation of O_2 inactivated apoFNR (demonstrated *in vitro*). Reactivation of [2Fe-2S]·FNR to [4Fe-4S]·FNR has not been demonstrated so far. (See text for more details).

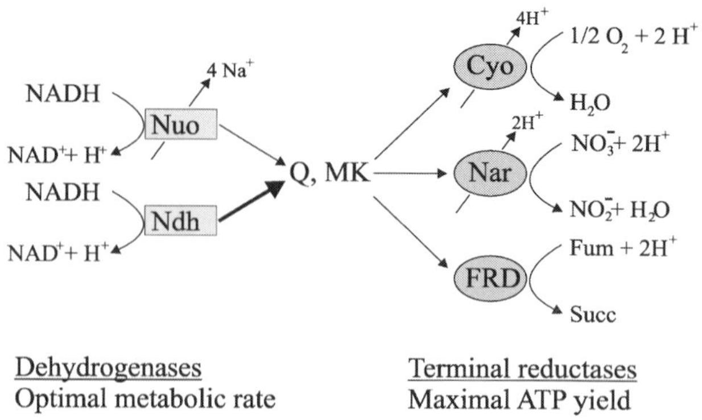

Figure 2. Regulation of respiratory enzyme synthesis in *E. coli*: high metabolic rates *versus* high efficiency (ATP yield). The figure gives the $H^+/2e^-$ ratios for the various enzymes. From the terminal reductases, nitrate reductase (Nar) synthesis is repressed by O_2, synthesis of fumarate reductase (Frd) by O_2 and nitrate. The synthesis of the terminal reductases ensures preferred use of electron acceptors with high H^+/e^- ratios and ATP yield (maximal ATP yield; efficiency control). The respiratory dehydrogenases (here shown for NADH dehydrogenase I or Nuo and NADH dehydrogenase II or Ndh) differ in coupling efficiency (4 Na^+ / 2 e^- versus 0 H^+ / 2 e^-) and activity (indicated by the thickness of the arrow). When possible (e.g. in aerobic respiration), the Ndh enzyme with high activity, but low efficiency is used (maximal rate; rate control). The synthesis of the dehydrogenases is regulated at the transcriptional level by electron acceptors (O_2, nitrate, fumarate), carbon source, and growth phase (Fis, IHF proteins). For details see text. Q, ubiquinone; MK, menaquinone.

FNR function, and how O_2 from the medium is able to react with FNR in the cytoplasm.

Regulatory Rationale for the Switch from Aerobic to Anaerobic Metabolism

Synthesis of terminal reductases: Significance of high ATP yield ('efficiency'). The switch from aerobic to anaerobic respiration in *E. coli* has been studied in detail for the terminal reductases (for reviews see Spiro and Guest, 1990; Gunsalus, 1992; Iuchi and Lin, 1993; Stewart, 1993; Unden *et al.*, 1995; Unden and Bongaerts, 1997). The genes encoding the terminal reductases are transcriptionally regulated by electron acceptors and other factors like growth phase and C-source. By this regulation the enzymes are synthesized in a hierarchical way ensuring preferential use of O_2, whereas under anaerobic conditions nitrate is preferred (Figure 2). When neither O_2 nor nitrate are available, fumarate reductase and the fermentative enzymes are synthesized. The switch from aerobic to nitrate and fumarate respiration or fermentation corresponds to a progressive decrease in ATP yields (Gunsalus, 1992; Unden and Bongaerts, 1997). This regulation ensures preferential use of electron acceptors with high ATP yields, and is effected by regulators responding to O_2, nitrate and fumarate. In addition, O_2 sequentially controls expression of metabolic pathways by using different regulatory O_2 tensions (Becker *et al.*, 1996; Becker *et al.*, 1997). Thus aerobic respiration, anaerobic respiration and fermentation are functional at decreasing O_2 tensions in the medium (> 5 μM O_2, < 5 μM O_2, < 1 μM O_2, respectively).

Synthesis of respiratory dehydrogenases: Significance of high metabolic rates. The synthesis and use of the respiratory dehydrogenases, on the other hand, is not optimized for maximal ATP yields (Unden and Bongaerts, 1997; Tran *et al.*, 1997; Wackwitz *et al.*, 1999; Unden and Schirawski, 1997). *E. coli* contains many different respiratory dehydrogenases which all use the quinones as electron acceptors. Likewise, for the same electron acceptor often more than one respiratory dehydrogenase is found. For example, NADH is oxidized by the NADH dehydrogenase I and II isoenzymes (NADH: quinone oxidoreductase) (Calhoun *et al.*, 1993; Tran *et al.*, 1997; Bongaerts *et al.*, 1995). NADH dehydrogenase I, or Nuo (*nuo* genes), couples the oxidation of NADH to H^+ or Na^+ translocation (Steuber *et al.*, 2000), whereas NADH dehydrogenase II, i.e. Ndh (*ndh* gene), does not (Figure 2). Thus the use of Nuo, but not of Ndh, generates a proton or Na^+ potential and drives phosphorylation of ADP. The enzyme activity of Ndh can be 3 times higher than that of Nuo and allows higher respiration rates.

The expression of the *nuo* and the *ndh* operons is regulated in response to electron acceptors (O_2, nitrate, fumarate), the C-source and the growth phase (Bongaerts *et al.*, 1995; Wackwitz *et al.*, 1999; Green and Guest, 1994; Green *et al.*, 1996; Tran *et al.*, 1997). The regulation is opposite for both operons. Thus the activity of the coupling enzyme (Nuo) is maximal in fumarate respiration, intermediate in aerobic respiration, and minimal in fermentation when it is not required.

The non-coupling Ndh, on the other hand, is found mainly in aerobic respiration. In NADH–fumarate respiration NADH dehydrogenase I represents the only coupling enzyme which explains its preferred use in this pathway. In aerobic respiration containing the coupling quinol oxidases Cyo or Cyd, however, Nuo is the minor enzyme compared to Ndh. As a consequence, H^+- or Na^+-translocation by NADH dehydrogenase can be estimated to be low in aerobic respiration (about 0.7 H^+ or Na^+/2e) and high in fumarate respiration (about 3.1 H^+ or Na^+/2e) (Tran *et al.*, 1997; Unden and Schirawski, 1997; Unden and Bongaerts, 1997).

Most of the other dehydrogenases used in aerobic respiration (e.g. aerobic glycerol-3-P dehydrogenase, pyruvate oxidase) apparently do not couple the redox reaction to H^+ translocation either, although for energetic reason H^+ translocation would be possible (for a review see Unden and Bongaerts, 1997). The dehydrogenases of anaerobic respiration (anaerobic glycerol-3-P dehydrogenase, hydrogenases, formate dehydrogenase) on the other hand, mostly seem to be coupling enzymes. As a consequence, in aerobic respiration the bacteria gain high metabolic flux rates at the expense of low efficiency or ATP yield. However, the efficiency of ATP yield has to be increased when the bacteria are limited for energy (i.e. anaerobic fumarate respiration). By the use of different regulatory principles for dehydrogenases (high metabolic rate) and terminal reductases (high ATP yield) the bacteria are able to balance high metabolic flux rates and high ATP yields for an optimal overall growth (Unden and Bongaerts, 1997; Unden and Schirawski, 1997).

Under specific conditions these regulatory principles can be cancelled. Thus in aerobic growth of *E. coli* (i.e. low efficiency of coupling), the synthesis of Ndh can be turned down and that of Nuo increased when higher ATP yields are desirable (Wackwitz *et al.*, 1999; Green *et al.*, 1996). This situation happens during the onset of exponential growth when high ATP yields are more important than speed of growth. The growth phase dependent transcriptional regulators Fis and IHF stimulate the expression of the *nuo* operon and repress *ndh*, and the resulting increase of Nuo at the expense of Ndh raises the ATP yield in early exponential aerobic growth (Wackwitz *et al.*, 1999). Thus the control of respiratory pathways is subject to a complicated transcriptional regulation in order to optimize growth with respect to efficiency (or energy conservation) and metabolic rates. This long-known phenomenon of bacterial growth and metabolism can now be explained to a great extent at a molecular level by transcriptional regulation.

O_2 as the Signal for FNR Inactivation

For reaction with FNR, O_2 has to diffuse from the medium to FNR in the cytoplasm, and this process represents an important part of the signalling pathway. O_2 is able to react directly with FNR *in vitro*, and also in the bacteria obviously no other mediators are required for the reaction (Jordan *et al.*, 1997). It has been shown, however, that ferricyanide is able *in vivo* and *in vitro* to promote inactivation of FNR function or [4Fe–4S]·FNR destruction in an artificial reaction (Unden *et al.*, 1990; Lazazzera *et al.*, 1996), indicating that inactivation of FNR can also be achieved by oxidizing agents without direct need for O_2. As shown below, the properties of O_2 diffusion to the cytoplasm are consistent with the presence of O_2 in the cytoplasm under aerobic and microaerobic conditions, which is a prerequisite for direct interaction between FNR and O_2 (Becker *et al.*, 1996; Unden *et al.*, 1997; Tseng *et al.*, 1996).

O_2 as the Regulatory Signal for FNR: The properties of O_2 as the regulatory signal for FNR were studied in growing bacteria in an oxystat by the use of FNR-dependent reporter gene fusions (Becker *et al.*, 1996). The oxystat allows growth of the bacteria at defined and constant O_2 tensions (pO_2). The expression of FNR-regulated genes decreases in a hyperbolic response with increasing O_2 tension in the medium. The O_2 tension corresponding to half-maximal induction or repression ($pO_{0.5}$) is in the range of 1 to 5 μM O_2 for most FNR regulated genes (equivalent to approx. 1 to 5 mbar, or 1 to $5 \cdot 10^2$ Pa of O_2, or 0.5 to 2.5% air saturation) (Fig. 3). Only for *nirB* encoding the soluble nitrite reductase a distinctly higher $pO_{0.5}$ was found (Tran *et al.*, 2000). Therefore at O_2 concentrations as low as 1 μM in the medium, sufficient O_2 should be present in the cytoplasm for FNR inactivation (Fig. 4). The $pO_{0.5}$ of *sdh* expression which is regulated by the ArcBA system in response to O_2 is in a similar range (about 5 μM O_2) (Becker *et al.*, 1996). The *cyd* and *cyo* operons encoding the quinol oxidases Cyo and Cyd which operate under aerobic or microaerobic conditions are subject to O_2 regulation by ArcA and FNR (Tseng *et al.*, 1996; Fu *et al.*, 1991). Expression of *cyd* is thus stimulated by ArcA under microaerobic conditions and repressed by FNR under anaerobic conditions, resulting in a maximal expression under microaerobic conditions. The function of fermentative metabolism started only at very low pO_2 values ($pO_{0.5} \leqslant 1$ μM O_2) (Becker *et al.*,

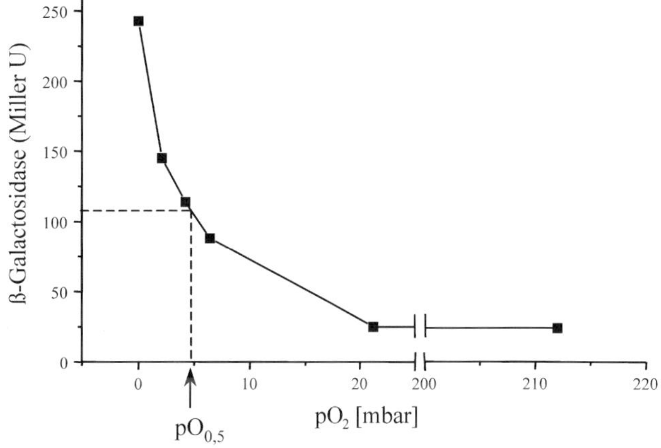

Figure 3. Expression of *frdA'*-*'lacZ* (fumarate reductase) of *E. coli* as a function of the pO_2 in the medium. Expression was measured in growing bacteria in an oxystat (Becker *et al.* 1996). The pO_2-value causing half-maximal expression ($pO_{0.5}$) is indicated.

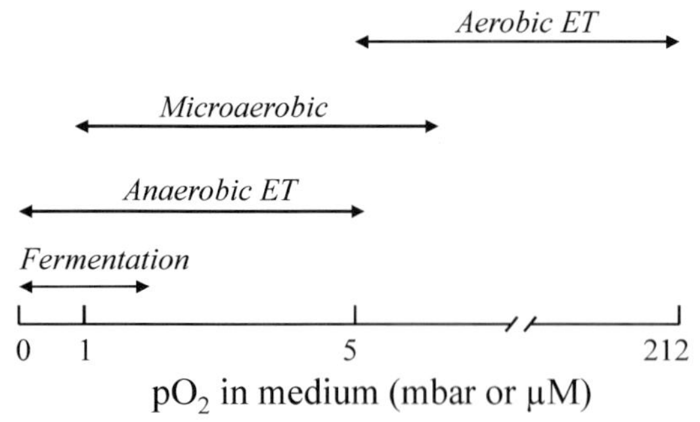

Figure 4. Relation of different types of catabolic reactions in *E. coli* to the oxygen tension (pO_2) in the medium. The figure gives approximate pO2 values for the synthesis and function of fermentation, anaerobic respiration or electron transport (ET), microaerobic respiration, and aerobic electron transport (Becker *et al.*, 1996; Becker *et al.*, 1997; Tseng *et al.* 1996; Fu *et al.*, 1991).

1997). Therefore aerobic, microaerobic and anaerobic respiration and fermentation take over with decreasing pO_2 sequentially. Aerobic respiration (mostly ArcBA regulated) is active at $pO_{0.5} > 5$ μM O_2, whereas anaerobic respiration (mostly FNR regulated) and fermentation function at $pO_{0.5} \leqslant 5$ μM O_2 and $pO_{0.5} \leqslant 1$ μM O_2, respectively.

Presence of O_2 in the Cytoplasm for Catabolic Purposes: Degradation of aromatic compounds like benzoate requires O_2 for ring cleavage by oxygenases. Due to the cytoplasmic location of the oxygenases, growth at the expense of aromatic compounds depends on the presence of O_2 in the cytoplasm. This is in contrast to the aerobic metabolism of glucose or succinate which requires O_2 for aerobic respiration by membraneous oxidases. The functioning of oxygenase dependent metabolic pathways, or growth on these substrates, therefore is indicative for the presence of O_2 in the cytoplasm (Arras *et al.*, 1998). *Pseudomonas putida* is able to grow at the expense of aromatic compounds under aerobic and microaerobic conditions. The oxygen tension which causes decrease of the maximal growth rate by 50% ($pO_{0.5}$) is about 8 μM O_2 for aromatic compounds compared to about 1.5 μM O_2 for glucose or succinate (Arras *et al.*, 1998). At the corresponding $pO_{0.5}$ value the oxygenases became limiting in catabolism, and the substrate of the oxygenase was excreted (Arras *et al.*, 1998). Thus under aerobic and microaerobic conditions, the diffusion of O_2 into the cytoplasm occurs at high rates which is even sufficient to support catabolic reactions with high O_2 requirement in the cytoplasm. The difference in the $pO_{0.5}$ values for FNR dependent regulation in *E. coli* ($pO_{0.5}$ = 1 to 5 μM O_2) and for the growth on aromatic compounds ($pO_{0.5}$ = 8 μM O_2) should reflect the low rates of cytoplasmic O_2 consumption in the former and the high rates of O_2 consumption in the latter case.

Stability and Availability of O_2 within the Bacteria under Microaerobic Conditions?: In *E. coli* thiols represent the major redox active compounds in the cytoplasm. For components like glutathione (GSH/GSSG couple), glutaredoxin or thioredoxin (reduced and oxidized forms) redox potentials in the range of −200 mV to −300 mV are assumed (Tuggle and Fuchs, 1985; Hwang *et al.*, 1992; Prinz *et al.*, 1997). Reaction of O_2 (E_0' = + 820 mV) with thiols is slow compared to the high diffusion rates. Therefore O_2 is rather stable in this environment for kinetic reasons and not in thermodynamic equilibrium with the thiol compounds. This explains why O_2 is present in the bacteria at the same time as the thiols even under microaerobic conditions and despite the large difference in redox potential.

The supply of O_2 to the cytoplasm by diffusion can be calculated from the diffusion coefficients for O_2 in water, phospholipid, and protein solution (cytoplasm), and the cell dimensions (Unden *et al.*, 1995; Becker *et al.*, 1996). In *E. coli* and *Pseudomonas* the rate of O_2 diffusion into the bacteria can be compared to the respiration rate, i.e. the consumption of O_2 at the cytoplasmic membrane (Becker *et al.*, 1996; Unden *et al.*, 1995). Under microaerobic conditions with 1 μM dissolved O_2 in the medium (corresponding to about 0.5% air saturation) the diffusion of O_2 into the bacteria amounts to about 3 μmol $O_2 \times min^{-1} \times mg^{-1}$ protein. The O_2 diffusion exceeds the O_2 consumption by respiration and other reactions (about 0.4 μmol $O_2 \times min^{-1} \times mg^{-1}$ protein). Thus the experiments and the calculations concurringly demonstrate that under microaerobic conditions O_2 can be predicted to be present and is found in the cytoplasm. This is in agreement with the measured $pO_{0.5}$ values for FNR regulation (1 to 5 μM O_2), oxygenase function ($pO_{0.5} \sim 8$ μM O_2) and the switch to fermentation ($pO_{0.5} \leqslant 1$ μM O_2). Only at distinctively lower pO_2 values the cytoplasm might become anoxic due to consumption of O_2 by respiration and reaction with cytoplasmic thiols.

Glutathione and the Formation of Active FNR

Formation of Active FNR: FNR protein is present in *E. coli* in roughly the same concentrations under aerobic and anaerobic growth conditions (Spiro and Guest, 1987; Unden and Duchêne, 1987). Therefore FNR function has to be controlled at the functional level, and it has been shown that inactive FNR from aerobically grown *E. coli* can be converted under anaerobic conditions *in vivo* to the functional state without need for protein synthesis (Engel *et al.*, 1991). This suggests a reactivation of O_2-inactivated FNR in *E. coli* under anaerobic conditions.

According to the scheme of Figure 1 three biochemically and functionally different forms of FNR have been identified. [4Fe–4S]·FNR and [2Fe–2S]·FNR can be differentiated by Mössbauer spectroscopy and have been demonstrated *in vitro*, but also *in vivo* (Khoroshilova *et al.*, 1997; Popescu *et al.*, 1998; Kiley and Beinert, 1999) whereas formation of apoFNR as a product of O_2-inactivation has been demonstrated *in vitro* (Lazazzera *et al.*, 1996; Tran *et al.*, 2000). The interconversion of the different forms in Figure 1 has been demonstrated so far mainly *in vitro*, only conversion of [4Fe–4S]·FNR to [2Fe–2S]·FNR has been followed *in vitro* and *in vivo* (Khoroshilova *et al.*, 1997; Popescu *et al.*, 1998). Production of apoFNR from

[2Fe–2S]·FNR has been shown only *in vitro* due to the lack of appropriate methods *in vivo* (Lazazzera *et al.*, 1996; Tran *et al.*, 2000). The generation of [4Fe–4S]·FNR has been studied mainly *in vitro*, but the involvement of the IscS desulfurase has been demonstrated *in vivo* by mutagenesis. The formation of [4Fe–4S]·FNR from apoFNR obviously is part of *de novo* synthesis of active FNR. The reaction requires cysteine desulfurase (Khoroshilova *et al.*, 1995, 1997; Green *et al.*, 1996; Tran *et al.*, 2000; Schwartz *et al.*, 2000; Zheng *et al.*, 1993) which catalyzes desulfuration of the cysteine providing HS^- (presumably via enzyme bound persulfide) for the FeS cluster formation. *E. coli* contains 3 genes homologous to the *Azotobacter vinelandii nifS* encoding cysteine desulfurase (NifS). The IscS isoenzyme (*iscS* gene) seems to be most important for formation of [4Fe–4S]·FNR *in vivo* (Schwartz *et al.*, 2000). In addition, IscU is required *in vivo* for efficient supply of Fe, and as a scaffold for FeS cluster formation (Urbina *et al.*, 2001; Tokumoto *et al.*, 2002). O_2 inactivated apoFNR can also be used efficiently for reconstitution of active FNR in a very similar reaction (Tran *et al.*, 2000). Therefore *in vitro* FNR can be inactivated by O_2 and reactivated in a cyclic process as shown in Figure 1, and in this way O_2 inactivation of FNR is reversible. It is not known, however, whether the reactivation of O_2-exposed apoFNR is of physiological relevance in *E. coli*. Direct reactivation of [4Fe–4S]·FNR from [2Fe–2S]·FNR has been demonstrated so far neither *in vitro* nor *in vivo*.

Glutathione as the Reducing Agent for the Formation of [4Fe–4S]·FNR: Mutants of *E. coli* deficient of glutathione (*gshA*) show a shift in the switch point for FNR dependent regulation to distinctly lower $pO_{0.5}$ values, suggesting a role for glutathione in the control of FNR function in the bacteria (Tran *et al.*, 2000). Complete loss of FNR function in the *gshA* mutant was not observed, suggesting that other cellular reductants can take over the function of glutathione in parts. This might apply in particular under anaerobic conditions when slow activation by the artificial reaction might be sufficient to retain FNR in the active state. *In vitro*, glutathione can be used as the reducing agent for the formation of active FNR ([4Fe–4S]·FNR) from apoFNR and from O_2-inactivated apoFNR (Tran *et al.*, 2000). Whether glutathione supports also the conversion of [2Fe–2S]·FNR to [4Fe–4S]·FNR is not known. *In vivo*, the increased sensitivity of FNR dependent regulation in the *gshA* mutant, could in part also be caused by a lower O_2 consumption in the cytoplasm, or by effects of toxic O_2-products formed in the glutathione deficient mutant at higher rates.

Control of the Functional State of FNR by O_2 and Glutathione

If the different forms of FNR can be interconverted as suggested in Figure 1, the functional state of FNR would depend on the ratio of active ([4Fe–4S]·FNR) and inactive FNR ([2Fe–2S]·FNR, apoFNR), or the rates of ([4Fe–4S]·FNR) formation and disassembly (Fig. 5).

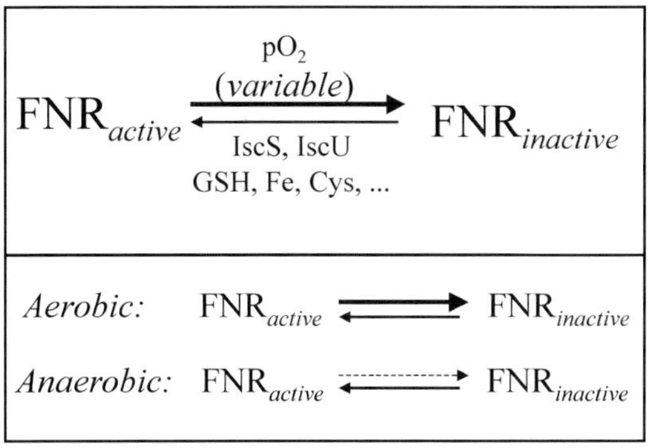

Figure 5. Factors suggested to control the interconversion between active FNR ([4Fe-4S]·FNR) and inactive forms FNR ([2Fe-2S]·FNR, apoFNR) *in vivo*. It is supposed that O_2 causes a rapid inactivation of FNR, whereas reactivation of FNR (compare Fig. 1) is a slow and constant process. As a consequence, under aerobic conditions, inactivation might prevail, under anaerobic conditions formation of active FNR. Thus the functional state of FNR is governed mainly by the pO_2 of the medium.

The rate of [4Fe-4S]·FNR inactivation is supposed to be variable depending on the pO_2 in the medium or cytoplasm (Becker *et al.*, 1996). It is not known, however, which of the factors required for the formation of active FNR (Fe^{2+}, cysteine, IscS, IscU, glutathione or other reducing agents) varies in concentration or activity significantly in the bacteria in response to O_2 availability. Assuming that these factors do not change largely in response to O_2 availability, the amounts of active and inactive FNR depend largely on the pO_2 in the bacteria due to variation of the FNR inactivation rate (Fig. X/Eq. 1). At high pO_2 the inactivation is supposed to predominate the formation of active FNR, resulting in the presence of high amounts of inactive FNR. Under conditions of low O_2 supply, or anoxic conditions, on the other hand, the rates of FNR inactivation would be low. Then even low rates of formation of active FNR would be sufficient to reverse the ratio of active and inactive FNR. As far as known, the factors required for FNR reconstitution, are present in similar concentrations or activies under aerobic and anaerobic conditions. This applies to the total FNR contents (Unden and Duchêne, 1987), and presumably also to the glutathione content and GSSG/GSH ratio (Hwang *et al.*, 1992; Prinz *et al.*, 1997), supporting the view that the formation of active FNR occurs at roughly constant rates. Understanding FNR function *in vivo* would require in particular more information on these factors which are required for the formation of active FNR from apoFNR and from [2Fe-2S]·FNR.

Acknowledgement

The work described from the authors laboratory was supported by a grant in the priority programme "Regulatory networks in bacteria" from the Deutsche Forschungsgemeinschaft and by the Fonds der

Chemischen Industrie. This contribution is an up-dated version of a publication by Unden *et al.* (2002).

References

Arras, T., Schirawski, J., and Unden, G. 1998. Availability of O_2 as a substrate in the cytoplasm of bacteria under oxic and micro-oxic conditions. J. Bacteriol. 180: 2133–2136.

Bauer, C.E., Elsen, S., and Bird, T.H. 1999. Mechanisms of redox control of gene expression. Annu. Rev. Microbiol. 53: 495–523.

Becker, S., Holighaus, G., Gabrielczyk, T., and Unden, G. 1996. O_2 as the regulatory signal for FNR-dependent gene regulation in *Escherichia coli.* J. Bacteriol. 178: 4515–4521.

Becker, S., Vlad, D., Schuster, S., Pfeiffer, P., and Unden, G. 1997. Regulatory O_2 tensions for the synthesis of fermentation products in *Escherichia coli* and relation to aerobic respiration. Arch. Microbiol. 168: 290–296.

Bongaerts, J., Zoske, S., Weidner, U., and Unden, G. 1995. Transcriptional regulation of the proton translocating NADH dehydrogenase genes (*nuoA-N*) of *Escherichia coli* by electron acceptors, electron donors and gene regulators. Mol. Microbiol. 16: 521–534.

Calhoun, M.W., Oden, K.L., Gennis, R.B., Teixeira de Mattos M.J., and Neijssel, O.M. 1993. Energetic efficiency of *Escherichia coli*: effects of mutations in components of the aerobic respiratory chain. J. Bacteriol. 175: 3020–3025.

Dixon, R. 1998. The oxygen-responsive NifL-NifA complex: a novel two-component regulatory system controlling nitrogenase synthesis in γ-proteobacteria. Arch. Microbiol. 371–380.

Engel, P., Trageser, M., and Unden, G. 1991. Reversible interconversion of the functional state of the gene regulator FNR from *Escherichia col in vivo* by O_2 and iron availability. Arch Microbiol. 156: 463–470.

Fu, H.A., Iuchi, S., and Lin. E.C.C. 1991. The requirement for ArcA and Fnr for peak expression of the *cyd* operon in *Escherichia coli* under microaerobic conditions. Mol. Gen. Genet 226: 209–213.

Georgellis, D., Kwon, O., and Lin, E.C. 2001. Quinones as the redox signal for the arc two-component system of bacteria. Science 292: 2314–2316.

Green, J., Bennett, B., Jordan, P., Ralph, E.T., Thompson, A.J., and Guest, J.R. 1996. Reconstitution of the [4Fe4S] cluster in FNR and demonstration of the aerobic-anaerobic transcription switch *in vitro*. Biochem. J. 316: 887–892.

Green, J., and Guest, J.R. 1994. Regulation of transcription at the *ndh* promoter of *Escherichia coli* by FNR and novel factors. Mol. Microbiol. 12: 433–444.

Green, J., Anjum, M.F., and Guest, J.R. 1996. The *ndh*-binding protein (Nbp) regulates the *ndh* gene of *Escherichia coli* in response to growth phase and is identical to Fis. Mol. Microbiol. 20: 1043–1055.

Gunsalus, R.P. 1992. Control of electron flow in *Escherichia coli*: coordinated transcription of respiratory pathway genes. J. Bacteriol. 174: 7069–7074.

Hwang, C., Sinskey, A.J., and Lodish, H.F. 1992 Oxidized redox state of glutathione in the endoplasmatic reticulum. Science 257: 1496–1502.

Iuchi, S., and Lin, E.C.C. 1993. Adaptation of *Escherichia coli* to redox environments by gene expression. Mol. Microbiol. 9: 9–15.

Jordan, P.A., Thomson, A.J., Ralph, E.T., Guest, J.R., and Green, J. 1997. FNR is a direct oxygen sensor having a biphasic response curve. FEBS Lett. 416: 349–352.

Kiley, P.J., and Beinert, H. 1999. Oxygen sensing by the global regulator FNR: the role of the iron-sulfur cluster. FEMS Microbiol. Rev. 22: 341–352.

Khoroshilova, N., Beinert, H., and Kiley, P.J. 1995. Association of a polynuclear iron-sulfur center with a mutant FNR protein enhances DNA binding. Proc. Natl. Acad. Sci. USA 92: 2499–2503.

Khoroshilova, N., Popescu, C., Münck, E., Beinert, H., and Kiley, P.J. 1997. Iron-sulfur cluster disassembly in the FNR protein of *Escherichia coli* by O_2: [4Fe4S] to [2Fe2S] conversion with loss of biological activity. Proc. Natl. Acad. Sci. USA 94: 6087–6092.

Klopprogge, K., and Schmitz, R.A. 1999. NifL of *Klebsiella pneumoniae*:redox characterization in relation to nitrogen source. Biochim. Biophys. Acta 1431: 462–470.

Lazazzera, B.A., Beinert, H., Khoroshilova, N., Kennedy, M.C., and Kiley, P.I. 1996. DNA-binding and dimerization of the Fe-S containing FNR protein from *Escherichia coli* are regulated by oxygen. J. Biol. Chem. 271: 2762–2768.

Oh, J.I, and Kaplan, S. (2000) Redox signalling: globalization of gene expression. EMBO J. 19: 4237–4247.

Patschkowski, T., Bates, D.M., and Kiley, P.J. (2000) Mechanisms for sensing and responding to oxygen deprivation. In 'Bacterial stress responses', G. Storz, R. Hengge-Aronis (eds.) ASM Press, Washington, pp 61–78.

Popescu, C.V., Bates, D.M., Beinert, H., Münck, E. and KIley, P.I. 1998. Mösbauer spectroscopy as a tool for the study of activation/inactivation of the transcription regulator FNR in whole cells of *Escherichia coli.* Proc. Natl. Acad. Sci. USA. 95: 13431–13435.

Prinz, W.A., Åslund, F., Holmgren A., and Beckwith, J. (1997) The role of the thioredoxin and glutaredoxin pathways in reducing protein disulfide bonds in the *Escherichia coli* cytoplasm. J. Biol. Chem. 272: 15661–15667

Schwartz, C.J., Djaman, O., Imlay, J.A., and Kiley, P.J. 2000. The cysteine desulfurase, IscS, has a major role in *in vivo* Fe-S cluster formation in *Escherichia coli.* Proc. Natl. Acad. Sci. USA 97: 9009–9014.

Spiro, S., and Guest, J.R. 1987. Regulation and over-expression of the *fnr* gene of *Escherichia coli.* J. Gen. Microbiol. 133: 3279–3288.

Spiro, S., and Guest, J.R. 1990. FNR and its role in oxygen-regulated gene expression in *Escherichia coli.* FEMS Microbiol. Rev. 75: 399–428.

Steuber, J., Schmid, C., Rufibach, M., and Dimroth, P. 2000. Na^+ translocation by complex I (NADH:quinone oxidoreductase) of *Escherichia coli.* Mol. Microbiol. 35: 428–434.

Stewart, V. 1993. Nitrate regulation of anaerobic respiratory gene expression in *Escherichia coli.* Mol. Microbiol. 9: 425–434.

Tokumoto, U., Nomura, S., Minami, Y., Mihara, H., Kato, S., Kurihara, T., Esaki, N., Kanazawa, H., Matsubara, H., and Takahashi, Y. (2002) Network of protein-protein interactions among iron-sulfur cluster assembly proteins of *Escherichia coli.* J. Biochem (Tokyo). 131:713–719.

Tran, Q.H., Arras, T., Becker, S., Holighaus, G., Ohlberger, G., and Unden, G. (2000) Role of glutathione in the formation of the active form of the oxygen sensor FNR ([4Fe4S]·FNR) and in the control of FNR function. Eur. J. Biochem. 267: 4817–4824.

Tran, Q.H., Bongaerts, J., Vlad, D., and Unden, G. 1997. Requirement for the proton-pumping NADH dehydrogenase I of *Escherichia coli* in NADH → fumarate respiration and bioenergetic implications. Eur. J. Biochem. 244: 155–160.

Tran, Q.H., and Unden, G. 1998. Changes in the proton potential and the cellular energetics of *Escherichia coli* during growth by aerobic or anaerobic respiration or by fermentation. Eur. J. Biochem. 251: 538–543.

Tseng, C.-P., Albrecht, J., and Gunsalus, R.P. 1996. Effect of microaerophilic cell growth conditions on expression of the aerobic (*cyoABCD* and *cydABC*) and anaerobic (*narGHJI, frdABCD*, and *dmsABC*) respiratory pathway genes in *E. coli.* J. Bacteriol. 178: 1094–1098.

Tuggle, C.K., and Fuchs, J.A. 1985. Glutathione reductase is not required for maintenance of reduced glutathione in *Escherichia coli* K-12. J. Bacteriol. 162: 448–450.

Unden, G., Bongaerts, J., Becker, S., Holighaus, G., Schirawski, J., and Six, S. 1995. O_2-sensing and O_2-dependent gene regulation in facultatively anaerobic bacteria. Arch. Microbiol. 164: 81–90.

Unden, G., and Bongaerts, J. 1997. Alternative respiratory pathways of *Escherichia coli*: Energetics and transcriptional regulation in response to electron acceptors. Biochim. Biophys. Acta. 1320: 217–234.

Unden, G., and Duchêne, A. 1987. On the role of cyclic AMP and the FNR protein in *Escherichia coli* growing anaerobically. Arch. Microbiol. 147: 1195–2000.

Unden, G., Trageser, M., and Duchéne, A. 1990. Effect of positive redox potentials on the expression of anaerobic respiratory enzymes in *Escherichia coli.* Mol. Microbiol. 4: 315–319.

Unden, G., and Schirawski, J. 1997. The O_2 responsive transcriptional regulator FNR of *Escherichia coli*: The search for signals and reactions. Mol. Microbiol. 25: 205–210.

Unden, G. 1998. Transcriptional regulation and energetics of alternative respiratory pathways in facultatively anaerobic bacteria. Biochim. Biophys. Acta. 1365: 220–224.

Unden, G. Achebach, S., Holighaus, G., Tran, H.Q., Wackwitz, B., and Zeuner, Y. 2002. Control of FNR function of *Escherichia coli* by O_2 and reducing conditions. J. Mol. Microbiol. Biotechnol. 4: 263–268.

Urbina, H.D., Silberg, J.J., Hoff, K.G., and Vickery, L.E. 2001. Transfer of sulfur from IscS to IscU during Fe/S cluster assembly. J. Biol. Chem. 176: 44521–44526.

Wackwitz, B., Bongaerts, J., Goodman, S.D., and Unden, G. 1999. Growth phase dependent regulation of *nuoA-N* expression of *Escherichia coli* by the Fis protein: DNA binding and bioenergetic significance. Mol. Gen. Genet. 262: 876–883; *Corrig.* Mol. Gen. Genet. 2000. 263: 1070.

Zheng, L., While, R.H., Cash, V.L., Jack, R.F., and Dean, D.R. 1993. Cysteine desulfurase activity indicates a role for NifS in metallocluster biosynthesis. Proc. Natl. Acad. Sci. USA. 90: 2754–2758.

Zientz, E., Bongaerts, J., and Unden, G. 1998. Fumarate regulation of gene expression in *Escherichia coli* by the DcuSR (*dcuSR* genes) two-component regulatory system. J. Bacteriol. 180: 5421–5425.

Zientz, E., Janausch, I.G., Six, S., and Unden, G. 1999. Function of DcuC as the C_4-dicarboxylate carrier during glucose fermentation by *Escherichia coli*. J. Bacteriol. 181: 3716–3720.

Chapter 17

The Molecular Biology of Formate Metabolism in Enterobacteria

Susanne Leonhartsberger, Ingrid Korsa, and August Böck

Abstract

Formate is the signature compound in the anaerobic metabolism of *Escherichia coli* and other enterobacteria. Its synthesis and degradation is integrated in a net-work of metabolic routes which are elegantly regulated to adjust the carbon flux to the metabolic needs. This review summarises the information on the biochemistry of synthesis and degradation of formate, on the genetics of the members of the regulon and on the mechanism underlying the regulation.

Formate is a key metabolite in the energy metabolism of many bacteria. Because of its low redox potential of −420 mV it can be oxidised not only through the aerobic respiratory chain but it can also serve as an electron donor for the anaerobic reduction of fumarate (Macy *et al.*, 1976) and of nitrate or nitrite (Wimpenny and Cole, 1967). The prominent role of formate in energy metabolism is particularily pronounced in the mixed-acid-fermentation of enterobacteria since one third of the sugar carbon degraded is converted into formate. Depending on the physiological condition, formate may be secreted, oxidised aerobically, used as a reductant in anaerobic respiration with a variety of terminal reductases, or converted into CO_2 and H_2 via the formate hydrogen-lyase reaction (Stephenson and Stickland, 1932; Ordal and Halvorson, 1939). In the latter physiological situation acidification is counteracted by the conversion of the formic acid into a neutral molecule, i.e. hydrogen, which possesses a redox potential comparable to that of formic acid. All these reactions are neatly tied together in a regulatory network, designated formate regulon, in which formate itself plays a decisive role as effector molecule (Rossmann *et al.*, 1991). In this review we shall deal first with an outline of the biochemical reactions involved in formate production and consumption and with the transcriptional units whose expression is under the control of formate. Finally, we shall concentrate on the regulation of formate metabolism and discuss the knowledge available on the main regulators involved, the FhlA and HycA proteins.

Formate Metabolism

The key reaction in the switch of aerobic carbohydrate degradation to fermentation is the cleavage of pyruvate (Figure 1). Aerobically and also under nitrate respiratory conditions, pyruvate is oxidised by pyruvate dehydrogenase (Kaiser and Sawers, 1994). Upon shift to anaerobic growth in the absence of external electron acceptors pyruvate is cleaved predominantly by pyruvate-formate-lyase (PFL) in a non-oxidative reaction thus yielding acetyl-CoA and formate. Both the activity and the synthesis of PFL are subject to complex regulation. In aerobically grown cells PFL is present in a basal amount but in an inactive form. Shift to anaerobiosis leads to activation of the enzyme by converting it into a radical form via abstraction of a hydrogen atom from glycyl residue 734. Flavodoxin is functioning as an electron donor and S-adenosylmethionine as cofactor (for review see Kessler and Knappe, 1997).

$$\begin{aligned}&\text{PFL} - \text{H} + \text{S} - \text{adenosylmethionine} + \text{flavodoxin}_{\text{red}} \rightarrow \\ &\quad \text{PFL}^{\circ} + 5' - \text{desoxyadenosine} \\ &\quad + \text{methionine} + \text{flavodoxin}_{\text{ox}}\end{aligned}$$

The three-dimensional structure of PFL was determined recently in complex with the substrate analogon oxamate; it suggested the involvement of two cysteine residues in the reaction mechanism whereby one, Cys418, serves as intermediary thiyl radical and Cys419 as H-atom donor (Becker *et al.*, 1999; Plaga *et al.*, 2000). PFL in its glycine radical form is oxygen sensitive; when it is exposed to molecular oxygen the carbon backbone of the polypeptide is cleaved at the site of the glycine radical. Activation of PFL is catalysed by a specific activase whose gene is located downstream of the *pfl* structural gene. When fermentatively growing cells of *E. coli* are shifted to aerobic conditions, the radical form of PFL is deactivated in a hitherto uncharacterised reaction catalysed by AdhE, the enzyme reducing acetyl-CoA to ethanol with acetaldehyde as an intermediate (Kessler and Knappe, 1997).

Concomitant with activation of the enzyme during a shift from aerobiosis to fermentative growth conditions an approximately 12 to 15-fold induction of PFL synthesis takes place. Besides elongation factor Tu PFL is thus the most abundant protein in fermenting cells. Factors involved in the induction process are the Fnr protein, the ArcAB two-component regulatory system and the integration host factor (Sawers and Watson, 1998; Alexeeva *et al.*, 2000).

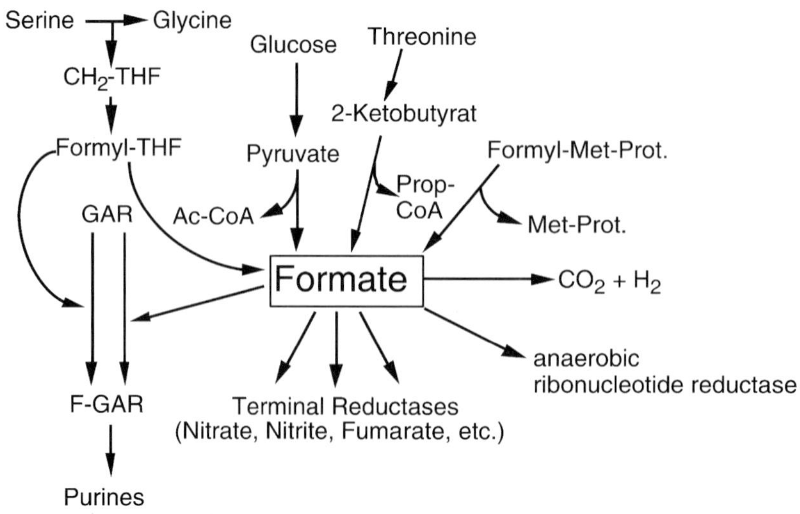

Figure 1. Synthesis and degradation of formate in the metabolism of *E. coli*. The pathways are shown schematically. THF: tetrahydrofolate, Ac-CoA: acetyl-CoA, Prop-CoA: propionyl-CoA, Met-Prot.: proteins with N-terminal methionine, GAR: glycinamide ribonucleotide, F-GAR: formyl glycinamide ribonucleotide. For details see text.

As a consequence of the high-level induction and the activation, actively fermenting *E. coli* cells accumulate formate levels up to 20 mM in their cytoplasm (cited in Sawers, 1994).

Recently, a second anaerobic ketoacid formate-lyase has been discovered in *E. coli*. It is involved in anaerobic threonine metabolism and cleaves 2-keto butyrate into propionyl-CoA and formate (Hesslinger *et al.*, 1998). Moderate contributions to the formate levels in *E. coli* are also provided by the posttranslational deformylation of N-formyl-methionine residues and during purine biosynthesis (Figure 1).

In addition, formate serves as the hydrogen donor for the ribonucleoside-triphosphate reductase responsible for the production of deoxyribonucleoside triphosphates during anaerobic growth of *E. coli* (Mulliez *et al.*, 1995).

Formate, therefore, is present both in aerobically and anaerobically grown cells; accordingly, both aerobic and anaerobic degradation routes have evolved (for detailed review see Sawers, 1994 and Unden and Bongaerts, 1997). Aerobic cells express a formate oxidoreductase (FDH-O) first described by Pinsent (1954); it appears to couple formate oxidation with ubiquinone reduction in the respiratory chain. The enzyme is also present under nitrate respiratory condition (Sawers *et al.*, 1991). The predominant enzyme in nitrate reducing cells, however, is FDH-N. The membrane-bound enzyme has been purified by Enoch and Lester (1975) and well characterised in different groups (see Sawers, 1994). Its synthesis is induced during anaerobiosis and by nitrate. FDH-H finally, is a component of the formate hydrogen-lyase complex of *E. coli*. It is loosely associated with the inner side of the cytoplasmic membrane and is readily detached during breakage of the cells. FDH-H was purified by Axley *et al.* (1991) and its three-dimensional structure has been resolved (Boyington *et al.*, 1997). All three formate dehydrogenases are selenoproteins carrying a molybdopterin dinucleotide cofactor. Of relevance for the discussion of the regulatory aspect is that FDH-N and FDH-H possess greatly different apparent affinities for formate, namely 0.12 mM for FDH-N and 26 mM for FDH-H (Enoch and Lester, 1975; Axley *et al.*, 1990; for detailed review see Unden and Bongaerts, 1997). Intracellularly generated formate can be readily exchanged with the environment. Two gene products were identified with a function in formate transport. They are FocA which is the product of the first gene in the *pfl* transcriptional unit (Suppmann and Sawers, 1994) and FocB, a product of one of the genes of the *hyf* operon from *E. coli* (Andrews *et al.*, 1997). Experimental evidence for such a transport function has been provided for FocA but not yet for FocB.

Genetics of Formate Metabolism

E. coli harbours four transcriptional units on its chromosome whose expression is determined by the intracellular concentration of formate (Figure 2A).

fdhF: The monocistronic *fdhF* gene codes for the selenopolypeptide of FDH-H. It is transcribed from a sigma-54 dependent promoter whose activation requires the presence of an upstream regulatory sequence (UAS) centered between −100 to −140 nucleotides upstream of the transcription start site (Birkmann and Böck, 1989; Birkmann *et al.*, 1987).

hyc genes: The *hycBCDEFGH* genes code for components of the formate hydrogen-lyase systems which accept electrons from FDH-H and reduce protons to molecular hydrogen (see Figure 2B) (Böhm *et al.*, 1990; Sauter *et al.*, 1992). HycB displays sequence similarity with the small subunit of formate dehydrogenases and could serve as the membrane docking target of FDH-H, HycC and HycD are

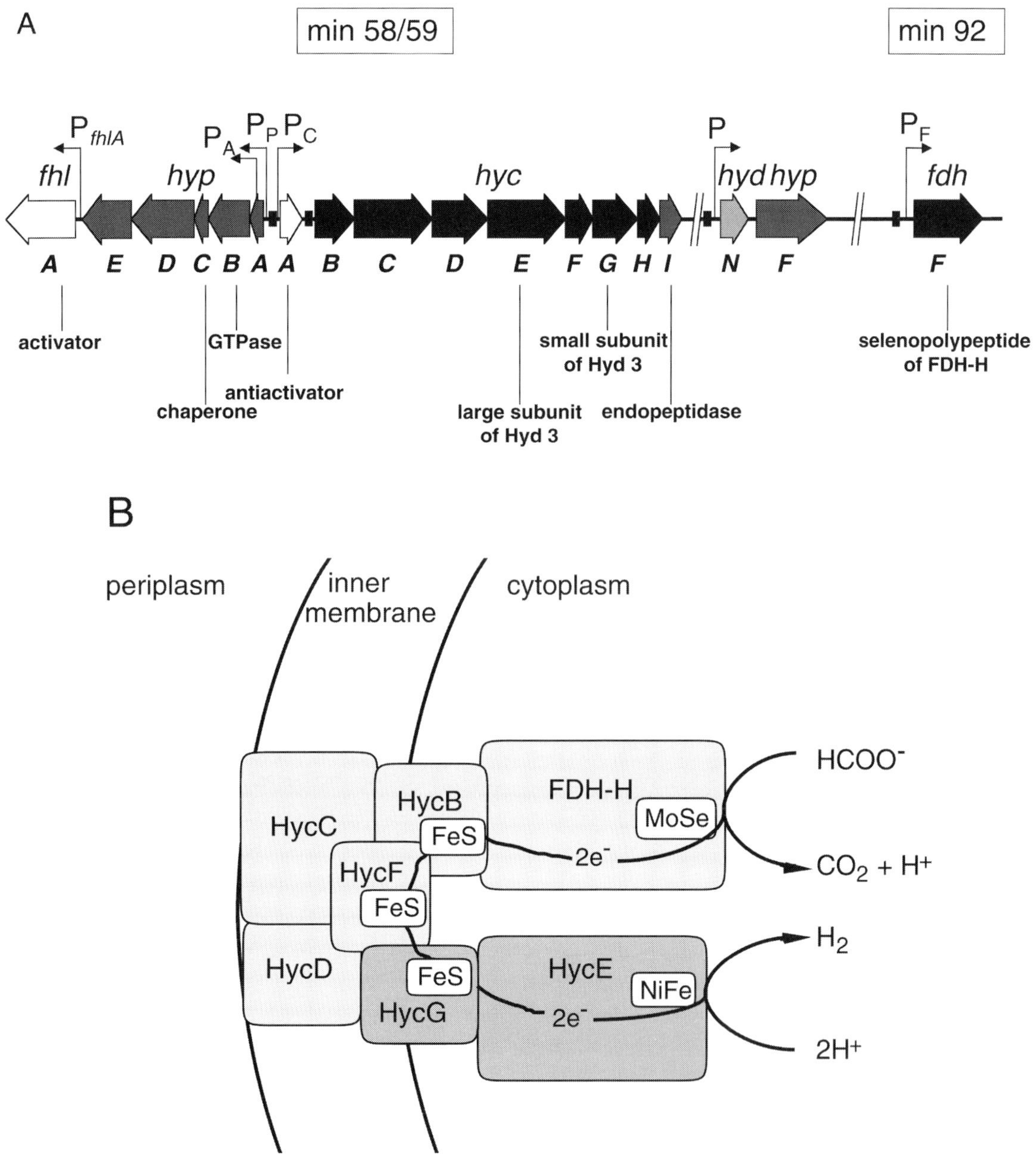

Figure 2. Organisation of the genes and structural model of the FHL complex in *E. coli*. A. The four transcriptional units of the formate regulon. Big arrows represent genes, small boxes UAS sequences, small arrows promoters. Genes coding for structural components of the FHL complex are shown in black, genes of accessory proteins in grey and regulatory genes in white. Hyd: hydrogenase. Functions of the gene products are shown where biochemical evidence was presented. B. Model of the FHL complex associated to the membrane and the functions of its components. Subunits of the hydrogenase 3 are shaded in dark grey.

membrane-integral proteins with a putative anchoring function and HycE and HycG are the large and small subunits of hydrogenase 3, respectively. Judged from sequence signatures, HycF appears to be an electron carrier but its position within the redox chain is unknown. The function of HycH which is required for generation of formate hydrogen-lyase activity is unknown. As discussed later, HycA is involved in regulation and HycI in the maturation of the complex. The *hyc* operon is expressed from a sigma-54 dependent promoter (P_c) located upstream of *hycA* co-operating with an UAS positioned in the center of the *hyc-hyp* intergenic region (Lutz *et al.*, 1990).

The *hyp* operon is inversely oriented to the *hyc* transcriptional unit; the five promoter-proximal genes of this operon code for proteins involved in the maturation of hydrogenase 3 (*hypA*, *hypC*) or of all three hydrogenases (*hypB*, *hypD*, *hypE*) (for review see Maier and Böck, 1996). The ultimate gene of this operon is *fhlA* which codes for the central regulator of the formate regulon (Sankar *et al.*, 1988; Schlensog and Böck, 1990). The regulation of the expression of

Table 1. Physiological conditions influencing the expression of the genes of the formate regulon and the experimental evidence for the involvement.

Factor	Experimental Evidence	Reference
Formate	no expression in Δ*pfl* strain, formate induces	Birkmann *et al.*, 1987
Molybdate	no expression in *mod* strain, MoO_4^{2-} and WO_4^{2-} induce	Schlensog *et al.*, 1989 Rosentel *et al.*, 1995
Oxygen or nitrate	external electron acceptors repress expression of *fdhF::lacZ* or *hyc::lacZ*	Pecher *et al.*, 1983
pH < 7	expression of *fdhF::lacZ* or *hyc::lacZ* correlates with gas formation only at acidic pH	Rossmann *et al.*, 1991

this operon implies the activity of three promoters. P_P is located 5′ to *hypA* and its activation is sigma-54 dependent. Interestingly, its corresponding UAS is positioned in the intergenic region between *hycA* and *hycB*, some 800 bp more upstream (Hopper *et al.*, 1994). A second promoter P_A is located within the *hypA* gene; it acts with sigma-70 and requires the Fnr protein for activation (Lutz *et al.*, 1991). Finally, there is a weak and constitutively active promoter in the intergenic region between *hypE* and *fhlA*.

A fourth transcriptional unit whose expression is responsive to formate is located downstream of the *hyc* operon and consists of two genes, *hydN* with a putative role in electron transfer and *hypF* with a function in the maturation of all three hydrogenases (Maier *et al.*, 1996).

Physiological Conditions Affecting Expression of the Formate Regulon

The expression of the genes of the formate regulon depends on a considerable number of physiological conditions (Peck and Gest, 1957; Wimpenny and Cole, 1967; Pecher *et al.*, 1983). They are summarised in Table 1, together with the experimental evidence. Expression is absolutely dependent on the presence of formate and molybdate and is counteracted by the terminal electron acceptors oxygen and nitrate. Fumarate exerts only a moderate effect. Finally and most importantly, the genes of the regulon are only induced at a pH in the medium lower than 7.0.

Rossmann *et al.* (1991) presented a model in which the effects of formate, oxygen, nitrate and pH are all integrated to represent the effect of a single effector molecule, namely formate (Figure 3). Under aerobic conditions, PFL is inactive and formate is not present at significant concentrations since FDH-O efficiently consumes the low amounts formed via the non-fermentative routes (see Figure 3). Upon shifting *E. coli* cells to fermentative growth conditions, PFL is activated and induced in its synthesis and formate is produced. At neutral pH the majority of it is excreted, possibly under involvement of the activity of the FocA protein. At more acidic pH, the scenario depends on the relative affinities of formate for FDH-N and FDH-H and for the central regulator of the formate regulon FhlA. When nitrate is present, FDH-N is induced and effectively oxidises formate because of its superior K_m. In the absence of nitrate, formate accumulates until it reaches a concentration sufficient to bind to FhlA, which has an apparent binding constant of 5 mM (Hopper and Böck, 1995). FhlA in complex with the ligand formate then binds to the UAS and activates transcription from the sigma-54 promoters. As a consequence, FDH-H and the other components of the formate hydrogen-lyase system are synthesized and their activity counterbalances a further increase of formate by converting it into CO_2 and H_2 (Rossmann *et al.*, 1991).

How does the pH signal tie in? Initially it was argued that the effect of acidification relies on the redistribution of formate from the extracellular space into the cytoplasm, possibly as a proton driven import (Rossmann *et al.*, 1991). However, in view of the pK_a value of formate of about 3.4 this possibility appears not plausible since full induction of the expression is reached already at a pH value of 6.0 The resulting increase in the protonated form of formate should be insufficient to increase the intracellular concentrations to a value above 1 mM.

Molybdate dependence of the expression of the genes coding for hydrogenase 3 seems to be mediated by different systems. The ModE protein has been shown to be a required component in the activation of the *hyc* operon to an optimal level in the presence of formate (Self *et al.*, 1999). Additionally, MoeA is involved in the activation of the *hyc* operon (Hasona *et al.*, 1998) and it has been suggested that it catalyses the production of an activated form of molybdate, which in turn might interact with the transcriptional activator FhlA (Self and Shanmugam, 2000).

Mechanism of Action of FhlA

FhlA belongs to the NtrC group of transcriptional regulators which consist of an N-terminal domain A, a conserved central domain C and a C-terminal domain D. Domain C – which contains an ATP binding site – is thought to interact with RNA polymerase x sigma-54 and domain D has been shown to interact with the UAS with its helix-turn-helix motif. The three domains are usually separated by short linkers (Drummond *et al.*, 1986).

FhlA belongs to a subclass of these regulators which share with the NtrC prototypes only the structural features of the C and D domains (Schlensog and Böck, 1990). This subclass which also includes NifA, DmpR and XylR lack an aspartyl residue at the A domain which is the site of phosphorylation in the classical NtrC-type regulators. Regulators of this type do not

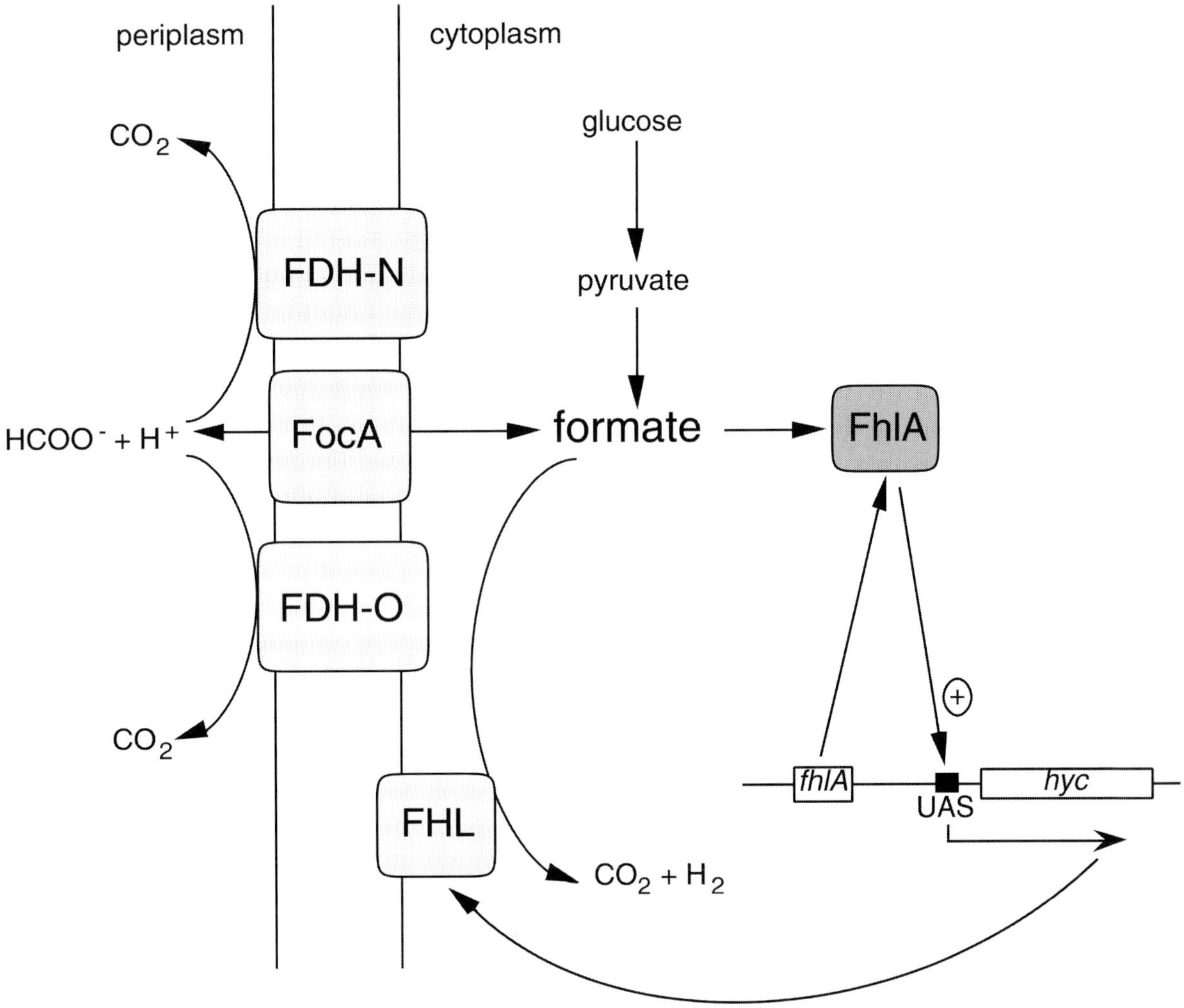

Figure 3. Different metabolic systems competing for the degradation of formate under different conditions. For details see text.

appear to require phosphorylation by a sensor kinase but liganding of a small molecule to be converted into the activated form. In the case of XylR and DmpR these effector molecules are aromatic compounds (Shingler and Moore, 1994), in the case of FhlA it is formate. Experimental proof that formate *per se* activates has been provided by *in vitro* transcription-translation assays and by analysing the ATPase activity of purified FhlA in response to formate and its UAS (Hopper *et al.*, 1994; Hopper and Böck, 1995). Moreover, azide, a structural analogon of formate, can mimick its effect. Mutants of FhlA have been selected that activate transcription in the absence of formate and it has been shown that the amino acid exchanges are located in domain A of FhlA (Korsa and Böck, 1997). Although a structure of FhlA with its effector is not yet available, all these results show that formate binds to FhlA most probably in the N-terminal domain A.

The predominant functional consequence of the binding of formate is a significant stimulation of the ATP hydrolysis activity due to a decrease of the K_M of FhlA for ATP. Interestingly, an increase of the ATP concentration resulted in ligand – independent ATP hydrolysis, which supports the conjecture that the modulation of the ATPase activity is the key task of the ligand. This assumption is supported by the observation that the FhlA variants which activate transcription in the absence of formate show the same kinetics of ATP hydrolysis as wild-type FhlA in the presence of the effector (Hopper and Böck, 1995; Korsa and Böck, 1997). When the complete domain A was removed, the truncated FhlA showed high and constitutive ATP hydrolysis as well as transcriptional activation whereas the domain A in *trans* reduced the activity of full-length FhlA, indicating (i) that domain A – like that in other regulators – modulates the binding and hydrolysis of ATP, and (ii) that domain A binds formate (Leonhartsberger *et al.*, 2000). It also supports the notion that FhlA does not need phosphorylation to be active. However, as the activity of the N-terminally truncated FhlA is still stimulated in the presence of UAS, activation by DNA and formate seem to be at least partially independent processes.

Purified FhlA interacts with both the specific DNA (carrying the UAS) and unspecific DNA (lacking the UAS) as judged by stimulation of ATPase activity.

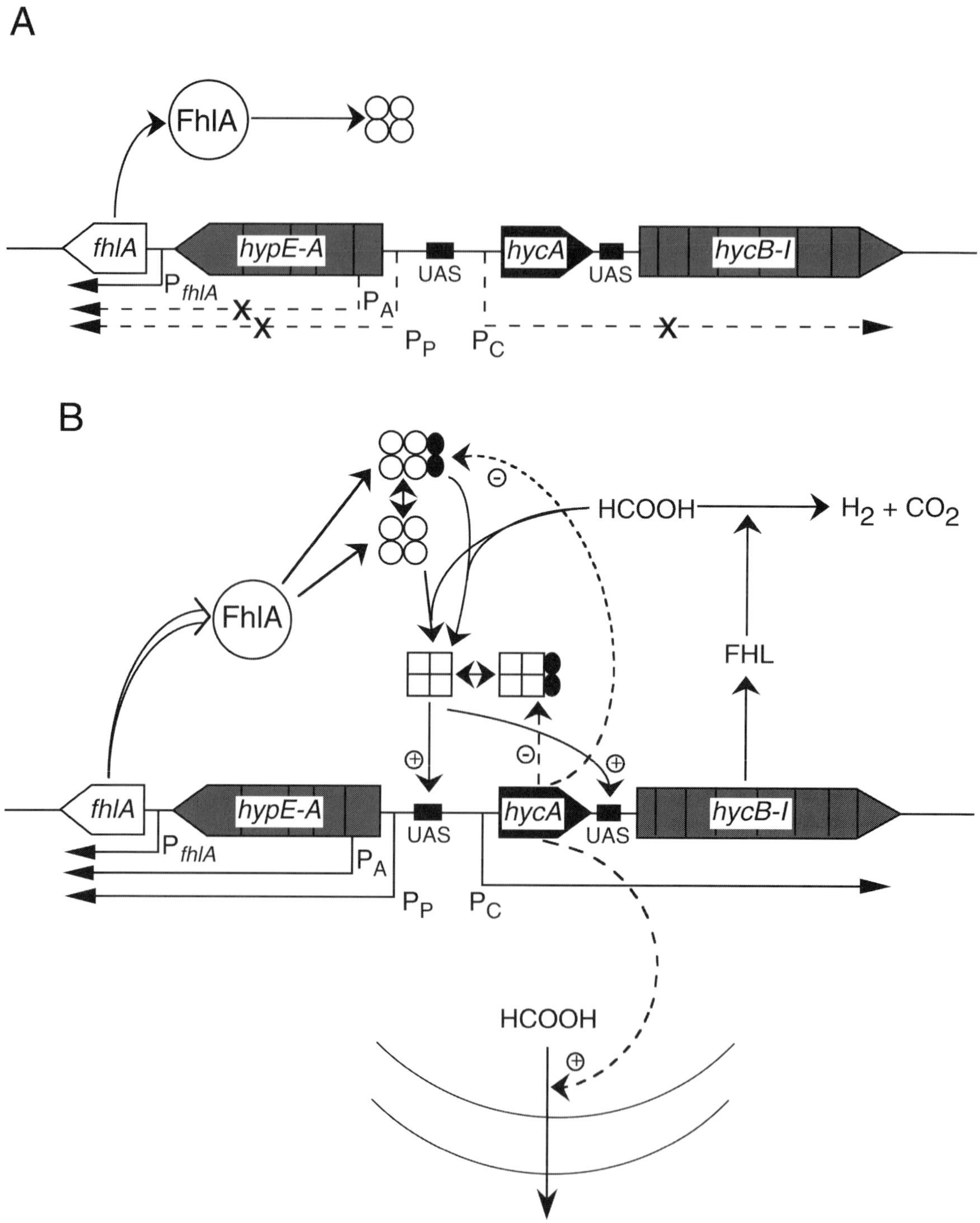

Figure 4. Regulation of the expression of the genes of the formate regulon under aerobic (A) and anaerobic (B) conditions. For details see text.

However, when both formate and DNA ligands are offered, only the specific DNA triggers a drastic stimulation of the kinetics of ATP hydrolysis (Hopper and Böck, 1995). It is at present unknown whether the increase in ATPase activity is a direct consequence of the binding of formate and/or DNA or whether the ligands change the oligomerisation state of the protein which leads to the observed changes in ATPase activity. Such an oligomerisation has been shown to be crucial for the activity of other regulators of the NtrC family (Weiss *et al.*, 1992). *In vitro* transcription analyses revealed features of FhlA that allowed the proposition of a model in which FhlA undergoes a change in its oligomeric state for transcription activation and that this oligomerisation is favored by high nucleoside triphosphate concentrations, by the effector formate and by the target DNA (Hopper *et al.*, 1996).

Recently, the DNA-binding site of FhlA was analysed in detail. Each UAS consists of two binding sites of 16 bp length with the consensus sequence 5′-CATTTCGTACGAAATG-3′ separated by a spacer region. Hydroxyradical footprinting analysis identified six sites on each strand within the UAS which are bound by FhlA (Leonhartsberger *et al.*, 2000). Formate and ATP had no detectable influence on the binding pattern. A model was proposed in which FhlA binds the DNA on only one side of the helix, probably as a tetramer. *In vivo* experiments with constructs in which insertions shifted the binding sites on different sites of the helix confirmed this model (Leonhartsberger *et al.*, 2000).

The Role of HycA as Antiactivator

When a mutation was introduced into *hycA* – the promoter-proximal gene of the *hyc* operon – the formate hydrogen-lyase activity was increased two-fold. When *hycA* was overexpressed *in trans*, the activity was decreased to undetectable levels (Sauter *et al.*, 1992). Clearly, this behavior indicates that HycA has a regulatory role counteracting the activity of FhlA. As *hycA* is part of the *hyc* operon the amount of the protein in the cell in turn is regulated by FhlA and formate. Theoretically, there are two levels at which HycA might act. First, HycA might interact with FhlA via protein-protein contact and down-modulate its activity; secondly, HycA could be involved in the control of the cellular formate concentration, e.g., by functioning as an export facilitator which assists in extruding formate. At present, the issue is unresolved. It is also unknown whether HycA directly senses some extra- or intracellular signal.

Related Systems

Klebsiella species ferment glucose essentially along the same routes as *E. coli* but they produce predominantly neutral end products like ethanol and butanediol. The three enzymes involved in the conversion of two moles of pyruvate into one mole of butanediol are organised in one operon (*budABC*) (for review see Magee and Kosaric, 1987). In *Klebsiella terrigena* expression of the *bud* operon is regulated by BudR which is encoded by a gene located in inverse orientation to *budABC* and separated from it by a 106 bp intergenic region (Mayer *et al.*, 1995). BudR is a LysR-type regulator and specifically responds to the acetate concentration in the medium. Acidic pH and anaerobiosis stimulate the effect of acetate. It appears that BudR coordinates the activity of the energy-conserving, non-reductive but acidifying acetate formation pathway with that of the non-energy-conserving and non-acidifying butanediol pathway (Mayer *et al.*, 1995).

Synposis

Figure 4 summarises the scenarioes of the expression of the genes of the formate regulon under aerobic (part A) and fermentative (B) conditions. Aerobically, *fhlA* is solely transcribed from its constitutive promoter; due to the lack of PFL activity the genes of the formate regulon are not induced. Under fermentative growth conditions there is an autoinduction of the expression of *fhlA* since transcription initiates at P_P and continues over *fhlA*. Promoter P_A is active under anaerobic conditions in the absence of high formate concentrations; its physiological relevance appears to reside in the provision of the *hyp* gene products for the maturation of hydrogenases 1 and 2 (Lutz *et al.*, 1990).

The autoinduction of *fhlA* expression is counteracted by the formation of HycA as an antiactivator and by the formation of formate hydrogen-lyase which decreases the formate concentration.

An intriguing additional issue is that *fhlA* translation is under the control of OxyS (Altuvia *et al.*, 1998). The production of this 109 nucleotide untranslated RNA is induced in response to oxidative stress in *E. coli*. It has been shown that OxyS RNA inhibits *fhlA* translation by pairing with a sequence overlapping the ribosome binding site and a short sequence within the coding region of *fhlA*. Kissing complex formation between OxyS and *fhlA* results in a stable antisense-target complex (Argaman and Altuvia, 2000). Oxidative stress, like a shift from fermentation to aerobic conditions, as characteristic for the life style of *E. coli*, thus can immediately shutdown synthesis of the regulator and block wasteful formation of the oxygen-sensitive formate hydrogen-lyase system.

Acknowledgement

The work in the authors laboratory summarized in this review has been continuously supported by grants from the Deutsche Forschungsgemeinschaft and the Fonds der Chemischen Industrie.

References

Alexeeva, S., de Kort, B., Sawers, G., Hellingwerf, K.J., and de Mattos, M.J. 2000. Effects of limited aeration and of the ArcAB system on intermediary pyruvate catabolism in *Escherichia coli*. J. Bacteriol. 17: 4934–4940.

Altuvia, S., Zhang, A., Argaman, L., Tiwari, A., and Storz, G. 1998. The *Escherichia coli* OxyS regulatory RNA represses *fhlA* translation by blocking ribosome binding. EMBO J. 17: 6069–6075.

Andrews, S.C., Berks, B.C., McClay, J., Ambler, A., Quail, M.A., Golby, P., and Guest, J.R. 1997. A 12-cistron *Escherichia coli* operon (*hyf*) encoding a putative proton-translocating formate hydrogenlyase system. Microbiology 143: 3633–3647.

Argaman, L., and Altuvia, S. 2000. *fhlA* repression by OxyS RNA: Kissing complex formation at two sites results in a stable antisense-target RNA complex. J. Mol. Biol. 300: 1101–1112.

Axley, M.J., Böck, A., and Stadtman, T.C. 1991. Catalytic properties of an *Escherichia coli* formate dehydrogenase mutant in which sulfur replaces selenium. Proc. Natl. Acad. Sci. USA 88: 8450–8454.

Axley, M.J., Grahame, D.A., and Stadtman, T.C. 1990. *Escherichia coli* formate-hydrogen lyase: purification and properties of the selenium-dependent formate dehydrogenase component. J. Biol. Chem. 265: 18213–18218.

Becker, A., Fritz-Wolf, K., Kabsch, W., Knappe, J., Schultz, S., and Wagner, A.F.V. 1999. Structure and mechanism of the glycyl radical enzyme pyruvate formate-lyase. Nat. Struct. Biol. 10: 969–975.

Birkmann, A., and Böck, A. 1989. Characterisation of a *cis* regulatory DNA element necessary for formate induction of the formate dehydrogenase gene (*fdhF*) of *Escherichia coli*. Mol. Microbiol. 3: 187–195.

Birkmann, A., Zinoni, F., Sawers, G., and Böck, A. 1987. Factors affecting transcriptional regulation of the formate-hydrogen-lyase pathway of *Escherichia coli*. Arch. Microbiol. 148: 44–51.

Böhm, R., Sauter, M., and Böck, A. 1990. Nucleotide sequence and expression of an operon in *Escherichia coli* coding for formate hydrogenlyase components. Mol. Microbiol. 4: 231–243.

Boyington, J.C., Gladyshev, V.N., Khangulov, S.V., Stadtman, T.C., and Sun, PD. 1997. Crystal structure of formate dehydrogenase H: catalysis involving Mo, molybdopterin, selenocysteine, and an Fe4S4 cluster. Science 275: 1305–1308.

Drummond, M., Whitty, P., and Wootton, J. 1986. Sequence and domain relationships of *ntrC* and *nifA* from *Klebsiella pneumoniae*: homologies to other regulatory proteins. EMBO J. 5: 441–447.

Enoch, H.G., and Lester, R.L. 1975. The purification and properties of formate dehydrogenase and nitrate reductase from *Escherichia coli*. J. Biol. Chem. 250: 6693–6705.

Hesslinger, C., Fairhurst, S.A., and Sawers, G. 1998. Novel keto acid formate-lyase and propionate kinase enzymes are components of an anaerobic pathway in *Escherichia coli* that degrades L-threonine to propionate. Mol. Microbiol. 27: 477–492.

Hopper, S., Babst, M., Schlensog, V., Fischer, H.-M., Hennecke, H., and Böck, A. 1994. Regulated expression *in vitro* of genes coding for formate hydrogenlyase components of *Escherichia coli*. J. Biol. Chem. 269: 19597–19604.

Hopper, S., and Böck, A. 1995. Effector-mediated stimulation of ATPase activity by the σ^{54}-dependent transcriptional activator FHLA from *Escherichia coli*. J. Bacteriol. 177: 2798–2803.

Hopper, S., Korsa, I., and Böck, A. 1996. The nucleotide concentration determines the specificity of *in vitro* transcription activation by the σ^{54}-dependent activator FhlA. J. Bacteriol. 178: 199–203.

Kaiser, M., and Sawers, G. 1994. Pyruvate formate-lyase is not essential for nitrate respiration by *Escherichia coli*. FEMS Microbiol. Lett. 117: 163–168.

Kessler, D., and Knappe, J. 1997. Anaerobic dissimilation of pyruvate. In: "Escherichia coli and Salmonella". F.C. Neidhardt, (ed.) ASM Press, Washington DC, p. 199–205.

Korsa, I., and Böck, A. 1997. Characterization of *fhlA* mutations resulting in ligand-independent transcriptional activation and ATP hydrolysis. J. Bacteriol. 179: 41–45.

Leonardo, M.R., Dailly, Y., and Clark, D.P. 1996. Role of NAD in regulating the *adhE* gene of *Escherichia coli*. J. Bacteriol. 178: 6013–6018.

Leonhartsberger, S., Ehrenreich, A., and Böck, A. 2000. Analysis of the domain structure and the DNA binding site of the transcriptional activator FhlA. Eur. J. Biochem. 267: 3672–3684.

Lutz, S., Böhm, R., Beier, A., and Böck, A. 1990. Characterisation of divergent NtrA-dependent promoters in the anaerobically expressed gene cluster coding for hydrogenase 3 components of *Escherichia coli*. Mol. Microbiol. 4: 13–20.

Lutz, S., Jacobi, A., Schlensog, V., Böhm, R., Sawers, G., and Böck, A. 1991. Molecular characterisation of an operon (*hyp*) necessary for the activity of the three hydrogenase isoenzymes in *Escherichia coli*. Mol. Microbiol. 5: 123–135.

Macy, J., Kulla, H., and Gottschalk, G. 1976. Hydrogen-dependent anaerobic growth of *Escherichia coli* on L-malate:succinate formation. J. Bacteriol. 125: 423–428.

Magee, R.J., and Kosaric, N. 1987. The microbial production of 2,3-butanediol. Adv. Appl. Microbiol. 32: 89–161.

Maier, T., and Böck, A. 1996. Nickel incorporation into hydrogenases. In: Advances in inorganic biochemistry: mechanisms of metallocenter assembly. R.P. Hausinger, G.L. Eichhorn, and L.G. Marzilli, (eds.) VCH Publishers Inc., NY, p. 173–192.

Maier, T., Binder, U., and Böck, A. 1996. Analysis of the *hydA* locus of *Escherichia coli*: two genes (*hydN* and *hypF*) involved in formate and hydrogen metabolism. Arch. Microbiol. 165: 333–341.

Mayer, D., Schlensog, V., and Böck, A. 1995. Identification of the transcriptional activator controlling the butanediol fermentation pathway in *Klebsiella terrigena*. J. Bacteriol. 177: 5261–5269.

Mulliez, E., Ollagnier, S., Fontecave, M., Eliasson, R., and Reichard, P. 1995. Formate is the hydrogen donor for the anaerobic ribonucleotide reductase from *Escherichia coli*. Proc. Natl. Acad. Sci. USA 92, 8759–8762.

Ordal, E.J., and Halvorson, H.O. 1939. A comparison of hydrogen production from sugars and formic acid by normal and variant strains of *Escherichia coli*. J. Bacteriol. 38: 199–220.

Pecher, A., Zinoni, F., Jatisatienr, C., Wirth, R., Hennecke, H., and Böck, A. 1983. On the redox control of synthesis of anaerobically induced enzymes in enterobacteriaceae. Arch. Microbiol. 136: 131–136.

Peck, H.D., and Gest, H. 1957. Formic dehydrogenase and the hydrogenlyase enzyme complex in coli-aerogenes bacteria. J. Bacteriol. 73: 706–721.

Pinsent, J. 1954. The need for selenite and molybdate in the formation of formic dehydrogenase by members of the coli-aerogenes group of bacteria. Biochem. J. 57: 10–16.

Plaga, W., Vielhaber, G., Wallach, J., and Knappe, J. 2000. Modification of Cys-418 of pyruvate formate-lyase by methacrylic acid, based on its radical mechanism. FEBS Lett. 466: 45–48.

Rosentel, J.K., Healy, F., Maupin-Furlow, J.A., Lee, J.H., and Shanmugam, K.T. 1995. Molybdate and regulation of *mod* (molybdate transport), *fdhF*, and *hyc* (formate hydrogenlyase) operons in *Escherichia coli*. J. Bacteriol. 177: 4857–4864.

Rossmann, R., Sawers, G., and Böck, A. 1991. Mechanism of regulation of the formate-hydrogenlyase pathway by oxygen, nitrat, and pH: definition of the formate regulon. Mol. Microbiol. 5: 2807–2814.

Sankar, P., Lee, J.H., and Shanmugam, K.T. 1988. Gene-product relationships of *fhlA* and *fdv* genes of *Escherichia coli*. J. Bacteriol. 170: 5440–5445.

Sauter, M., Böhm, R., and Böck, A. 1992. Mutational analysis of the operon (*hyc*) determining hydrogenase 3 formation in *Escherichia coli*. Mol. Microbiol. 6: 1523–1532.

Sawers, G. 1994. The hydrogenases and formate dehydrogenases of *Escherichia coli*. Antonie van Leeuwenhoek. 66: 57–88.

Sawers, G., and Watson, G. 1998. A glycyl radical solution: oxygen-dependent interconversion of pyruvate formate-lyase. Mol. Microbiol. 29: 945–954.

Sawers, G., Heider, J., Zehelein, E., and Böck, A. 1991. Expression and operon structure of the *sel* genes of *Escherichia coli* and identification of a third selenium-containing formate dehydrogenase isoenzyme. J. Bacteriol. 173: 4983–4993.

Schlensog, V., Birkmann, A., and Böck, A. 1989. Mutations in *trans* which affect the anaerobic expression of a formate dehydrogenase (*fdhF*) structural gene. Arch. Microbiol. 152: 83–89.

Schlensog, V., and Böck, A. 1990. Identification and sequence analysis of the gene encoding the transcriptional activator of the formate hydrogenlyase system of *Escherichia coli*. Mol. Microbiol. 4: 1319–1327.

Schlensog, V., Lutz, S., and Böck, A. 1994. Purification and DNA-binding properties of FHLA, the transcriptional activator of the formate hydrogenlyase system from *Escherichia coli*. J. Biol. Chem. 269: 19590–19596.

Self, W.T., Grunden, A.M., Hasona, A., and Shanmugam, K.T. 1999. Transcriptional regulation of molyboenzyme synthesis in *Escherichia coli* in response to molybdenum: ModE-molybdate, a repressor of the *modABCD* (molybdate transport) operon is a secondary transcriptional activator for the *hyc* and *nar*operons. Microbiology 145: 41–55.

Self, W.T., and Shanmugam, K.T. 2000. Isolation and characterization of mutated FhlA proteins which activate transcription of the *hyc* operon (formate hydrogenlyase) of *Escherichia coli* in the absence of molybdate. FEMS Microbiol. Lett. 184: 47–52.

Shingler, V., and Moore, T. 1994. Sensing of aromatic compounds by the DmpR transcriptional activator of phenol-catabolizing *Pseudomonas* sp. strain CF600. J. Bacteriol. 176: 1555–1560.

Stephenson, M., and Stickland, L.H. 1932. Hydrogenlyases. Bacterial enzymes liberating molecular hydrogen. Biochem. J. 26: 712–724.

Suppmann, B., and Sawers, G. 1994. Isolation and characterization of hypophosphite-resistant mutants of *Escherichia coli*: identification of the FocA protein, encoded by the *pfl* operon as a putative formate transporter. Mol. Microbiol. 11: 965–982.

Unden, G., and Bongaerts, J. 1997. Alternative respiratory pathways of *Escherichia coli*: energetics and transcriptional regulation in response to electron acceptors. Biochim. Biophys. Acta 1320: 217–234.

Weiss, V., Claverie-Martin, F., and Magasanik, B. 1992. Phosphorylation of nitrogen regulator I of *Escherichia coli* induces strong cooperative binding to DNA essential for activation of transcription. Proc. Natl. Acad. Sci. USA 89: 5088–5092.

Wimpenny, J.W.T., and Cole, J.A. 1967. The regulation of metabolism in facultative bacteria. III. The effect of nitrate. Biochim. Biophys. Acta 148: 133–242.

Chapter 18

Nitric Oxide Signaling and NO Dependent Transcriptional Control in Bacterial Denitrification by Members of the FNR-CRP Regulator Family

Walter G. Zumft

Abstract

Bacterial denitrification transforms nitrate to dinitrogen. The process is expressed facultatively in response to environmental conditions. Around 50 components make up the denitrification apparatus and its assembly pathways. We are beginning to understand how exogenous signals provided by oxygen and N oxides are processed for activating the underlying gene programs. Key signals are provided by nitrate, nitric oxide, and a low oxygen tension. In the genus *Pseudomonas* the nitrate signal is processed by a two component regulatory system which activates the *nar* operon encoding respiratory nitrate reductase. Nitric oxide is not only an essential respiratory substrate of the denitrifying cell but constitutes in nanomolar concentrations also a key signal for the expression of nitrite reductase and NO reductase which control cellular NO homeostasis. The signal pathway in the genera *Pseudomonas, Paracoccus* and *Rhodobacter* involves regulators of the FNR family of transcription factors, which cluster phylogenetically in a separate subgroup. In contrast, *Ralstonia eutropha* requires a sigma-54 dependent regulator of the NtrC family for the expression of its quinol-dependent NO reductase. Important questions are directed currently at the mechanism(s) of activation of these transcription factors by NO, and avoidance of crosstalk with FNR factors at target promoters operating with identical recognition motifs.

Introduction

Nitric oxide (NO) is generated and reduced by bacterial denitrification. "Monoxide d'azote" (nitrogen monoxide) was described to originate from a biological process in one of the first studies on denitrification, more than one hundred years ago (Gayon and Dupetit, 1886). Cell-free NO formation by bacteria was shown by Mori and coworkers (Iwasaki *et al.*, 1956) and several lines of evidence secured NO the biological function of an obligate intermediate in nitrite denitrification. Denitrification is a respiratory mode of energy conservation based on the reduction of N oxides. The complete denitrification process consists thus in the consecutive action of four respiratory oxidoreductases acting on nitrate, nitrite, nitric oxide, and nitrous oxide (N_2O) to yield dinitrogen. The overall reaction is essential for biogeochemical N-cycling since dinitrogen fixation needs to be balanced by a reverse reaction to prevent nitrate accumulation at detrimental levels in the biosphere.

Regulatory studies of how the facultatively functioning denitrification system is expressed have been directed at a select group of species representing distinct metabolic themes. The best-studied organotrophic denitrifiers harbor the cytochrome cd_1 nitrite reductase; in contrast, *Rhodobacter sphaeroides*, a photodenitrifier, depends on the Cu-containing nitrite reductase. Extensive biochemical and genetic characterizations of the denitrification apparatus of the heterotrophic denitrifiers *Pseudomonas stutzeri* and *Pseudomonas aeruginosa* have revealed the modifications of the system within a single genus. The intensively studied aerobic modes of respiration of *Paracoccus denitrificans* (Baker *et al.*, 1998) provide the background with which to compare the anaerobic system of denitrification. *Ralstonia eutropha*, a hydrogen bacterium and cytochrome cd_1-dependent denitrifier, harbors a novel quinol-dependent NO reductase and shows unprecedented organizational and regulatory modifications of a in part megaplasmid-encoded denitrification system.

Reviews covering the multiple facets of the denitrification process have been published elsewhere, such as a description of the participating enzymes (Berks *et al.*, 1995), the biochemistry of nitrite reductase and NO generation (Silvestrini *et al.*, 1994; Cutruzzolà, 1999), NO metabolism and properties of NO reductases (Zumft, 1993; Hendriks *et al.*, 2000), and the systematics of denitrifying bacteria (Zumft, 1992). For a coverage of the cell biology, molecular genetics and biochemistry of the denitrification process the reader is referred to a comprehensive review (Zumft, 1997). Here, I describe the current status of our understanding of how the expression of the denitrification apparatus is regulated with emphasis on NO as the signal molecule and signal transduction pathways involving regulators of the FNR family. Nitrate and oxygen signaling which are necessary for establishing the overall process and are interlaced with NO dependent regulation will also be summarized. Variations to the theme will be described and the mechanism of NO sensing discussed in light of the available experimental evidence.

Nitric Oxide, an Essential Nutrient

Whenever a bacterium encounters a high NO concentration over a prolonged time, the toxic aspect of NO will be dominating irrespective of the organism being a denitrifier or not. NO at 0.5 mM completely prevents the transcription of nitrite reductase and NO reductase genes in *P. stutzeri* (Vollack and Zumft, 2001). Even in denitrifiers the loss of NO reductase is a lethal event due to NO accumulation, whereas a *nir-nor* double mutant, that is unable to reduce nitrite, is viable (Braun and Zumft, 1991). Thus, hardly a report on nitric oxide metabolism fails to state the toxicity of NO, giving the impression that a cell would do all to avoid this compound. However, the view manifest in such statements needs differentiation.

In bacteria, NO is reaction product and substrate of denitrification at the same time, i.e., an ordinary cell metabolite. The physiologically relevant NO concentration is in the low nanomolar range. Toward this end NO can be compared to a trace metal, whose deleterious or beneficial properties depend on the concentration to which a bacterium is exposed. Ample evidence for the metabolic role of NO stems from several complementary approaches. Central findings comprise the identification and purification of the NO reductase complex (Heiss *et al.*, 1989), isolation and mutational inactivation of the reductase structural genes, *norC* and *norB* (Braun and Zumft, 1991), demonstration of the built-up of a steady state concentration of NO during denitrification (Goretski *et al.*, 1990), proof for the occurrence of NO in the environment (Conrad, 1996), and evidence for a role of NO reductase in the evolution of aerobic respiration (Giuffrè *et al.*, 1999). As will be shown, NO has also the function of a signal molecule in activating gene programs for its own metabolism besides its role in cellular respiration.

Enzymology of NO Homeostasis

The steady state concentrations of free NO during denitrification was found in the range of 20 to 30 nM and 50 nM, by direct measurement of NO (Goretski *et al.*, 1990) or by gas stripping (Zafiriou *et al.*, 1989), respectively. Nitrate-denitrifying *P. aeruginosa* develops a steady state concentration of 1 to 2 nM NO (Kalkowski and Conrad, 1991). The apparent K_m value for NO consumption by denitrifying bacteria and soil samples ranges from about 1 to 70 nM (for review, see Conrad, 1996). Thus, the bacterial NO-metabolizing system is fully operative at an extremely low NO concentration.

Evolutionary Convergence for NO Generation by Heme- or Cu-Containing Nitrite Reductases

NO homeostasis is determined by NO production and sink reactions. The major enzymatic NO generators of denitrifiers are two types of respiratory nitrite reductases (Cutruzzolà, 1999; Watmough *et al.*, 1999). They are found on an either/or basis without recognizable taxonomic pattern in denitrifying bacteria suggestive of multiple horizontal gene transfer events. The *nirS* gene encodes the tetraheme cytochrome cd_1 nitrite reductase in *P. aeruginosa*, *P. stutzeri*, *Pa. denitrificans*, *Paracoccus pantotrophus* (formerly *Thiosphaera pantotropha*), *R. eutropha*, and several other denitrifiers. The mixed heme protein carries a heme C in the electron transfer domain and an unusual porphyrinedione, referred to as heme D_1, at the catalytic site. The second type of nitrite reductase contains Cu, and is encoded by the *nirK* gene. Representative species harboring this enzyme are *R. sphaeroides*, *Achromobacter cycloclastes*, *Alcaligenes xylosoxidans*, *Alcaligenes faecalis* S-6, *Pseudomonas* sp. G-179, *Bacillus halodenitrificans*, and also the archaeon *Haloferax denitrificans* (Inatomi and Hochstein, 1996). The crystal structure of both types of nitrite reductases has been determined. Although cytochrome cd_1 and Cu-containing nitrite reductases both exhibit some oxygen reductase activity, no evidence exists that this property would attribute them a dual physiological function in anaerobic and aerobic respiratory metabolism. Primary and secondary structures of the two types of nitrite reductase are without relationship indicating independent evolutionary trajectories. Thus, the biochemistry of NO generation provides us with a fine example of evolutionary convergence, where the cell employs in distinct proteins different metals and cofactors to achieve the same ends for catalysis and cellular bioenergetics.

Diversity of Electron Donation to a Conserved Binuclear Heme-Nonheme Iron Site in Nitric Oxide Reductases

It is NO reductase that reverses dinitrogen fixation and catalyzes N-N bond formation by reductively transforming NO to N_2O. The enzyme appears to belong to an evolutionary ancient lineage of respiratory complexes (Saraste and Castresana, 1994; van der Oost *et al.*, 1994). An X-ray structure is not yet available, but the catalytic subunit, NorB, has been modeled onto the subunit II of cytochrome *c* oxidase (Kannt *et al.*, 1998). The model is thought to come close to the real situation due to the tight correspondence in the transmembrane helices, putative histidine ligands, and limited extra membrane domains allowing little structural variability. Within the concept of an evolutionary linkage between the two types of enzymes it is striking that certain NO reductases have oxygen reductase activity (Fujiwara and Fukumori, 1996), whereas the cytochrome oxidases of *Thermus thermophilus* have significant NO reductase activity (Giuffrè *et al.*, 1999).

The *R. eutropha* NO reductase is distant in a phylogenetic comparison to the *Pseudomonas* NO reductase. This is due to an N-terminal sequence extension of 280 amino acids, which increases the mass of the catalytic subunit from about 53 to 75 kDa (Cramm *et al.*, 1997). The extension comprises two putative transmembrane helices (to raise the total number from 12 to 14) and a periplasmic domain extending between helices one and two. Different from the situation of *P. stutzeri*, *P. aeruginosa*, *Pa. denitrificans* and other closely studied denitrifiers, the *Ralstonia norB* gene is not associated with a second gene, encoding the cytochrome *c* subunit of NO reductase. The purified enzyme has no cytochrome *c*-oxidizing activity but shows NO dependent quinol-oxidation (Cramm *et al.*, 1999). Homologues of the *Ralstonia*-type enzyme are encoded in the genomes of *Neisseria* sp., *Synechocystis* sp., *Bacillus stearothermophilus*, *Staphylococcus aureus*, *Mycobacterium avium* and *Corynebacterium diphtheriae*. It has been proposed to refer to this enzyme as 'qNOR' to differentiate it from the cytochrome dependent 'cNOR' (Hendriks *et al.*, 2000). A third variant of NO reductase has been found recently in the grampositive denitrifier *Bacillus azotoformans*. This dimeric enzyme is quinol dependent but carries also a Cu_A center (Suharti *et al.*, 2001), known otherwise only from N_2O reductase and cytochrome *c* oxidase. The electron paramagnetic resonance spectrum shows a well resolved hyperfine splitting due to the binuclear Cu_A center, whereas the optical spectrum is characteristic of the heme B content.

Chemical analysis, electron paramagnetic resonance and optical spectroscopy for all purified NO reductases indicate that the catalytic subunit of the enzyme has low-spin and high-spin hemes and nonheme iron, as established initially for the NO reductases of *P. stutzeri* (Cheesman *et al.*, 1998; Kannt *et al.*, 1998) and *Pa. denitrificans* (Hendriks *et al.*, 1998; Moënne-Loccoz and de Vries, 1998). Heme and nonheme iron form the binuclear catalytic site where NO is reduced to N_2O (a center homologous to the binuclear oxygen-reducing site of cytochrome *c* oxidase and other heme-copper oxidases). The catalytic subunit of NO reductase has conserved its architecture and is apparently interlaced during evolution with that of the active site of the heme-copper oxidases and their electron donation pathways. The later vary in both the NO reductases and terminal oxidases among cytochrome *c*, quinone, and Cu_A redox centers.

Protective Mechanisms

Within a limited concentration range bacteria are able to cope with NO, and some seem to have developed protective measures to extend the range of NO tolerance or counteract a particular NO susceptibility. The best studied examples of protective proteins, other than the NO reductases themselves, are flavohemoglobin and cytochrome *c'*. Both proteins cannot be presumed to belong to the obligatory inventory of a denitrifier; yet, where they occur, they act as alternative sinks and take part in NO homeostasis. Flavohemoglobins have a wide distribution among bacteria. This includes the denitrifiers *R. eutropha* (Cramm *et al.*, 1994), *P. stutzeri* (Vollack and Zumft, 2001), and *P. aeruginosa* (Stover *et al.*, 2000). Flavohemoglobin binds heme B and FAD. The flavoprotein domain is structurally similar to that of ferredoxin-NADP-reductase. Flavohemoglobin is involved in NO detoxification and depending on the availability of oxygen, it eliminates NO in an NADH dependent reaction either by its oxygenase activity, generating mainly nitrate, or by reducing NO to N_2O (for review, see Poole and Hughes, 2000). The synthesis of flavohemoglobin in *Escherichia coli* is induced by 0.2 mM sodium nitroprusside (SNP), *S*-nitrosothiols, or nitrite. Cytochrome *c'* is a periplasmic heme protein that has its heme C-binding motif located C-terminally. The heme iron is high-spin. It binds NO and CO and confers distinct optical properties to this protein. Cytochrome *c'* is found in several denitrifiers and is thought to function as an NO scavenger or buffer for NO (Yoshimura *et al.*, 1988, 1993; Moir, 1999; Cross *et al.*, 2000). Recent evidence suggests that cytochrome *c'* indeed functions as an NO reductase (Cross *et al.*, 2001).

Nitric Oxide as a Signal Molecule

The discovery of the function of NO as a regulatory signal in bacterial denitrification has its roots in observations with nitrite reductase mutants. Inactivation of *nirS*, the structural gene for cytochrome cd_1, not only causes loss of nitrite reduction but also decreases NO reduction and leads to a low level of NO reductase expression (Ye *et al.*, 1992; de Boer *et al.*, 1994; Palmedo *et al.*, 1995; Arai *et al.*, 1999). This effect is also observed when nitrite reduction is impeded other than by inactivating *nirS*, for instance genes required for heme D_1 biosynthesis. A crucial observation to explain these results came from interspecies exchange of *nirK*. In spite of the different biochemical nature of the two types of nitrite reductases, it is possible to express *nirK* in active form in a $NirS^-$ background (Glockner *et al.*, 1993). This relieves the low level expression of NO reductase. Since the common denominator of NirK and NirS proteins is the generation of NO, we proposed NO as inducer of its own reductase and existence of an NO-signaling mechanism (Zumft *et al.*, 1994; Palmedo *et al.*, 1995).

Subsequent work has provided supportive evidence for this concept. The preferred experimental approach consists in the use of an artificial NO donor (mostly the nitrosating agent SNP) for whole cells. The signal activity of NO is monitored with *lacZ* reporter gene fusions of putative target promoters provided in *trans* or introduced chromosomally. Such studies with the *nirK* promoter of *R. sphaeroides* (Kwiatkowski and Shapleigh, 1996; Tosques *et al.*, 1996a) and the *nirS and norC* promoters of *Pa. denitrificans* (van Spanning *et al.*, 1999; Hutchings and Spiro, 2000) have shown that SNP serves as an activator. The effective concentration of NO as an inducer was determined with *P. stutzeri* by adding NO gas to a cell suspension and following RNA transcripts. Five to 50 nM NO were sufficient for a maximal transcription of the *nirSTB* and *norCB* operons (Vollack and Zumft, 2001). Interestingly, the effective NO concentration is close to that building up during steady state denitrification.

The *nir* and *nor* genes of *Pseudomonas* sp. and *Pa. denitrificans* are activated by millimolar concentrations of nitrite (Arai *et al.*, 1991, 1999; Härtig and Zumft, 1999; Murai *et al.*, 2000). Expression of *nirS* from *P. aeruginosa* shows a linear response up to 5 mM nitrite. The principal nitrite response can now be explained by the generation of NO from nitrite. However, it is open whether the nitrite effect is entirely due to NO generation or whether there are genuine regulatory responses triggered by nitrite (or also nitrate), and acting on *nir* and *nor* genes. With regard to NO as the signal molecule, the question has to be answered where this signal comes from during the transition from aerobiosis to denitrification. There may be alternative NO sources in the cell. Nitrate respiration has been brought into the context of NO formation (Ji and Hollocher, 1989; Kalkowski and Conrad, 1991), but this requires further experimental scrutiny with regard to gene regulation.

Nitrate and Oxygen Signaling

When nitrate is added to anoxic cells it induces first respiratory nitrate reductase and generates nitrite (Körner and Zumft, 1989; Körner, 1993; Baumann *et al.*, 1996). This, in turn, may trigger the NO-specific signal transduction pathway by forming NO enzymatically or nonenzymatically. The activation of the *narGHJI* operon, encoding respiratory nitrate reductase is under the control of NarXL, a nitrate (to some extent also nitrite) responsive two-component system. The genes for the nitrate sensor, *narX*, and the response regulator, *narL*, are located upstream of the *nar* operon in *P. stutzeri* (Härtig *et al.*, 1999), *P. aeruginosa* (Rompf *et al.*, 1998b), and *Pseudomonas fluorescens* (Philippot *et al.*, 2001). Knockout mutagenesis of *narXL* prevents *nar* expression in *P. stutzeri*, but the other reductases of denitrification are not affected. The *Pseudomonas* NarXL regulatory system is structurally and functionally homologous to that of *E. coli* (Darwin and Stewart, 1996). Induction of denitrification in *Pa. denitrificans* shows a well established nitrate response (Baumann *et al.*, 1996), but it is not clear whether this is mediated by a NarXL system. Full expression of the *nar* operon of *Pa. pantotrophus* requires besides nitrate also NarR, a regulator with similarity to FNR factors without the cysteine motif (see below) (Wood *et al.*, 2001).

A nitrite sensitive two-component system has been identified in *Neisseria gonorrhoeae*, which activates *aniA*, encoding the Cu-containing dissimilatory nitrite reductase (Lissenden *et al.*, 2000). Because of the effector specificity, the system is being referred to as NarPQ, akin to the *E. coli* two-component system with preference for nitrite. Nitrite denitrification is a general trait of the genus *Neisseria*. *N. gonorrhoeae* reduces low concentrations of nitrite, but not nitrate, and generates N_2O. The reaction involves a nitrite reductase, located unusually in the outer membrane, and a quinol-dependent NO reductase. The nitrite reductase is essential for establishing anaerobic growth. The qNOR of nondenitrifying bacteria is often found associated with a nitrite reductase (Hendriks *et al.*, 2000). This may confer a selective advantage to a bacterium and perhaps is important also for establishing pathogenicity. A bacterium that is exposed in its habitat to NO under microaerophilic or intermittently to oxic and anoxic conditions can eliminate NO and nitrite (the latter formed from NO oxidation) by reduction to nonreactive N_2O, provided that it disposes over both nitrite reductase and NO reductase.

Expression of the *nar* operon depends also on a signal generated by low oxygen tension or full anaerobiosis. This response can be mediated by an FNR-type transcription factor such as ANR in *P. aeruginosa* (Ye *et al.*, 1995) or FnrP in *Pa. denitrificans* (van Spanning *et al.*, 1997). In *N. gonorrhoeae* the expression of the *aniA* gene is dependent also on an FNR-type regulator. The anaerobic regulator of *P. stutzeri* acting on the *nar* operon has not been found yet. Although this bacterium harbors with FnrA a canonical FNR-type regulator, its inactivation does not affect nitrate respiration (Cuypers and Zumft, 1993). Although, the overall picture for activating the *narGHJI* operon and *aniA* seems to involve regulators of the NarXL and the FNR families, mediating nitrate/nitrite and oxygen response, respectively, this dual signal transduction pathway has to be demonstrated yet for any one of the model denitrifiers. If we consider denitrification a rather ancient respiratory mode, several modifications may have been realized. In *Bradyrhizobium japonicum* oxygen is sensed by the heme protein FixL of the FixLJ two-component regulatory system. The response regulator FixJ activates *fixK*$_2$ encoding an FNR-related transcription factor activating *nar* (Nellen-Anthamatten *et al.*, 1998). It is not known whether a further signal originating from an N oxide is fed into the regulatory cascade.

The DNR Group of FNR-Related Transcription Factors

On sequencing the regions downstream of the *nor* operons of *P. aeruginosa* (Arai *et al.*, 1995), *Pa. denitrificans* (van Spanning *et al.*, 1995), *R. sphaeroides* (Tosques *et al.*, 1996b) and *P. stutzeri* (Zumft, 1997; Vollack *et al.*, 1999) open reading frames were found, whose products showed similarity to the *E. coli* transcription factor FNR. Knockout mutagenesis of these factors revealed in each case that they are necessary for the transcriptional control of nitrite reductase and NO reductase genes, i.e., *nirS*, *nirK*, and *norCB*. Their action assures the coordinate expression of the principal enzymes for NO homeostasis. The factors were termed DNR for 'dissimilatory nitrate respiration regulator' in *P. aeruginosa*, and NNR or NnrR for 'nitrite, nitric oxide reductase regulator' in *Pa. denitrificans* and *R. sphaeroides*, respectively. In *P. stutzeri*, where at least four FNR-related transcription factors coexist, the factor acting on the *nirSTB* and *norCB* operons was termed DnrD (formerly FnrD), to indicate its involvement in denitrification.

The denitrification regulators are characterized by lack of the cysteine motif necessary for binding a [4Fe-4S] cluster that mediates the oxygen response of FNR-type factors. Initially, DNR and NNR proteins were considered as FixK-like factors of the FNR family (Zumft, 1997). FixK transcription factors also lack the FeS cluster and are functionally involved in nitrogen fixation. Examination of the relationship of the denitrification regulators with the FixK-like factors, however, show them to group in an own branch in the phylogenetic tree of the FNR family (Vollack *et al.*, 1999) and are referred to collectively here as DNR group. DNR factors, except that from *Synechocystis* sp., share the same recognition helix for DNA binding, which is a conserved ExxSR motif and makes contacts to TA and GC base pairs (Spiro, 1994).

The glutamate residue in the recognition helix is absent from NnrR of *R. sphaeroides*, *Pseudomonas* sp. G179, and *A. faecalis* S-6. A number of other putative NnrR-related factors found in genome projects share overall sequence conservation but not (providing the absence of sequence uncertainties) the conserved recognition motif. The current status shows a correlation of NnrR-related factors targeting *nirK* genes for Cu-containing nitrite reductase on one side (*R. sphaeroides* 2.4.3 and 2.4.1, *R. sphaeroides* f. sp. *denitrificans*, *Pseudomonas* sp. G-179, and *A. faecalis* S-6), and DNR-related factors acting on *nirS* genes for cytochrome cd_1 nitrite reductase on the other side (*P. aeruginosa*, *P. stutzeri*, and *Pa. denitrificans*) (unpublished data). If the structural and functional divide consolidates, it may be appropriate to consider separate subgroups for DNR and NnrR factors and adopt labels according to phylogenetic placement.

Other than the FNR-type factors, the DNR-NnrR factors are not involved in oxygen sensing. Mutation of the single cysteine of NNR does not cause loss of NNR activity (Hutchings *et al.*, 2000). This distinguishes this protein from the oxygen-sensing Flp, an FNR-like factor of *Lactobacillus* (Gostick *et al.*, 1998). Flp mediates oxygen response by the redox reaction of (–SH groups of cysteines on neighboring subunits. Several other proteins are similar to DNR factors. Of those, only the function of HbaR from *Rhodopseudomonas palustris* has been investigated. HbaR acts on *hbaA* encoding 4-hydroxybenzoate-CoA ligase of the pathway for degradation of aromatic compounds (Egland and Harwood, 2000). It is thought to become activated as a transcription factor by interaction with 4-hydroxybenzoate. Such a function is supported from genetic evidence, but demonstration of HbaR binding directly to the target promoter in response to the effector molecule is still necessary.

A Sigma Factor σ^{54} Dependent NtrC-Related Transcription Factor in *Ralstonia*

In *R. eutropha* the gene encoding a quinol-dependent NO reductase is duplicated on the chromosome and the megaplasmid. Either reductase is functional and a mutation in both structural genes is required to abolish the ability of the bacterium to reduce NO (Cramm *et al.*, 1997). The NO reductase gene, *norB,* is organized as an operon with an upstream located *norA* gene (Pohlmann *et al.*, 2000). NorA is a homologue of DnrN and not an NO reductase gene proper. Homologues of DnrD and NorA are ScdA of *Staphylococcus aureus*, YtfE of *E. coli* (accession number P39313) and *Haemophilus influenzae* (accession number P45312), and a slightly smaller protein (mol. mass 18 kDa) of *Neisseria meningitidis* (accession numbers AAF41739 and CAB84804) (Vollack and Zumft, 2001). The precise function of these proteins is unknown but their existence in different bacteria attributes them a broader significance. Mutagenesis of *dnrN* affects transcript pattern and mRNA stability of the *nirS* operon in *P. stutzeri.*

The regulatory situation in *R. eutropha* is quite different from that of other denitrifiers studied thus far. Upstream of *norA* and transcribed in the opposite direction, the *norR* gene encodes a regulator of the NtrC family. An identical, duplicated gene arrangement is found on the chromosome and the megaplasmid (Pohlmann *et al.*, 2000). The C-terminus of the NorR proteins consists in a putative DNA-binding domain; the central domains have conserved ATP-binding motifs of response regulators interacting with RpoN. Expression of *norR* is sigma-54 dependent, explaining the previous observation of an RpoN requirement for denitrification in this bacterium. *R. eutropha* responds to nitrate with the activation of the promoters of *nirS*, *norA*, and *nosZ*. *nirS* and *nosZ* are activated in a co-ordinate manner, whereas *norA* activation is sequential in relation to the other two genes. The *norA* operon responds also to SNP. In a *nirS* mutant SNP still acts as a strong activating signal, but activation by both nitrate and nitrite is reduced to about one third of wild-type level. This indicates that nitrate and/or nitrite metabolism generate the NO signal, but might also provide an independent further signal.

Expression of qNOR of *N. gonorrhoeae* is induced by the NO donor spermine-NO in an *aniA* null background suppressing NO formation from nitrite (Householder *et al.*, 2000). Induction is possible both aerobically and anaerobically. Gene expression is not dependent on FNR or NarP but may require a DNR or NtrC factor yet to be identified. In the case of *Synechocystis* sp. PCC6803 it is remarkable that the gene for qNOR is next to an open reading frame whose gene product clearly belongs into the DNR branch of regulators. This opens the possibility that the regulation of NO reductase in this cyanobacterium is DNR dependent, a potential variation for qNOR regulation versus the situation realized in *R. eutropha.*

NO Responsive Promoters

DNR factors serve as activators of *nir* and *nor* genes but it has to be established that this is a direct dependence of the respective promoters. In as much as sequences of *nir* and *nor* genes became available promoter regions could be searched for consensus motifs with a putative regulatory function. In all *nirS* and *norC* promoters partially palindromic motifs were found with a high similarity to the binding motif for FNR, here referred to as the DNR box (Figure 1). The center of these motifs is located 40.5 to 43.5 nucleotides upstream of the transcript start, which makes the promoters to belong to class II. In this class the activator makes contact promoter-proximal to the β-subunit of RNA polymerase, overlapping the −35 promoter element, and promoter-distal with the C-terminal domain of the α-subunit (Busby and

nirS **promoter**

TTGATtgccGTCAA	**-43.5**	***P. stutzeri***
TTGATtccgGTCAA		***P. aeruginosa***
TTaAcaatgGTCAA	**-41.5**	***Pa. pantotrophus***
TTaAcaaagGTCAA	**-41.5**	***Pa. denitrificans***

norC **promoter**

TTGATtgccATCAA	**-40.5**	***P. stutzeri***
TTGATtgccATCAA		***P. aeruginosa***
TTGActttcATCAA	**-43.5**	***Pa. denitrificans***

Figure 1. Recognition sites for DNR and NNR regulators in the promoters of denitrification genes. Where known, the distance of the center of the motif from the transcript start is indicated.

Ebright, 1999). Direct binding of a DNR-NNR factor to the DNR box is suggested from indirect evidence but has not been shown experimentally yet. The DNR boxes of the *nirS* promoter (Saunders *et al.*, 1999) and the *norC* promoter (Hutchings and Spiro, 2000) of *Pa. denitrificans* have been altered in several nucleotides. Such altered promoters fused to *lacZ* reporter constructs have no β-galactosidase activity under anaerobic conditions in the wild-type background, indicating that NNR binds to the palindromic sequence in this bacterium. Thus, the loss of *nir* and *nor* expression in *dnr* or *nnr* mutants is due to DNR or NNR dependent promoters.

The *nor* genes of *Pa. denitrificans* are organized as a six-gene operon, *norCBQDEF*, yielding transcripts of 5.4 and 2.2 kb. The former covers the entire operon, whereas the short transcript corresponds to *norCB* and may be the product of mRNA processing (Murai *et al.*, 2000). A function of the additional genes of the operon has not been clearly established yet. The *norC* promoter of *Pa. denitrificans* has two transcription-start sites (Hutchings and Spiro, 2000). The major transcript initiates 43.5 nucleotides downstream of the DNR box. A minor transcript is found also in aerobically grown cells. However, it is absent in an *nnr* mutant, which suggests that NNR plays also a role in the *aerobic* expression of the *nnr* gene. Deletion of the DNA region upstream of this promoter does not cause a loss of anaerobic promoter activity, underlining that the DNR box is the major *cis*-acting element for *nor* operon expression.

The *nirS* gene of *P. stutzeri* is transcribed as an operon to yield a *nirSTB* transcript together with a monocistronic *nirS* mRNA. *nirT* was the first member of a now sizable gene family encoding tetraheme cytochromes. Several of them are involved in electron donation to enzymes, for instance NapC to periplasmic nitrate reductase. The half-life of the *nirS* mRNA is about 13 min, which is also the case for the *norCB* and *nosZ* mRNAs (Härtig and Zumft, 1999). The *nirS* gene of *Pa. pantotrophus* is not part of an operon (Saunders *et al.*, 2000). In anaerobically grown cells a transcript of about 1800 nucleotides is found, corresponding to the size of the structural gene, which is absent from aerobically grown cells. This is important in the context whether this bacterium is indeed an aerobic denitrifier. The molecular analysis supports now what was already in doubt since aerobic cells have no detectable cytochrome cd_1 (Moir *et al.*, 1995).

The *nirS* promoter of *Pa. denitrificans* is activated by NNR, however, it requires also the *nirI* gene product, encoded just upstream of *nirS* and in opposite direction (Saunders *et al.*, 1999). The role of NirI is puzzling. It exerts a positional effect on *nirS* transcription, since it is required upstream of *nirS* and cannot be supplied in *trans*. A *nirI* strain has no detectable *nirS* promoter activity. NirI is homologous to NosR, a membrane-bound putative regulator of *P. stutzeri* required for the transcription of *nosZ*, the structural gene for nitrous oxide reductase. *nosZ* is not expressed in a *nosR* mutant, but NosR has no effect on *nirS* (Cuypers *et al.*, 1992). *nosR* is always found associated with *nosZ*, but *nirI* is not a constant component of *nir* gene clusters in the genetically analyzed denitrifiers; if essential, it must be located elsewhere. The principal topology of NosR has been determined (Cuypers *et al.*, 1995; Zumft, 1997). It consists of an N-terminal transmembrane anchor followed by a large periplasmic domain which is followed by five transmembrane helices and ends in a C-terminal cytoplasmic domain with one or two putative FeS centers. NirI shows the same topology. The precise function of these interesting membrane proteins in the transcriptional activation of denitrification genes remains to be explored in more detail.

Considerable interest is focusing now on the question of how members of the DNR group are activated by NO. Constitutive overexpression of *dnrD*, presumably leading also to a higher DnrD level in the cell, is insufficient to activate DnrD dependent target promoters unless the N oxide signal is also provided (Vollack and Zumft, 2001). This is a strong argument for the necessity of DnrD activation by NO and the existence of interconvertible active and inactive states. The pathway seems to be mechanistically not tied to denitrification to allow for the finding that NNR expressed in *E. coli* can be activated by SNP (Hutchings *et al.*, 2000). Figure 2 depicts various hypotheses of activating the regulator by NO, which are discussed below. It is feasible that there is more than one signal transduction pathways involving the NO-dependent transcription factors of denitrification. Remarkably, the *norA* promoter of *R. eutropha*, which is dependent on the NtrC-related regulator NorR, is linearly responsive toward SNP up to 2 mM (Pohlmann *et al.*, 2000).

Interaction of NO with a transcription factor may be of a direct or an indirect nature. NO may be sensed by and influence an enzymatic activity, for instance a nucleotide cyclase, and regulate DNR activity by providing a cofactor. DNR could also be modified covalently by an interacting protein or a cascade of

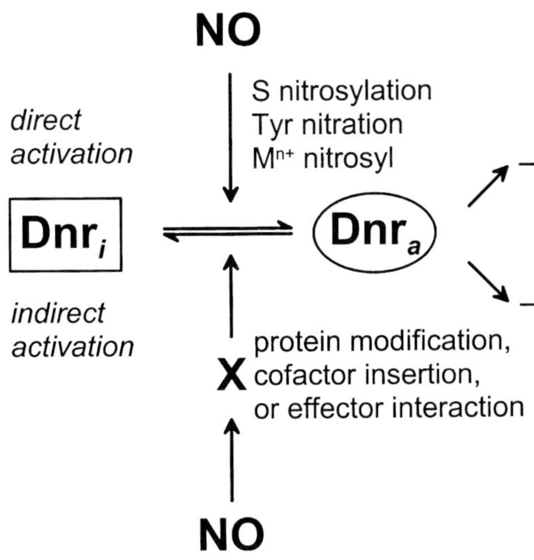

Figure 2. Model of interlacing NO and Dnr-mediated signal transduction. Dnr factors (the same holds for Nnr factors) are assumed to exist in active and inactive conformations, determining DNA binding to the recognition sequence, or DNR box, located approximately 40 bp upstream of the transcription start. Various mechanisms, direct or indirect and all of them still hypothetical, allow for the interconversion of the two Dnr forms. Details are discussed in the text. Activation has to be reversible but the requirements of the reverse reaction remain open at this moment.

proteins, one of which acts as an NO sensor. Supportive evidence for such a complex pathway is not available. More findings seem to exclude currently certain possibilities rather than they would clearly support a particular mechanism. NO is a radical and the aspect of NO toxicity has led to the identification of a number of NO-dependent reactions, which potentially can provide a mechanism for signal transduction by altering the binding activity of a DNR factor toward its target promoter. Thus, a cellular reactant with NO may also be considered a candidate effector for activation of DNR. Reaction of NO with superoxide leads to formation of peroxynitrite, which in turn could lead to nitration of a tyrosine. A conserved tyrosine found in NNR and other factors activating *nir* and *nor* genes has been mutated to phenylalanine, and although NNR activity was reduced, this tyrosine residue is not essential (Hutchings *et al.*, 2000). Under anaerobic conditions of the N_2O-grown cell, DnrD is active, which does exclude the possibility of tyrosine nitration since peroxynitrite cannot be formed (Vollack and Zumft, 2001).

NNR has a single cysteine whose mutation to serine has even a lesser effect on NNR activity than the tyrosine-phenylalanine exchange. This argues against an –SH nitrosylation mechanism. From structural considerations, we have excluded above the possibility that DNR and NNR factors are FeS proteins. A mechanism of a direct nitrosylation of an FeS cluster, as shown for SoxR of the *sox* regulon involved in the NO stress response of *E. coli* (Ding and Demple, 2000), or nitrosylation of the FeS cluster of the iron regulatory protein 1 of the eukaryotic cell (Wardrop *et al.*, 2000), is not feasible thus. This leaves us currently with the challenging and intriguing problem of how DNR, NNR and NorR are converted in the cell by NO to their active state as transcription factors.

Eluding Crosstalk with Other FNR-Type Regulators

Denitrification is a facultative process and its gene program is expressed under conditions of low oxygen tension. The [4Fe-4S] protein FNR of *E. coli* is the paradigm of an oxygen responsive transcriptional regulator (Kiley and Beinert, 1999; Green *et al.*, 2001). The first of an FNR-type regulator in denitrifying bacteria, ANR, was discovered in the context of studying arginine and pyruvate catabolism of *P. aeruginosa*. Several denitrifying bacteria are now known where FNR-type regulators, mediating oxygen response, and DNR factors, mediating NO response, coexist. Since DNR and FNR-type regulators recognize the same binding site in their respective target promoters it leads to the question of discriminatory elements (van Spanning *et al.*, 1997; Zumft, 1997). Binding of these regulators could be highly discriminatory, recognizing a single nucleotide which, however, seems not to be the case. There is no distinct modification in the DNR box that would provide a discriminatory element compared with the FNR box. DNR boxes of the *nor* promoters correspond to the TTGAT-N_4-ATCAA consensus of the FNR box (Figure 1). Intriguingly, a half site of the DNR box on *nirS* promoters is GTCAA rather than ATCAA. However, since the same regulator is assumed to bind to both the *nirS* and *norC* promoters it is clear that there is no discriminatory element in a GTCAA versus an ATCAA half site. A promoter analysis addressing the influence of neighboring sequences on a specific recognition event is not yet available. It is conceivable that further proteins have to bind at the promoter. The *nirS* promoter of *P. stutzeri* has a clear −10 sigma factor σ^{70} element (Härtig and Zumft, 1999), but such sequence is not evident in the *Pa. pantotrophus* pendant (Saunders *et al.*, 2000). It is likely that the fine tuning of promoters with FNR and DNR boxes is not uniformly throughout the different genera.

ANR serves not only as positive anaerobic regulator, but also as an *aerobic* regulator for nitrite reductase (Ka *et al.*, 1997) and the *hemN* gene of *P. aeruginosa*. The latter encodes the oxygen independent coproporphyrinogen III oxidase (Rompf *et al.*, 1998a). Under denitrifying conditions both ANR and DNR are required for *hemN* expression. On the other hand, ANR and FnrP function also as repressors for certain genes (Ray and Williams, 1997; van Spanning *et al.*, 1997). NNR (van Spanning *et al.*, 1997) and DnrD (Vollack *et al.*, 1999) downregulate *nar*. Nitrate reductase activity of *Pa. denitrificans* is increased in an *nnr* mutant; in *P. stutzeri* inactivation of *dnrD* causes an increase in the activity and level of the enzyme as well as the amount of *nar* mRNA (unpublished data). DNR factors have been found to operate sometimes also in a hierarchical relationship to FNR-type factors to form a two-step regulatory cascade. The ANR→DNR hierarchy in *P. aeruginosa* (Arai *et al.*, 1997) can explain the integration of oxygen and substrate signals for the expression of denitrification genes. Other examples, not related to denitrification, are FnrA→DnrS of *P. stutzeri* targeting anaerobic ribonucleotide reductase (Vollack *et al.*, 1999), and AadR→HbaR of *R. palustris* regulating degradation of aromatics (Egland and Harwood, 2000).

In conclusion, during the past few years we have seen the uncovering of a remarkable versatility of sensory mechanisms integrated into the FNR scaffold. The best understood process is that of oxygen-sensing via the [4Fe-4S] cluster transformation and reassembly mechanism in the canonical FNR-type proteins (Kiley and Beinert, 1999). Then there is the alternative oxygen-sensing mechanism via (SH redox chemistry by FLP proteins (Gostick *et al.*, 1998). Further, we have NO-sensing by DNR and NnrR-related factors as principal facet of regulating denitrification, as described here, and there is the possibility of 4-hydroxybenzoate-sensing by HbaR (Egland and Harwood, 2000). Finally, CO-sensing has been integrated into the FNR-CRP family by developing a hemoprotein variant (Lanzilotta *et al.*, 2000).

Acknowledgments

The work was supported by the Deutsche Forschungsgemeinschaft and Fonds der Chemischen Industrie.

References

Arai, H., Igarashi, Y., and Kodama, T. 1991. Nitrite activates the transcription of the *Pseudomonas aeruginosa* nitrite reductase and cytochrome *c*-551 operon under anaerobic conditions. FEBS Lett. 288: 227–228.

Arai, H., Igarashi, Y., and Kodama, T. 1995. Expression of the *nir* and *nor* genes for denitrification of *Pseudomonas aeruginosa* requires a novel CRP/FNR-related transcriptional regulator, DNR, in addition to ANR. FEBS Lett. 371: 73–76.

Arai, H., Kodama, T., and Igarashi, Y. 1997. Cascade regulation of the two CRP/FNR-related transcriptional regulators (ANR and DNR) and the denitrification enzymes in *Pseudomonas aeruginosa*. Mol. Microbiol. 25: 1141–1148.

Arai, H., Kodama, T., and Igarashi, Y. 1999. Effect of nitrogen oxides on expression of the *nir* and *nor* genes for denitrification in *Pseudomonas aeruginosa*. FEMS Microbiol. Lett. 170: 19–24.

Baker, S.C., Ferguson, S.J., Ludwig, B., Page, M.D., Richter, O.-M.H., and van Spanning, R.J.M. 1998. Molecular genetics of the genus *Paracoccus*: metabolic versatile bacteria with bioenergetic flexibility. Microbiol. Mol. Biol. Rev. 62: 1046–1078.

Baumann, B., Snozzi, M., Zehnder, A.J.B., and van der Meer, J.R. 1996. Dynamics of denitrification activity of *Paracoccus denitrificans* in continuous culture during aerobic-anaerobic changes. J. Bacteriol. 178: 4367–4374.

Berks, B.C., Ferguson, S.J., Moir, J.W.B., and Richardson, D.J. 1995. Enzymes and associated electron transport systems that catalyse the respiratory reduction of nitrogen oxides and oxyanions. Biochim. Biophys. Acta 1232: 97–173.

Braun, C., and Zumft, W.G. 1991. Marker exchange of the structural genes for nitric oxide reductase blocks the denitrification pathway of *Pseudomonas stutzeri* at nitric oxide. J. Biol. Chem. 266: 22785–22788.

Busby, S., and Ebright, R.H. 1999. Transcription activation by catabolite activator protein (CAP). J. Mol. Biol. 293: 199–213.

Cheesman, M.R., Zumft, W.G., and Thomson, A.J. 1998. The MCD and EPR of the heme centers of nitric oxide reductase from *Pseudomonas stutzeri*: evidence that the enzyme is structurally related to the heme-copper oxidases. Biochemistry 37: 3994–4000.

Conrad, R. 1996. Soil microorganisms as controllers of atmospheric trace gases (H_2, CO, CH_4, OCS, N_2O, and NO). Microbiol. Rev. 60: 609–640.

Cramm, R., Siddiqui, R.A., and Friedrich, B. 1994. Primary sequence and evidence for a physiological function of the flavohemoprotein of *Alcaligenes eutrophus*. J. Biol. Chem. 269: 7349–7354.

Cramm, R., Siddiqui, R.A., and Friedrich, B. 1997. Two isofunctional nitric oxide reductases in *Alcaligenes eutrophus* H16. J. Bacteriol. 179: 6769–6777.

Cramm, R., Pohlmann, A., and Friedrich, B. 1999. Purification and characterization of the single-component nitric oxide reductase from *Ralstonia eutropha* H16. FEBS Lett. 460: 6–10.

Cross, R., Aish, J., Paston, S.J., Poole, R.K., and Moir, J.W.B. 2000. Cytochrome *c'* from *Rhodobacter capsulatus* confers increased resistance to nitric oxide. J. Bacteriol. 182: 1442–1447.

Cross, R., Lloyd, D., Poole, R.K., and Moir, J.W.B. 2001. Enzymatic removal of nitric oxide catalyzed by cytochrome *c'* in *Rhodobacter capsulatus*. J. Bacteriol. 183: 3050–3054.

Cutruzzolà, F. 1999. Bacterial nitric oxide synthesis. Biochim. Biophys. Acta 1411: 231–249.

Cuypers, H., and Zumft, W.G. 1993. Anaerobic control of denitrification in *Pseudomonas stutzeri* escapes mutagenesis of an *fnr*-like gene. J. Bacteriol. 175: 7236–7246.

Cuypers, H., Viebrock-Sambale, A., and Zumft, W.G. 1992. NosR, a membrane-bound regulatory component necessary for expression of nitrous oxide reductase in denitrifying *Pseudomonas stutzeri*. J. Bacteriol. 174: 5332–5339.

Cuypers, H., Berghöfer, J., and Zumft, W.G. 1995. Multiple *nosZ* promoters and anaerobic expression of *nos* genes necessary for *Pseudomonas stutzeri* nitrous oxide reductase and assembly of its copper centers. Biochim. Biophys. Acta 1264: 183–190.

Darwin, A.J., and Stewart, V. 1996. The NAR modulon systems: nitrate and nitrite regulation of anaerobic gene expression. In: Regulation of gene expression in *Escherichia coli*. E.C.C. Lin, and A.S. Lynch, (eds.), Chapman & Hall, New York, NY. p. 343–359.

de Boer, A.P.N., Reijnders, W.N.M., Kuenen, J.G., Stouthamer, A.H., and van Spanning, R.J.M. 1994. Isolation, sequencing and mutational analysis of a gene cluster involved in nitrite reduction in *Paracoccus denitrificans*. Antonie Leeuwenhoek 66: 111–127.

Ding, H., and Demple, B. 2000. Direct nitric oxide signal transduction via nitrosylation of iron-sulfur centers in the SoxR transcription activator. Proc. Natl. Acad. Sci. USA 97: 5146–5150.

Egland, P.G., and Harwood, C.S. 2000. HbaR, a 4-hydroxybenzoate sensor and FNR-CRP superfamily member, regulates anaerobic 4-hydroxybenzoate degradation by *Rhodopseudomonas palustris*. J. Bacteriol. 182: 100–106.

Fujiwara, T., and Fukumori, Y. 1996. Cytochrome *cb*-type nitric oxide reductase with cytochrome *c* oxidase activity from *Paracoccus denitrificans* ATCC 35512. J. Bacteriol. 178: 1866–1871.

Gayon, U., and Dupetit, G. 1886. Recherches sur la réduction des nitrates par les infiniment petits. Mem. Soc. Sci. Phys. Nat. (Bordeaux) Ser. 3 (2): 201–307.

Giuffrè, A., Stubauer, G., Sarti, P., Brunori, M., Zumft, W.G., Buse, G., and Soulimane, T. 1999. The heme-copper oxidases of *Thermus thermophilus* catalyze the reduction of nitric oxide: evolutionary implications. Proc. Natl. Acad. Sci. USA 96: 14718–14723.

Glockner, A.B., Jüngst, A., and Zumft, W.G. 1993. Copper-containing nitrite reductase from *Pseudomonas aureofaciens* is functional in a mutationally cytochrome cd_1-free background (NirS$^-$) of *Pseudomonas stutzeri*. Arch. Microbiol. 160: 18–26.

Goretski, J., Zafiriou, O.C., and Hollocher, T.C. 1990. Steady-state nitric oxide concentrations during denitrification. J. Biol. Chem. 265: 11535–11538.

Gostick, D.O., Green, J., Irvine, A.S., Gasson, M.J., and Guest, J.R. 1998. A novel regulatory switch mediated by the FNR-like protein of *Lactobacillus casei*. Microbiology 144: 705–717.

Green, J., Scott, C., and Guest, J. 2001. Functional versatility in the CRP-FNR superfamily of transcription factors: FNR and FLP. Adv. Microb. Physiol. 44: 1–34.

Härtig, E., and Zumft, W.G. 1999. Kinetics of *nirS* expression (cytochrome cd_1 nitrite reductase) in *Pseudomonas stutzeri* during the transition from aerobic respiration to denitrification: evidence for a denitrification-specific nitrate- and nitrite-responsive regulatory system. J. Bacteriol. 181: 161–166.

Härtig, E., Schiek, U., Vollack, K.-U., and Zumft, W.G. 1999. Nitrate and nitrite control of respiratory nitrate reduction in denitrifying *Pseudomonas stutzeri* by a two-component regulatory system homologous to NarXL of *Escherichia coli*. J. Bacteriol. 181: 3658–3665.

Heiss, B., Frunzke, K., and Zumft, W.G. 1989. Formation of the N–N bond from nitric oxide by a membrane-bound cytochrome *bc* complex of nitrate-respiring (denitrifying) *Pseudomonas stutzeri*. J. Bacteriol. 171: 3288–3297.

Hendriks, J., Warne, A., Gohlke, U., Haltia, T., Ludovici, C., Lübben, M., and Saraste, M. 1998. The active site of bacterial nitric oxide reductase is a dinuclear iron center. Biochemistry 37: 13102–13109.

Hendriks, J., Oubrie, A., Castresana, J., Urbani, A., Gemeinhardt, S., and Saraste, M. 2000. Nitric oxide reductases in bacteria. Biochim. Biophys. Acta 1459: 266–273.

Householder, T.C., Fozo, E.M., Cardinale, J.A., and Clark, V.L. 2000. Gonococcal nitric oxide reductase is encoded by a single gene, *norB*, which is required for anaerobic growth and is induced by nitric oxide. Infect. Immun. 68: 5241–5246.

Hutchings, M.I., and Spiro, S. 2000. The nitric oxide regulated *nor* promoter of *Paracoccus denitrificans*. Microbiology 146: 2635–2641.

Hutchings, M.I., Shearer, N., Wastell, S., van Spanning, R.J.M., and Spiro, S. 2000. Heterologous NNR-mediated nitric oxide signaling in *Escherichia coli*. J. Bacteriol. 182: 6434–6439.

Inatomi, K.-I., and Hochstein, L.I. 1996. The purification and properties of a copper nitrite reductase from *Haloferax denitrificans*. Curr. Microbiol. 32: 72–76.

Iwasaki, H., Matsubayashi, R., and Mori, T. 1956. Studies on denitrification. II. Production of nitric oxide and its utilization in the N-N-linkage formation by denitrifying bacteria. J. Biochem. 43: 295–305.

Ji, X.-B., and Hollocher, T.C. 1989. Nitrate reductase of *Escherichia coli* as a NO-producing nitrite reductase. Biochem. Arch. 5: 61–66.

Ka, J.-O., Urbance, J., Ye, R.W., Ahn, T.-Y., and Tiedje, J.M. 1997. Diversity of oxygen and N-oxide regulation of nitrite reductases in denitrifying bacteria. FEMS Microbiol. Lett. 156: 55–60.

Kalkowski, I., and Conrad, R. 1991. Metabolism of nitric oxide in denitrifying *Pseudomonas aeruginosa* and nitrate-respiring *Bacillus cereus*. FEMS Microbiol. Lett. 82: 107–111.

Kannt, A., Michel, H., Cheesman, M.R., Thomson, A.J., Dreusch, A.B., Körner, H., and Zumft, W.G. 1998. The electron transfer centers of nitric oxide reductase: homology with the heme-copper oxidase family. In: Biological electron-transfer chains: genetics, composition and mode of operation. G.W. Canters, and E. Vijgenboom, (eds.), Kluwer Academic Publishers, Dordrecht, The Netherlands. p. 279–291.

Kiley, P.J., and Beinert, H. 1999. Oxygen sensing by the global regulator FNR: the role of the iron-sulfur cluster. FEMS Microbiol. Rev. 22: 341–352.

Körner, H. 1993. Anaerobic expression of nitric oxide reductase from denitrifying *Pseudomonas stutzeri*. Arch. Microbiol. 159: 410–416.

Körner, H., and Zumft, W.G. 1989. Expression of denitrification enzymes in response to the dissolved oxygen level and respiratory substrate in continuous culture of *Pseudomonas stutzeri*. Appl. Environ. Microbiol. 55: 1670–1676.

Kwiatkowski, A.V., and Shapleigh, J.P. 1996. Requirement of nitric oxide for induction of genes whose products are involved in nitric oxide metabolism in *Rhodobacter sphaeroides* 2.4.3. J. Biol. Chem. 271: 24382–24388.

Lanzilotta, W.N., Schuller, D.J., Thorsteinsson, M.V., Kerby, R.L., Roberts, G.P., and Poulos, T.L. 2000. Structure of the CO sensing transcription activator CooA. Nature Struct. Biol. 7: 876–880.

Lissenden, S., Mohan, S., Overton, T., Regan, T., Crooke, H., Cardinale, J.A., Householder, T.C., Adams, P., O'Conner, C.D., Clark, V.L., Smith, H., and Cole, J.A. 2000. Identification of transcription activators that regulate gonococcal adaptation from aerobic to anaerobic or oxygen-limited growth. Mol. Microbiol. 37: 839–855.

Moënne-Loccoz, P., and de Vries, S. 1998. Structural characterization of the catalytic high-spin heme *b* of nitric oxide reductase: a resonance Raman study. J. Am. Chem. Soc. 120: 5147–5152.

Moir, J.W.B. 1999. Cytochrome *c'* from *Paracoccus denitrificans*: spectroscopic studies consistent with a role for the protein in nitric oxide metabolism. Biochim. Biophys. Acta 1430: 65–72.

Moir, J.W.B., Richardson, D.J., and Ferguson, S.J. 1995. The expression of redox proteins of denitrification in *Thiosphaera pantotropha* grown with oxygen, nitrate, and nitrous oxide as electron acceptors. Arch. Microbiol. 164: 43–49.

Murai, K., Miyake, K., Andoh, J., and Iijima, S. 2000. Transcriptional regulation of *nir* and *nor* operons of *Paracoccus denitrificans*. J. Biosci. Bioeng. 89: 384–387.

Nellen-Anthamatten, D., Rossi, P., Preisig, O., Kullik, I., Babst, M., Fischer, H.M., and Hennecke, H. 1998. *Bradyrhizobium japonicum* $FixK_2$, a crucial distributor in the FixLJ-dependent regulatory cascade for control of genes inducible by low oxygen levels. J. Bacteriol. 180: 5251–5255.

Palmedo, G., Seither, P., Körner, H., Matthews, J.C., Burkhalter, R.S., Timkovich, R., and Zumft, W.G. 1995. Resolution of the *nirD* locus for heme d_1 synthesis of cytochrome cd_1 (respiratory nitrite reductase) from *Pseudomonas stutzeri*. Eur. J. Biochem. 232: 737–746.

Philippot, L., Mirleau, P., Mazurier, S., Siblot, S., Hartmann, A., Lemanceau, P., and Germon, J.C. 2001. Characterization and transcriptional analysis of *Pseudomonas fluorescens* denitrifying clusters containing the *nar*, *nir*, *nor*,and *nos* genes. Biochim. Biophys. Acta 1517: 436–440.

Pohlmann, A., Cramm, R., Schmelz, K., and Friedrich, B. 2000. A novel NO-responding regulator controls the reduction of nitric oxide in *Ralstonia eutropha*. Mol. Microbiol. 38: 626–638.

Poole, R.K., and Hughes, M.N. 2000. New functions for the ancient globin family: bacterial responses to nitric oxide and nitrosative stress. Mol. Microbiol. 36: 775–783.

Ray, A., and Williams, H.D. 1997. The effects of mutation of the *anr* gene on the aerobic respiratory chain of *Pseudomonas aeruginosa*. FEMS Microbiol. Lett. 156: 227–232.

Rompf, A., Hungerer, C., Hoffmann, T., Lindenmeyer, M., Römling, U., Grob, U., Doss, M.O., Arai, H., Igarashi, Y., and Jahn, D. 1998a. Regulation of *Pseudomonas aeruginosa hemF* and *hemN* by the dual action of the redox response regulator Anr and Dnr. Mol. Microbiol. 29: 985–997.

Rompf, A., Hyland, S., Hoffmann, T., and Jahn, D. 1998b. Regulation of *Pseudomonas aeruginosa* nitrate respiration by NarX/L and the ANR/DNR system. EMBL accession no. Y15252.

Saraste, M., and Castresana, J. 1994. Cytochrome oxidase evolved by tinkering with denitrification enzymes. FEBS Lett. 341: 1–4.

Saunders, N.F.W., Houben, E.N.G., Koefoed, S., de Weert, S., Reijnders, W.N.M., Westerhoff, H.V., de Boer, A.P.N., and van Spanning, R.J.M. 1999. Transcription regulation of the *nir* gene cluster encoding nitrite reductase of *Paracoccus denitrificans* involves NNR and NirI, a novel type of membrane protein. Mol. Microbiol. 34: 24–36.

Saunders, N.F.W., Ferguson, S.J., and Baker, S.C. 2000. Transcriptional analysis of the *nirS* gene, encoding cytochrome cd_1 nitrite reductase, of *Paracoccus pantotrophus* LMD 92.63. Microbiology 146: 509–516.

Silvestrini, M.C., Falcinelli, S., Ciabatti, I., Cutruzzolà, F., and Brunori, M. 1994. *Pseudomonas aeruginosa* nitrite reductase (or cytochrome oxidase): an overview. Biochimie 76: 641–654.

Spiro, S. 1994. The FNR family of transcriptional regulators. Antonie Leeuwenhoek 66: 23–36.

Stover, C.K., Pham, X.Q., Erwin, A.L., Mizoguchi, S.D., Warrener, P., Hickey, M.J., Brinkman, F.S.L., Hufnagle, W.O., Kowalik, D.J., Lagrou, M., Garber, R.L., Goltry, L., Tolentino, E., Westbrock-Wadman, S., Yuan, Y., Brody, L.L., Coulter, S.N., Folger, K.R., Kas, A., Larbig, K., Lim, R., Smith, K., Spencer, D., Wong, G.K.-S., Wu, Z., Paulsen, I.T., Reizer, J., Saier, M.H., Hancock, R.E.W., Lory, S., and Olson, M.V. 2000. Complete genome sequence of *Pseudomonas aeruginosa* PAO1, an opportunistic pathogen. Nature 406: 959–964.

Suharti, Strampraad, M.J.F., Schröder, I., and de Vries, S. 2001. A novel copper A containing menaquinol NO reductase from *Bacillus azotoformans*. Biochemistry 40: 2632–2639.

Tosques, I.E., Kwiatkowski, A.V., Shi, J., and Shapleigh, J.P. 1996a. Characterization and regulation of the gene encoding nitrite reductase in *Rhodobacter sphaeroides* 2.4.3. J. Bacteriol. 179: 1090–1095.

Tosques, I.E., Shi, J., and Shapleigh, J.P. 1996b. Cloning and characterization of *nnrR*, whose product is required for the expression of proteins involved in nitric oxide metabolism in *Rhodobacter sphaeroides* 2.4.3. J. Bacteriol. 178: 4958–4964.

van der Oost, J., de Boer, A.P.N., de Gier, J.-W.L., Zumft, W.G., Stouthamer, A.H., and van Spanning, R.J.M. 1994. The heme-copper oxidase family consists of three distinct types of terminal oxidases and is related to nitric oxide reductase. FEMS Microbiol. Lett. 121: 1–10.

van Spanning, R.J.M., de Boer, A.P.N., Reijnders, W.N.M., Spiro, S., Westerhoff, H.V., Stouthamer, A.H., and van der Oost, J. 1995. Nitrite and nitric oxide reduction in *Paracoccus denitrificans* is under the control of NNR, a regulatory protein that belongs to the FNR family of transcriptional activators. FEBS Lett. 360: 151–154.

van Spanning, R.J.M., de Boer, A.P.N., Reijnders, W.N.M., Westerhoff, H.V., Stouthamer, A.H., and van der Oost, J. 1997. FnrP and NNR of *Paracoccus denitrificans* are both members of the FNR family of transcriptional activators but have distinct roles in respiratory adaptation in response to oxygen limitation. Mol. Microbiol. 23: 893–907.

van Spanning, R.J.M., Houben, E., Reijnders, W.N.M., Spiro, S., Westerhoff, H.V., and Saunders, N. 1999. Nitric oxide is a signal for NNR-mediated transcription activation in *Paracoccus denitrificans*. J. Bacteriol. 181: 4129–4132.

Vollack, K.-U., and Zumft, W.G. 2001. Nitric oxide signaling and transcriptional control of denitrification genes in *Pseudomonas stutzeri*. J. Bacteriol. 183: 2516–2526.

Vollack, K.-U., Härtig, E., Körner, H., and Zumft, W.G. 1999. Multiple transcription factors of the FNR family in denitrifying *Pseudomonas stutzeri*: characterization of four *fnr*-like genes, regulatory responses and cognate metabolic processes. Mol. Microbiol. 31: 1681–1694.

Wardrop, S.L., Watts, R.N., and Richardson, D.R. 2000. Nitrogen monoxide activates iron regulatory protein 1 RNA-binding activity by two possible mechanisms: effect on the [4Fe–4S] cluster and iron mobilization from cells. Biochemistry 39: 2748–2758.

Watmough, N.J., Butland, G., Cheesman, M.R., Moir, J.W.B., Richardson, D.J., and Spiro, S. 1999. Nitric oxide in bacteria: synthesis and consumption. Biochim. Biophys. Acta 1411: 456–474.

Wood, N.J., Alizadeh, T., Bennett, S., Pearce, J., Ferguson, S.J., Richardson, D.J., and Moir, J.W.B. 2001. Maximal expression of membrane-bound nitrate reductase in *Paracoccus* is induced by nitrate via a third FNR-like regulator named NarR. J. Bacteriol., 183: 3606–3613.

Ye, R.W., Arunakumari, A., Averill, B.A., and Tiedje, J.M. 1992. Mutants of *Pseudomonas fluorescens* deficient in dissimilatory nitrite reduction are also altered in nitric oxide reduction. J. Bacteriol. 174: 2560–2564.

Ye, R.W., Haas, D., Ka, J.-O., Krishnapillai, V., Zimmermann, A., Baird, C., and Tiedje, J.M. 1995. Anaerobic activation of the entire denitrification pathway in *Pseudomonas aeruginosa* requires Anr, an analog of Fnr. J. Bacteriol. 177: 3606–3609.

Yoshimura, T., Iwasaki, H., Shidara, S., Suzuki, S., Nakahara, A., and Matsubara, T. 1988. Nitric oxide complex of cytochrome *c'* in cells of denitrifying bacteria. J. Biochem. 103: 1016–1019.

Yoshimura, T., Shidara, S., Ozaki, T., and Kamada, H. 1993. Five coordinated nitrosylhemoprotein in the whole cells of denitrifying bacterium, *Achromobacter xylosoxidans* NCIB 11015. Arch. Microbiol. 160: 498–500.

Zafiriou, O.C., Hanley, Q.S., and Snyder, G. 1989. Nitric oxide and nitrous oxide production and cycling during dissimilatory nitrite reduction by *Pseudomonas perfectomarina*. J. Biol. Chem. 264: 5694–5699.

Zumft, W.G. 1992. The denitrifying prokaryotes. In: The prokaryotes. A handbook on the biology of bacteria: ecophysiology, isolation, identification, applications. A. Balows, H.G. Trüper, M. Dworkin, W. Harder, and K.-H. Schleifer, (eds.), Springer-Verlag, New York, NY. p. 554–582.

Zumft, W.G. 1993. The biological role of nitric oxide in bacteria. Arch. Microbiol. 160: 253–264.

Zumft, W.G. 1997. Cell biology and molecular basis of denitrification. Microbiol. Mol. Biol. Rev. 61: 533–616.

Zumft, W.G., Braun, C., and Cuypers, H. 1994. Nitric oxide reductase from *Pseudomonas stutzeri*: primary structure and gene organization of a novel bacterial cytochrome *bc* complex. Eur. J. Biochem. 219: 481–490.

Chapter 19

Regulation of Heme Biosynthesis in Non-Phototrophic Bacteria

Max Schobert and Dieter Jahn

Abstract

The biosynthesis of tetrapyrroles like hemes and chlorophylls is essential for most living organisms. In bacteria hemes are integral parts of energy conserving electron transport chains and cofactors of various enzymes. Changes of environmental conditions usually lead to an adaption of the bacterial energy metabolism and often coincide with significant changes of cellular heme levels. This review focuses on the known regulatory mechanisms in non-phototrophic bacteria involved in the control of the formation of the heme biosynthetic apparatus. Species specific differences in the mode of energy generation result in various regulatory strategies. Focusing on the well investigated bacteria *Bacillus subtilis, Escherichia coli, Pseudomonas aeruginosa,* and *Salmonella typhimurium* the involved environmental stimuli, employed transcriptional regulators and promoter structures as well as the role of protein stability are described. Broad variations of the used regulatory principles were observed.

Introduction

Biosynthetic pathways for the formation of tetrapyrroles are found in most known prokaryotes. They are essential for the production of porphyrins like hemes, chlorophylls and bacteriochlorophylls (Figure 1). The central biosynthetic pathway branches to produce porphinoids like corrinoids (vitamin B_{12}), siroheme or coenzyme F_{430} (Dailey, 1990; Jordan, 1991; Friedmann and Thauer, 1992; Beale, 1996). The first committed step in tetrapyrrole biosynthesis is the production of 5-aminolevulinic acid (ALA). Two different routes exist in nature to generate this key intermediate. The C_4- or Shemin pathway, in which ALA is formed by the condensation of succinyl-CoA and glycine in a single enzymatic reaction catalyzed by ALA synthase is limited to yeast, mammalian cells and the α-group of proteobacteria (Avissar *et al.,* 1989; Suzuki *et al.*, 1997; Ferreira, 1999). During the C_5-pathway ALA is formed from the C_5-skeleton of glutamate. This pathway is found in all other bacteria, in archaea, plants and algae (Jahn *et al.*, 1992; Avissar and Moberg, 1995; Beale, 1996). During its initial step glutamyl-tRNA is reduced to glutamate-1-semialdehyde by glutamyl-tRNA reductase (HemA) (Avissar and Beale, 1989; Verkamp *et al.,* 1992; Moser *et al.,* 1999). In a second step a glutamate-1-semialdehyde-2,1-aminomutase (HemL) converts glutamate-1-semialdehyde to ALA (Ilag and Jahn, 1992; Jahn *et al.*, 1992). Further seven enzymatic steps are required for the formation of protoheme IX. The involved enzymes and the nature of the chemical intermediates are remarkably conserved among all organisms (Figure 1).

Regulation of Heme Biosynthesis in Bacteria

The heme demand of bacterial cells changes drastically in dependence of its mode of energy generation. In *E. coli* and *P. aeruginosa* significantly varying heme levels were detected for aerobically and anaerobically growing cells (Doss and Philipp-Dornston, 1971; Jacobs *et al.*, 1972; Philipp-Dornston and Doss, 1973; Hino and Ishida, 1973). Low heme levels were found in general, when these bacteria were subjected to fermentation, where no heme-containing respiratory enzymes are required. These physiological experiments suggested oxygen, nitrate and the carbon source as direct or indirect signals for heme biosynthesis regulation (Doss and Philipp-Dornston, 1971; Jacobs *et al.*, 1972; Jacobs *et al.*, 1973). Moreover, the accumulation of free heme as well as some of the biosynthetic intermediates of the pathway are toxic to the cells (Nakahigashi *et al.*, 1991). However, when heme binding proteins like hemoglobin, cytochromes or catalases are overproduced in *E. coli,* heme production increases in parallel up to 20-fold, indicating that heme is a strong regulator of its own biosynthesis (Woodard and Dailey, 1995; Verderber *et al.*, 1997).

There is substantial information available about tetrapyrrole biosynthesis and its regulation in yeast and higher organisms (Jordan, 1991; Reinbothe and Reinbothe, 1996; Suzuki *et al.*, 1997; Ferreira, 1999). However, most information about the regulation of heme biosynthesis via the C_5-route in bacteria is limited to *B. subtilis, E. coli, P. aeruginosa* and *S. typhimurium.* Physiological investigations identified two steps of the heme biosynthesis pathway as key regulatory points (Jacobs *et al.*, 1972; Philipp-Dornston and Doss, 1973). One major regulatory target in these organisms is the formation of ALA. The rate limiting character of the initial step of the pathway was convincingly documented by ALA-feeding experiments which led to increased heme formation and the accumulation and excretion of coproporphyrin III (Philipp-Dornston and Doss, 1973). Therefore, the second obvious regulatory point is

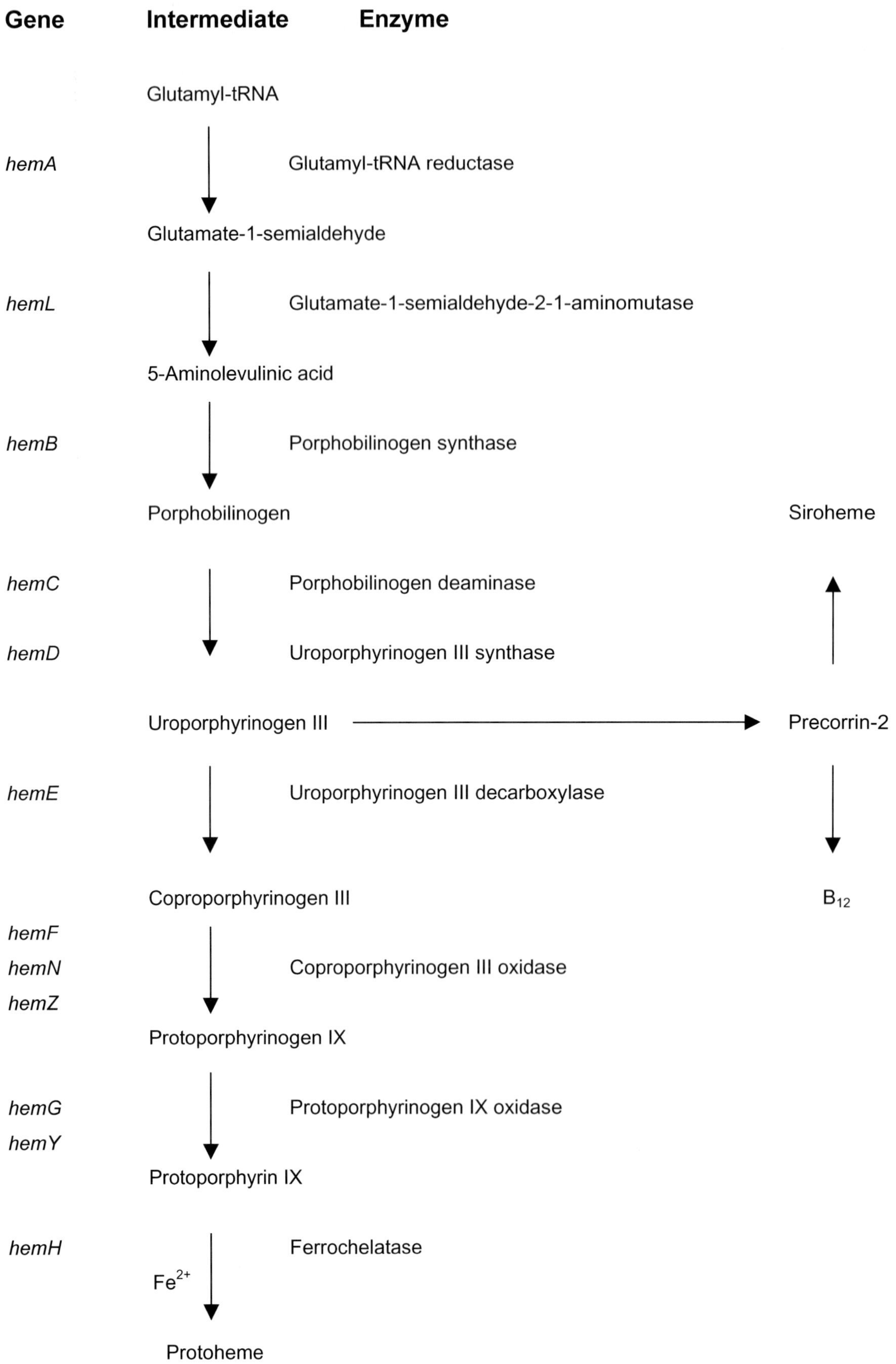

Figure 1. Pathway for the biosynthesis of tetrapyrroles in bacteria. Involved genes and their corresponding gene products as well as the chemical intermediates are listed.

the synthesis of protoporphyrinogen IX from coproporphyrinogen III by various coproporphyrinogen III oxidases named HemF, HemN and HemZ. One of these enzymes, HemF, requires molecular oxygen for catalysis and therefore cannot operate during anaerobic conditions where it is substituted by a second type of oxygen-independent enzyme (HemN, HemZ) (Tait, 1969; Tait, 1972; Seehra *et al.*, 1982; Xu *et al.*, 1992; Rompf *et al.*, 1998). The regulation of the late steps of heme biosynthesis is believed to be necessary to prevent the accumulation of potentially phototoxic byproducts or intermediates.

Interestingly, in plant-associated *rhizobia* the key regulatory step is located at the level of ALA utilization by porphobilinogen synthase (HemB). To control the flow of plant supplied ALA into the bacterial pathway *hemB* expression is strictly controlled (O'Brian, 1996).

Transcriptional Regulation of *hemA* Encoding Glutamyl-tRNA Reductase

A number of independent experiments demonstrate, that the first step in heme biosynthesis is strictly regulated. Several reports state that *E. coli* overproducing homologous or heterologous glutamyl-tRNA reductase accumulate ALA and subsequently heme (Chen *et al.*, 1994; Verderber *et al.*, 1997). These findings suggested a central regulatory role of the *hemA* gene. In *E. coli, S. typhimurium* and *P. aeruginosa* the *hemA* gene forms an operon with the *prfA* gene encoding protein release factor 1 (Elliott, 1989; Drolet *et al.*, 1989; Li *et al.*, 1989; Verkamp and Chelm, 1989; Hungerer *et al.*, 1995a). Upstream of the *hemA* gene of *E. coli* and transcribed in the opposite direction a gene was identified and previously named *hemM,* due to its potential involvement in heme biosynthesis (Ikemi *et al.*, 1992). However, several independent investigations using different bacterial species excluded its participation in tetrapyrrole biosynthesis and left it as an open reading frame of unknown function (Hungerer *et al.,* 1995a; McNicholas *et al.,*1997; Verderber *et al.*, 1997).

Regulation of *hemA* induction was first studied in *E. coli* and *S. typhimurium* using reporter gene fusions in *hemA* mutant strains (Choi *et al.*, 1996; Darie and Gunsalus, 1994; McNicholas *et al.*, 1997). These *hemA* mutants grow normally when ALA is supplied exogenously, but face a severe heme limitation after cells are transferred into a medium without ALA. The presence of the fermentable substrate pyruvate is essential under these conditions to allow slow exponential growth. Surprisingly, only moderate transcriptional control of the *hemA* gene was observed. Under these heme limiting conditions only a 2- to 3-fold increase of *hemA* transcription in *S. typhimurium* and a 4- to 6-fold increase in *E. coli* was observed (Choi *et al.*, 1996). *E. coli* forms significant different heme amounts during aerobic versus anaerobic growth. However, only a 1.5-fold anaerobic increase of a *hemA*-transcription was observed in anaerobically grown *E. coli* (McNicholas *et al.*, 1997). The presence of alternate electron acceptors like nitrate under anaerobic conditions did not influence *E. coli hemA* expression (Darie and Gunsalus, 1994).

Already Verkamp and Chelm (1989) identified two transcriptional start sites and two potential σ^{70}-dependent promoters for *E. coli hemA*. Promoter 2 contributed only 10–15% to the total *hemA* transcriptional activity, while 85–90% of *hemA* mRNA formation initiated at promoter 1. Interestingly, a Fnr-binding half site of the consensus was detected in promoter 1 in a position usually found in Fnr-repressed promoters (Melville and Gunsalus, 1996). Promoter activity is slightly elevated in wild type cells during anaerobic growth but repressed under anaerobic heme limiting conditions (McNicholas *et al.*, 1997) (Figure 2).

Transcriptional regulation of *P. aeruginosa hemA* was studied initially by quantitation of mRNA. The *hemA* mRNA levels in *P. aeruginosa* increase under anaerobic denitrifying conditions compared to aerobic growth. They are very low under fermentative conditions where no heme-containing cytochromes are required for respiration (Hungerer *et al.*, 1995a). Two transcriptional start sites were found upstream of *hemA*. Both transcripts are present during both aerobic and anaerobic growth conditions. A *P. aeruginosa* mutant with a defect in the oxygen regulator Anr (an *E. coli* Fnr homolog) exhibited similar overall increased *hemA* mRNA levels as the wild type under anaerobic denitrifying conditions, however the distribution of *hemA* mRNA from the two promoters changed significant. Almost no transcription was detected from promoter 1 while the induction of the more distant promoter 2 was drastically elevated. One half of a binding site of Anr was detected at promoter 1, while promoter 2 lacked any binding site motif (Hungerer *et al.*, 1995a). An cooperative Anr- and Dnr-dependence of promoter 1 was verified using reporter gene fusions. Dnr is a second Fnr-like transcriptional regulator of *P. aeruginosa*. However, Dnr is lacking the conserved cysteine residues for iron-sulfur cluster formation. In contrast to in *E. coli* and *S. typhimurium, P. aeruginosa hemA* expression was induced under anaerobic conditions by the presence of nitrate mediated by the two component regulatory system encoded by *narXL*. IHF was found important for overall *hemA* transcription (Krieger *et al.*, 2002) (Figure 3). Expression of *P. aeruginosa hemL* encoding glutamate-1-semialdehyde-2-1-aminomutase was induced by oxygen stress and anaerobic denitrifying conditions (Hungerer *et al.,* 1995b).

HemA Activity is Regulated by Proteolysis

While heme has no obvious inhibitory effect on the enzymatic activity of the purified HemA protein (Jahn *et al.*, 1991), glutamyl-tRNA reductase activity is reduced when heme is added to cell extracts of *E. coli* (Javor and Febre, 1992). This implies, that additional factors in combination with heme are required for the regulation of HemA activity.

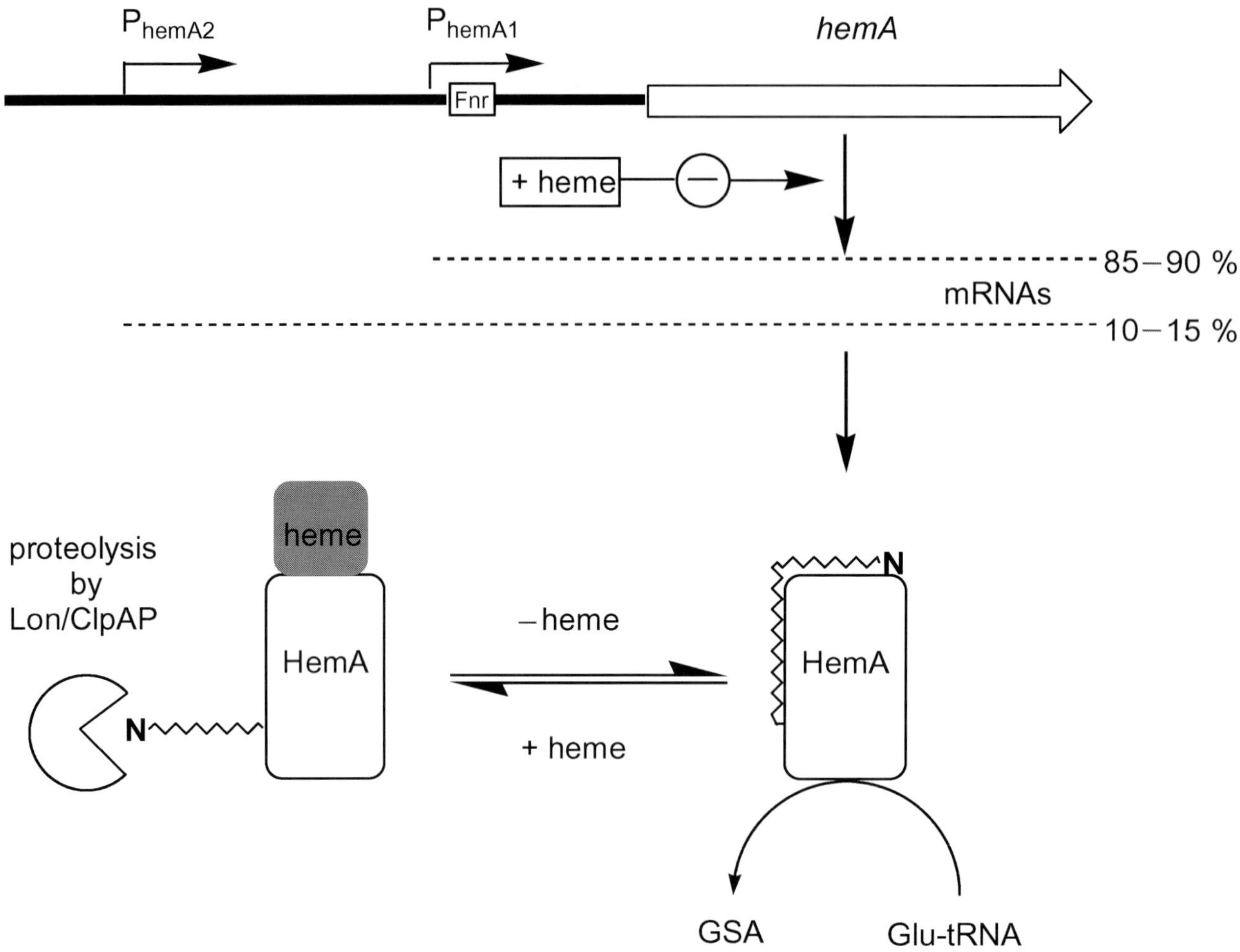

Figure 2. Model for the regulation of glutamyl-tRNA reductase (HemA) formation in *Escherichia coli* and *Salmonella typhimurium*. The *hemA* gene with two upstream promoters, the corresponding ratio of the transcription from both promoters as well as the Fnr binding half site in promoter 1 are indicated. Regulation of HemA stability by the proteases Lon and ClpAP in response to heme is displayed underneath. Abbreviations: Glu-tRNA, glutamyl-tRNA; GSA, glutamate-1-semialdehyde.

Elliott and coworkers first discovered a striking difference between *hemA-lacZ* expression and enzymatic activity of glutamyl-tRNA reductase in various *S. typhimurium* heme mutants after starvation for heme (Wang *et al.*, 1997). While *hemA-lacZ* expression increased only 1.5- to 2-fold in heme-deficient *hemB* or *hemL* mutants, the enzymatic activity of glutamyl-tRNA reductase increased 20- to 25-fold proportional to the parallel increase of HemA protein. Puls-chase experiments of HemA turnover in *hemL* mutants of *S. typhimurium* revealed a half-life of 20 min in the presence of ALA which increased to more than 300 min in heme-limited cells (Wang *et al.*, 1999a). Two ATP-depending proteases Lon and ClpP as well as the ClpA chaperone were identified to be involved in HemA turnover. In an *E. coli lon clpP* double mutant the HemA protein showed a significantly increased half life. These results clearly demonstrated, that cellular HemA concentration is regulated by conditional protein stability (Wang *et al.*, 1999a). A HemA mutant protein with a positive charge introduced close to the N-terminus was significantly stabilized and confirmed, that the N-terminus is important for proteolysis by the Lon and ClpP protease systems (Wang *et al.*, 1999b). The proposed model for the heme dependence of the proteolytic process suggested, that heme binding to HemA induces a conformational change of the protein and the exposition of a protease sensitive domain (Wang *et al.*, 1999a) (Figure 2).

Regulation of HemF and HemN

Early experiments with crude extracts of *Rhodobacter spheroides* revealed the presence of two enzyme activities for the conversion of coproporphyrinogen III to protoporphyrinogen IX under aerobic and anaerobic conditions (Tait, 1969; Tait, 1972). Genetic approaches identified two loci involved in coproporphyrinogen III oxidase activity and led to their cloning from various bacteria (Coomber *et al.*, 1992; Xu and Elliott, 1993; Troup *et al.*, 1994; Troup *et al.*, 1995; Xu and Elliott, 1994; Zeilstra-Ryalls, 1995; de Gier *et al.*, 1996; Homuth *et al.*, 1996; Hippler *et al.*, 1997;

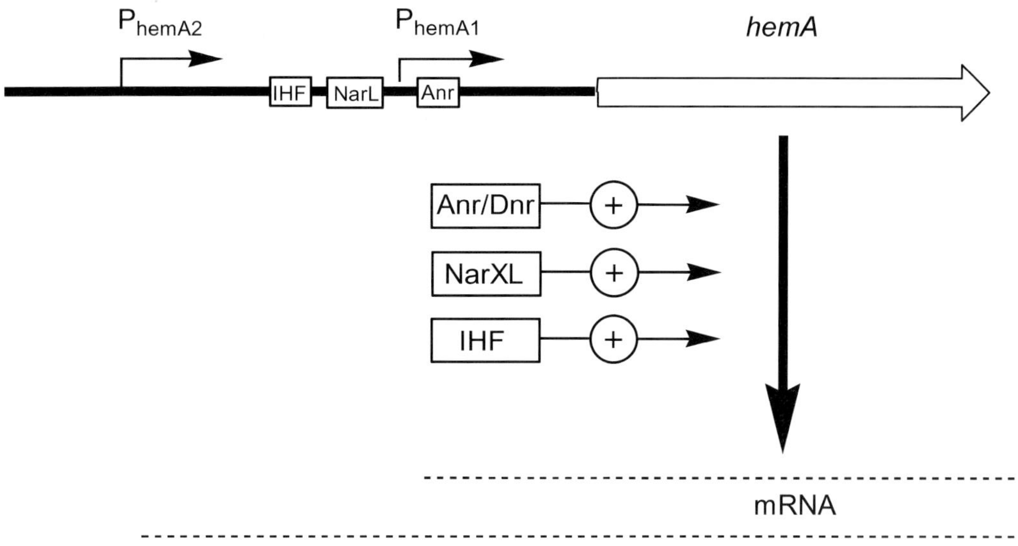

Figure 3. Regulation of *hemA* transcription in *Pseudomonas aeruginosa*.The *hemA* gene with two upstream promoters, the Anr binding half site in promoter 1 and IHF and NarL binding sites upstream of promoter 1 is shown.

Lieb *et al.*, 1998). The oxygen-dependent coproporphyrinogen III oxidase is encoded by *hemF*. HemF shows no obvious similarity to the oxygen-independent coproporphyrinogen III oxidase encoded by *hemN* except for a stretch of 21 amino acids with 40% identical residues. HemN contains a [4Fe–4S]-cluster and requires besides various other cofactors S-adenosyl methionine for enzyme activity (Layer *et al.*, 2002). The enzyme is active under anaerobic as well as aerobic conditions. Surprisingly, in *P. aeruginosa* a *hemF hemN* double mutant still synthesized heme under aerobic and anaerobic conditions, indicating the existence of a third oxygen-independent enzyme of the HemN-type named HemZ (Rompf *et al.*, 1998).

Regulation of *hemF* and *hemN* was found not as simple as expected from their oxygen-related enzymatic functions. In *P. aeruginosa* both *hemF* and *hemN* were induced under anaerobic growth conditions. The 7- to 10-fold anaerobic induction was only expected in the case of the *hemN* gene encoding the oxygen-independent enzyme. Surprisingly, also *hemF* encoding the oxygen-dependent enzyme was found 20-fold induced under anaerobic conditions (Rompf *et al.*, 1998).

However, anaerobic induction of the *hemF* analogue *HEM13* gene from *Saccharomyces cerevisiae* was observed before (Zagorec *et al.*, 1988). The authors proposed that *HEM13* is increasingly expressed under oxygen limiting conditions to maintain heme production. The enzyme possesses a high affinity for oxygen and catalyses a rate limiting step in yeast heme biosynthesis (Zagorec *et al.*, 1988). This explanation seems unlikely for the case of the anaerobic induction of the *P. aeruginosa hemF* gene due to the availability of the oxygen-independent HemN and HemZ. Moreover, a *P. aeruginosa hemF* mutant showed significantly reduced anaerobic growth indicating additional anaerobic functions for the HemF protein.

The promoter regions of *hemF* and *hemN* in *P. aeruginosa* contain a putative Anr/Dnr binding motif centered 44.5 bp and 41.5 bp upstream of the transcriptional start site, respectively (Rompf *et al.*, 1998). The regulatory proteins Anr and Dnr have been reported to form a regulatory cascade for the induction of denitrification genes under anaerobic conditions. During this cascade Anr, the Fnr analogue, induces expression of *dnr*. Dnr in turn induces transcription of various target genes (Arai *et al.*, 1997). Investigation of *P. aeruginosa hemF* and *hemN* transcription using *anr* or *dnr* mutant strains and co-expression of the *anr* and *dnr* genes *in trans* revealed dual action of Anr and Dnr during the transcriptional induction process. Surprisingly, Anr is essential for aerobic expression of *hemN* (Rompf *et al.*, 1998). Moreover, observed anaerobic *hemF* and *hemN* expression requires in parallel nitrate induction via the NarXL system and DNA-bending via integration host factor (IHF)(Figure 4).

In contrast to the situation in *P. aeruginosa* the *E. coli* and *S. typhimurium hemF* genes form an operon with an open reading frame coding for a protein with strong similarity to a cell wall amidase of *B. subtilis*. Xu and Elliot (1993) demonstrated that this gene is not involved in heme biosynthesis. Two *hemF* promoters were identified in *E. coli*. But transcription from these promoters does not vary significantly in dependence of changing oxygen tension and the presence of various alternative electron acceptors (Troup *et al.*, 1994). However, an OxyR-dependent low level induction of *E. coli hemF* in response to hydrogen peroxide was observed (Mukhopadhyay and Schellhorn, 1997). Increased heme synthesis is required to maintain the heme cofactor concentration for the hydrogen peroxide detoxifying catalases, which are induced in parallel under condition of oxygen stress. Experiments using reporter gene fusions showed an 3-fold anaerobic

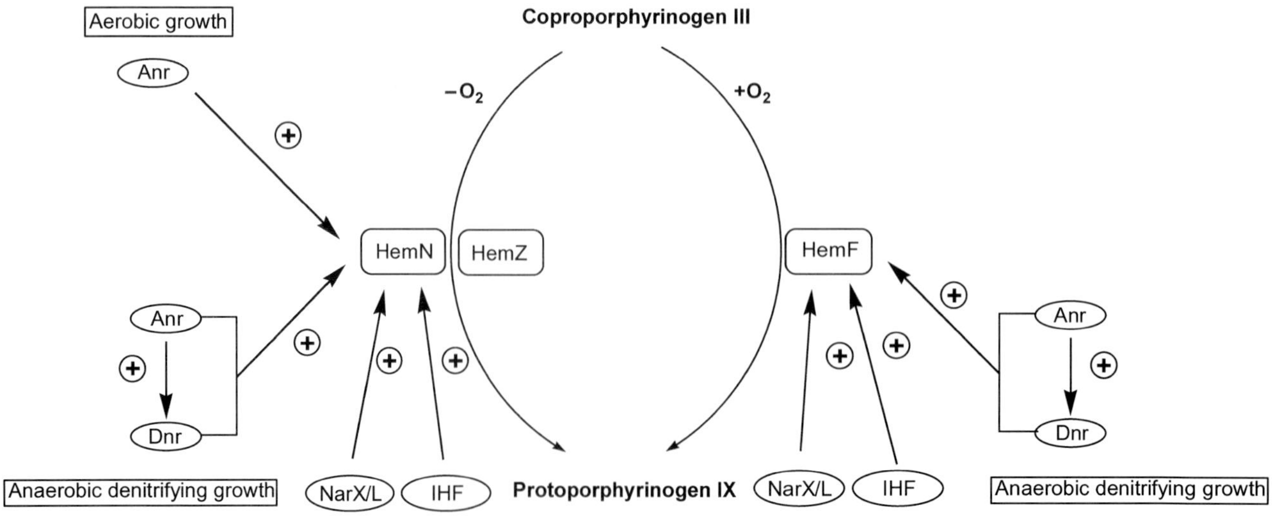

Figure 4. Model of *hemF* and *hemN* regulation in *Pseudomonas aeruginosa* by the dual action of Anr and Dnr, by NarX/L and IHF. The signal for Anr is a change in oxygen tension. The signal for NarX/L is the presence of nitrate. IHF is a DNA bending protein. The labels $-O_2$ and $+O_2$ indicate the oxygen-dependence of the enzymatic reaction catalyzed by the coproporphyrinogen III oxidases.

induction of *E. coli hemN* which was found independent of the presence of nitrate. Iron depletion by addition of a chelator led to a 4-fold decrease of *hemN* transcription (Troup *et al.*, 1995).

For *B. subtilis* the existence of two oxygen-independent coproporphyrinogen III oxidases, HemN and HemZ, has been reported (Hippler *et al.*, 1997; Homuth *et al.*, 1996; Homuth *et al.*, 1999). There is evidence for a third oxygen-independent enzyme since a *hemN hemZ* double mutant has no obvious growth phenotype (Homuth *et al.*, 1999). *B. subtilis* contains no HemF (Hippler *et al.*, 1997; Kunst *et al.*, 1997). Regulation of *B. subtilis hemN* and *hemZ* was studied using reporter gene fusions. Transcription of both genes is induced 5-fold under anaerobic conditions in the presence of nitrate. In agreement with the low heme requirement during fermentative growth no induction of those genes was observed under these conditions (Homuth *et al.*, 1999). To investigate which regulators participate in anaerobic induction of *B. subtilis hemN* and *hemZ,* reporter gene fusions of these genes were tested in *fnr, resDE* and *ywiD* mutant strains. These mutants are defective in previously identified redox regulatory genes (Nakano and Zuber, 1998). While aerobic expression of *hemN* and *hemZ* remained unaffected, anaerobic expression was drastically reduced in all three mutant strains (Homuth *et al.*, 1999). In *B. subtilis* the regulators encoded by *resDE, fnr* and *ywiD* form a redox regulatory cascade (Cruz Ramos *et al.,* 2000; Marino *et al.,* 2000). The two component regulatory system *resDE* is required for strong anaerobic induction of *fnr* expression (Nakano *et al.*, 1996). Fnr in turn binds to the highly conserved Fnr box upstream of *ywiD* and induces its anaerobic transcription. Finally YwiD co-activates transcription of various genes involved in anaerobic metabolism including *hemN* and *hemZ*. The *resDE* locus participates in the regulation of several tetrapyrrole-associated genetic loci including *qcrABC* (cytochrome *bc* complex), *ctaA* (required for heme A synthesis) and several cytochrome c biogenesis genes (Sun *et al.*, 1996). A fivefold induction of *hemZ* transcription was observed after peroxide stress, while the transcription of *hemN* remained unchanged. However, no typical binding site for the alternative stress sigma factor σ^B and no *per* box were detected in the promoter region of *hemZ*. This *per* box motif was previously found upstream of other genes which are subject to oxygen stress regulation like the *B. subtilis hemAXCDBL* operon encoding all enzymes for uroporphyrinogen III formation (Hansson *et al.,* 1991; Chen *et al.*, 1995) (Figure 5). Moreover

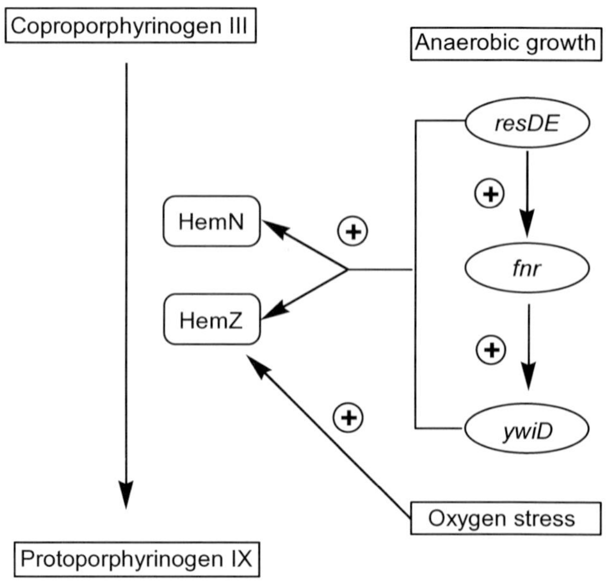

Figure 5. Model for the oxygen tension-dependent control of *hemN* and *hemZ* in *Bacillus subtilis* by the regulatory cascade of *resDE, fnr and ywiD,* and for the induction of *hemZ* expression by oxygen stress.

the *hemX* gene product was shown to affect the levels of glutamyl-tRNA reductase (Schroder *et al.*, 1994).

Recently a genomics approach discovered the anaerobic activation of the *B. subtilis hemEHY operon* (Hansson and Hederstedt, 1992; Hansson and Hederstedt, 1994; Ye *et al.,* 2000)

Regulation of Other Heme Biosynthetic Genes

Only little information is available about regulation of the remaining heme biosynthesis genes in bacteria. McNicholas *et al.* (1997) investigated transcriptional regulation of the genes encoding porphobilinogen deaminase (*hemC*), uroporphyrinogen III synthase (*hemD*), and the final enzyme of the tetrapyrrole biosynthesis, ferrochelatase (*hemH*) in *E. coli.* Only a 2-fold induction was observed for these genes under heme limiting conditions indicating a moderate transcriptional regulation.

Acknowledgements

We thank R. Krieger, A. Rompf, C. Hungerer, B. Troup, M. Lindenmeyer and K. Trunk for helpful discussions. This work was supported by the Deutsche Forschungsgemeinschaft, the Max-Planck-Gesellschaft, the Fonds der Chemischen Industrie and the Graduiertenkolleg "Biochemie der Enzyme". We thank R.K. Thauer (Max-Planck-Institute, Marburg, Germany) for continued support.

References

Arai, H., Kodama, T., and Igarashi, Y. 1997. Cascade regulation of the two CRP/FNR-related transcriptional regulators (ANR and DNR) and the denitrification enzymes in *Pseudomonas aeruginosa*. Mol. Microbiol. 25: 1141–1148.

Avissar, Y.J., and Beale, S.I. 1989. Identification of the enzymatic basis for delta-aminolevulinic acid auxotrophy in a *hemA* mutant of *Escherichia coli.* J. Bacteriol. 171: 2919–2924.

Avissar, Y.J., Ormerod, J.G., and Beale, S.I. 1989. Distribution of delta-aminolevulinic acid biosynthetic pathways among phototrophic bacterial groups. Arch. Microbiol. 151: 513–519.

Avissar, Y.J., and Moberg, P.A. 1995. The common origins of the pigments of life – early steps of chlorophyll biosynthesis. Photosynthesis Research 44: 221–242.

Beale, S.I. 1996. Biosynthesis of hemes. In: *Escherichia coli* and Salmonella: cellular and molecular biology. Frederick C. Neidhardt ed., ASM press, Washington, DC. pp. 731–748.

Chen, L., Keramati, L., and Helmann, J.D. 1995. Coordinate regulation of *Bacillus subtilis* peroxide stress genes by hydrogen peroxide and metal ions. Proc Natl Acad Sci USA 92: 8190–8194.

Chen, W., Russell, C.S., Murooka, Y., and Cosloy, S.D. 1994. 5-Aminolevulinic acid synthesis in *Escherichia coli* requires expression of *hemA*. J. Bacteriol. 176: 2743–2746.

Choi, P., Wang, L., Archer, C.D., and Elliott, T. 1996. Transcription of the glutamyl-tRNA reductase (*hemA*) gene in *Salmonella typhimurium* and *Escherichia coli*: role of the hemA P1 promoter and the arcA gene product. J. Bacteriol. 178: 638–646.

Coomber, S.A., Jones, R.M., Jordan, P.M., and Hunter, C.N. 1992. A putative anaerobic coproporphyrinogen III oxidase in *Rhodobacter sphaeroides*. I. Molecular cloning, transposon mutagenesis and sequence analysis of the gene. Mol. Microbiol. 6: 3159–3169.

Cruz Ramos, H., Hoffmann, T., Marino, M., Nedjari, H., Presecan-Siedel, E., Dreesen, O., Glaser, P., and Jahn, D. 2000. Fermentative Metabolism of *Bacillus subtilis*: physiology and regulation of gene expression. J. Bacteriol. 182: 3072–3080.

Dailey, H.A. (ed.), 1990 Biosynthesis of heme and chlorophylls. Mc Graw-Hill Publishing Co., Hightstown, N.J.

Darie, S., and Gunsalus, R.P. 1994. Effect of heme and oxygen availability on *hemA* gene expression in *Escherichia coli*: role of the *fnr, arcA*, and *himA* gene products. J. Bacteriol. 176: 5270–5276.

de Gier, J.W., Schepper, M., Reijnders, W.N., van Dyck, S.J., Slotboom, D.J., Warne, A., Saraste, M., Krab, K., Finel, M., Stouthamer, A.H., van Spanning, R.J., and van der Oost, J. 1996. Structural and functional analysis of aa3-type and cbb3-type cytochrome c oxidases of *Paracoccus denitrificans* reveals significant differences in proton-pump design. Mol. Microbiol. 20: 1247–1260.

Doss, M.O., and Philipp-Dornston, W.K. 1971. Porhyrin and heme biosynthesis from endogenous and exogenous δ-aminolevulinic acid in *Escherichia coli, Pseudomonas aeruginosa* and *Achromobacter metalcaligenes*. Hoppe-Seylers's Z Physiol. Chem. 352: 725–733.

Drolet, M., Peloquin, L., Echelard, Y., Cousineau, L., and Sasarman, A. 1989. Isolation and nucleotide sequence of the *hemA* gene of *Escherichia coli* K12. Mol. Gen. Genet. 216: 347–352.

Elliott, T. 1989. Cloning, genetic characterization, and nucleotide sequence of the *hemA-prfA* operon of *Salmonella typhimurium*. J. Bacteriol. 171: 3948–3960.

Ferreira, G.C. 1999. 5-Aminolevulinate synthase and mammalian heme biosynthesis. In: Iron Metabolism: Inorganic biochemistry and regulatory mechanisms. ed. by Gloria C. Ferreira, José J. G. Moura, and Ricardo Franco. Weinheim; New York; Chichester; Brisbane; Singapore; Toronto: Wiley-VCH, 1999. p. 15–34.

Friedmann, C.H., and Thauer, R.K. 1992. Macrocyclic tetrapyrrole biosynthesis in bacteria. In: Encyclopedia of Microbiology, Volume 3, Academic press. p 1–19.

Hansson, M., and Hederstedt, L. 1994. *Bacillus subtilis* HemY is a peripheral membrane protein essential for protoheme IX synthesis which can oxidize coproporphyrinogen III and protoporphyrinogen IX. J. Bacteriol. 176: 5962–5970.

Hansson, M., and Hederstedt, L. 1992. Cloning and characterization of the *Bacillus subtilis hemEHY* gene cluster, which encodes protoheme IX biosynthetic enzymes. J. Bacteriol. 174: 8081–8093.

Hansson, M., Rutberg, L., Schroder, I., and Hederstedt, L. 1991. The *Bacillus subtilis hemAXCDBL* gene cluster, which encodes enzymes of the biosynthetic pathway from glutamate to uroporphyrinogen III. J. Bacteriol. 173: 2590–2599.

Hino, S., and Ishida, A., 1973. Effect of oxygen on heme and cytochrome content in some facultative bacteria. Enzyme 16: 42–49.

Hippler, B., Homuth, G., Hoffmann, T., Hungerer, C., Schumann, W., and Jahn, D. 1997. Characterization of *Bacillus subtilis hemN*. J. Bacteriol. 179: 7181–7185.

Homuth, G., Heinemann, M., Zuber, U., and Schumann, W. 1996. The genes of *lepA* and *hemN* form a bicistronic operon in *Bacillus subtilis*. Microbiology 142: 1641–1649.

Homuth, G., Rompf, A., Schumann, W., and Jahn, D. 1999. Transcriptional control of *Bacillus subtilis hemN* and *hemZ*. J. Bacteriol. 181: 5922–5929.

Hungerer, C., Troup, B., Romling, U., and Jahn, D. 1995a. Regulation of the *hemA* gene during 5-aminolevulinic acid formation in *Pseudomonas aeruginosa*. J. Bacteriol. 177: 1435–1443.

Hungerer, C., Troup, B., Romling, U., and Jahn, D. 1995b. Cloning, mapping and characterization of the *Pseudomonas aeruginosa hemL* gene. Mol. Gen. Genet. 248: 375–380.

Ikemi, M., Murakami, K., Hashimoto, M., and Murooka, Y. 1992. Cloning and characterization of genes involved in the biosynthesis of delta-aminolevulinic acid in *Escherichia coli.* Gene 121: 127–132.

Ilag, L.L., and Jahn, D. 1992. Activity and spectroscopic properties of the *Escherichia coli* glutamate 1-semialdehyde aminotransferase and the putative active site mutant K265R. Biochemistry 31: 7143–7151.

Jacobs, N.J., Jacobs, J.M., and Morgan, H.E.J. 1972. Comparative effect of oxygen and nitrate on protoporphyrin and heme synthesis from delta-amino levulinic acid in bacterial cultures. J. Bacteriol. 112: 1444–1445.

Jacobs, N.J., Jacobs, J.M., and Mills, B.A. 1973. Role of oxygen in the late steps of heme synthesis in *pseudomonads* and *Escherichia coli.* Enzyme 16: 50–56.

Jahn, D., Michelsen, U., and Soll, D. 1991. Two glutamyl-tRNA reductase activities in *Escherichia coli.* J Biol Chem 266: 2542–2548.

Jahn, D., Verkamp, E., and Soll, D. 1992. Glutamyl-transfer RNA: a precursor of heme and chlorophyll biosynthesis. Trends Biochem Sci 17: 215–218.

Javor, G.T., and Febre, E.F. 1992. Enzymatic basis of thiol-stimulated secretion of porphyrins by *Escherichia coli.* J. Bacteriol. 174: 1072–1075.

Jordan, P.M. 1991. Biosynthesis of tetrapyrroles. In: A. Neuberger and L.L.M. van Deenen (ed.), New comprehensive biochemistry, Vol. 19. Elsevier, Amsterdam.

Krieger, R., Rompf, A., Schobert, M., and Jahn, D. 2002. The *Pseudomonas aeruginosa hemA* promoter is regulated by Anr, Dnr, NarL and Integration Host Factor. Mol. Genet. Genomics. 267: 409–417.

Kunst, F., Ogasawara, N., Moszer, I., Albertini, A.M., *et al.* 1997. The complete genome sequence of the gram-positive bacterium *Bacillus subtilis.* Nature 390: 249–256.

Layer, G., Verfurth, K., Mahlitz, E., and Jahn, D. 2002. Oxygen-independent coproporphyrinogen-III oxidase HemN from *Escherichia Coli.* J. Biol. Chem. 277: 34136–34142.

Li, J.M., Brathwaite, O., Cosloy, S.D., and Russell, C.S. 1989. 5-Aminolevulinic acid synthesis in *Escherichia coli.* J. Bacteriol. 171: 2547–2552.

Lieb, C., Siddiqui, R.A., Hippler, B., Jahn, D., and Friedrich, B. 1998. The *Alcaligenes eutrophus hemN* gene encoding the oxygen-independent coproporphyrinogen III oxidase, is required for heme biosynthesis during anaerobic growth. Arch. Microbiol. 169: 52–60.

Marino, M., Hoffmann, T., Schmid, R., Mobitz, H., and Jahn, D. 2000. Changes in protein synthesis during the adaptation of *Bacillus subtilis* to anaerobic growth conditions. Microbiology 146 (Pt 1): 97–105.

McNicholas, P.M., Javor, G., Darie, S., and Gunsalus, R.P. 1997. Expression of the heme biosynthetic pathway genes *hemCD, hemH, hemM,* and *hemA* of *Escherichia coli.* FEMS Microbiol Lett 146: 143–148.

Melville, S.B., and Gunsalus, R.P. 1996. Isolation of an oxygen-sensitive FNR protein of *Escherichia coli*: interaction at activator and repressor sites of FNR-controlled genes. Proc Natl Acad Sci USA 93: 1226–1231.

Moser, J., Lorenz, S., Hubschwerlen, C., Rompf, A., and Jahn, D. 1999. *Methanopyrus kandleri* glutamyl-tRNA reductase. J. Biol. Chem. 274: 30679–30685.

Mukhopadhyay, S., and Schellhorn, H.E. 1997. Identification and characterization of hydrogen peroxide-sensitive mutants of *Escherichia coli*: genes that require OxyR for expression. J. Bacteriol. 179: 330–338.

Nakahigashi, K., Nishimura, K., Miyamoto, K., and Inokuchi, H. 1991. Photosensitivity of a protoporphyrin-accumulating, light-sensitive mutant (visA) of *Escherichia coli* K-12. Proc. Natl. Acad. Sci. USA 88: 10520–10524.

Nakano, M.M., Zuber, P., Glaser, P., Danchin, A., and Hulett, F.M. 1996. Two-component regulatory proteins ResD-ResE are required for transcriptional activation of fnr upon oxygen limitation in *Bacillus subtilis.* J. Bacteriol. 178: 3796–3802.

Nakano, M.M., and Zuber, P. 1998. Anaerobic growth of a "strict anaerobe" (*Bacillus subtilis).* Annu Rev Microbiol 52: 165–190.

O'Brian, M.R. 1996. Heme synthesis in the *rhizobium-legume* symbiosis: a palette for bacterial and eukaryotic pigments.J. Bacteriol. 178: 2471–2478.

Philipp-Dornston, W.K., and Doss, M.O. 1973. Comparison of porphyrin and heme in various heterotrophic bacteria. Enzyme 16: 57–64.

Reinbothe, S., and Reinbothe, C., 1996. The regulation of enzymes involved in chlorophyll biosynthesis. Eur J Biochem 237: 323–343.

Rompf, A., Hungerer, C., Hoffmann, T., Lindenmeyer, M., Romling, U., Gross, U., Doss, M.O., Arai, H., Igarashi, Y., and Jahn, D. 1998. Regulation of *Pseudomonas aeruginosa hemF* and *hemN* by the dual action of the redox response regulators Anr and Dnr. Mol. Microbiol. 29: 985–997.

Schroder, I., Johansson, P., Rutberg, L., and Hederstedt, L. 1994. The hemX gene of the Bacillus subtilis hemAXCDBL operon encodes a membraneprotein, negatively affecting the steady-state cellular concentration of HemA(glutamyl-tRNA reductase). Microbiology 140: 731–740.

Seehra, J.S., Jordan, P.M., and Akhtar, M. 1982. Anaerobic and aerobic coproporphyrinogen III oxidases of *Rhodopseudomonas sphaeroides.* Biochem. J . 269: 709–718.

Sun, G., Sharkova, E., Chesnut, R., Birkey, S., Duggan, M.F., Sorokin, A., Pujic, P., Ehrlich, S.D., and Hulett, F.M. 1996. Regulators of aerobic and anaerobic respiration in *Bacillus subtilis.* J. Bacteriol. 178: 1374–1385.

Suzuki, J.Y., Bollivar, D.W., and Bauer, C.E. 1997. Genetic analysis of chlorophyll biosynthesis. Annu. Rev. Genet. 31: 61–89.

Tait, G.H. 1969. Coproporphyrinogenase activity in extracts from *Rhodopseudomonas sphaeroides.* Biochem Biophys Res Commun 37: 116–122.

Tait, G.H. 1972. Coproporphyrinogenase activities in extracts of *Rhodopseudomonas spheroides* and *Chromatium* strain D. Biochem. J. 128: 1159–1169.

Troup, B., Jahn, M., Hungerer, C., and Jahn, D. 1994. Isolation of the *hemF* operon containing the gene for the *Escherichia coli* aerobic coproporphyrinogen III oxidase by in vivo complementation of a *yeast* HEM13 mutant. J. Bacteriol. 176: 673–680.

Troup, B., Hungerer, C., and Jahn, D. 1995. Cloning and characterization of the *Escherichia coli hemN* gene encoding the oxygen-independent coproporphyrinogen III oxidase. J. Bacteriol. 177: 3326–3331.

Verderber, E., Lucast, L.J., Van Dehy, J.A., Cozart, P., Etter, J.B., and Best, E.A. 1997. Role of the *hemA* gene product and delta-aminolevulinic acid in regulation of *Escherichia coli* heme synthesis. J. Bacteriol. 179: 4583–4590.

Verkamp, E., and Chelm, B.K. 1989. Isolation, nucleotide sequence, and preliminary characterization of the *Escherichia coli* K-12 *hemA* gene. J. Bacteriol. 171: 4728–4735.

Verkamp, E., Jahn, M., Jahn, D., Kumar, A.M., and Soll, D. 1992. Glutamyl-tRNA reductase from *Escherichia coli* and *Synechocystis* 6803. Gene structure and expression. J. Biol. Chem. 267: 8275–8280.

Wang, L.Y., Brown, L., Elliott, M., and Elliott, T. 1997. Regulation of heme biosynthesis in *Salmonella typhimurium*: activity of glutamyl-tRNA reductase (HemA) is greatly elevated during heme limitation by a mechanism which increases abundance of the protein. J. Bacteriol. 179: 2907–2914.

Wang, L., Elliott, M., and Elliott, T. 1999a. Conditional stability of the HemA protein (glutamyl-tRNA reductase) regulates heme biosynthesis in *Salmonella typhimurium.* J. Bacteriol. 181: 1211–1219.

Wang, L., Wilson, S., and Elliott, T. 1999b. A mutant HemA protein with positive charge close to the N terminus is stabilized against heme-regulated proteolysis in *Salmonella typhimurium.* J. Bacteriol. 181: 6033–6041.

Woodard, S.I., and Dailey, H.A. 1995. Regulation of heme biosynthesis in *Escherichia coli.* Arch. Biochem. Biophys. 316: 110–115.

Xu, K., Delling, J., and Elliott, T. 1992. The genes required for heme synthesis in *Salmonella typhimurium* include those encoding alternative functions for aerobic and anaerobic coproporphyrinogen oxidation. J. Bacteriol. 174: 3953–3963.

Xu, K., and Elliott, T. 1993. An oxygen-dependent coproporphyrinogen oxidase encoded by the *hemF* gene of *Salmonella typhimurium.* J. Bacteriol. 175: 4990–4999.

Xu, K., and Elliott, T. 1994. Cloning, DNA sequence, and complementation analysis of the *Salmonella typhimurium hemN* gene encoding a putative oxygen-independent coproporphyrinogen III oxidase. J. Bacteriol. 176: 3196–3203.

Ye, R.W., Tao, W., Bedzyk, L., Young, T., Chen, M., and Li, L. 2000. Global gene expression profiles of *Bacillus subtilis* grown under anaerobic conditions. J. Bacteriol. 182: 4458–4465.

Zagorec, M., Buhler, J.M., Treich, I., Keng, T., Guarente, L., and Labbe-Bois, R. 1988. Isolation, sequence, and regulation by oxygen of the *yeast HEM13* gene coding for coproporphyrinogen oxidase. J. Biol. Chem. 263: 9718–9724.

Zeilstra-Ryalls, J.H., and Kaplan, S. 1995. Anaerobic and aerobic regulation in *Rhodobacter sphaeroides* 2.4.1: the role of the *fnrL* gene. J. Bacteriol. 177: 6422–6431.

Chapter 20

Multiple Levels of Regulation of Solventogenesis in *Clostridium acetobutylicum*

Brigitte Zickner, Michael Böhringer, Stephan Nakotte, Steffen Schaffer, Kai Thormann, and Peter Dürre

Abstract

Solvent synthesis in *Clostridium acetobutylicum* is induced in concert with sporulation to counteract the dangerous effects of produced butyric and acetic acids and to provide the cell with sufficient time to complete endospore formation. Cardinal transcription units for butanol and acetone production are the *sol* and *adc* operons encoding butyraldehyde/butanol dehydrogenase and coenzyme A transferase as well as acetoacetate decarboxylase. Induction is achieved by a decreased level of DNA supercoiling and the transcription factor Spo0A, possibly in cooperation with other regulatory proteins. A number of other operons is also turned on during this metabolic switch, whose physiological relevance, however, is only partly understood. The recent completion of *C. acetobutylicum* genome sequencing will pave the way for transcriptional profiling and thus allow comprehension of the coherent regulatory networks of solventogenesis and sporulation.

Introduction

Regulation of acetone and butanol formation in *Clostridium acetobutylicum* is embedded in a complex network, including sporulation, pH, and redox control. This obligate anaerobic bacterium degrades sugars or starch via a typical butyrate fermentation. Since high substrate concentrations are tolerated, large amounts of acetic and mainly butyric acid are formed that endanger survival by dissipation of the proton gradient over the cytoplasmic membrane. As a consequence, the organism ceases to make acids and switches to the production of neutral solvents for which even part of the excreted acids is taken up again and converted into butanol and acetone in a ratio of app. 2 to 1 (Figure 1). Concomitantly, the formation of endospores is started which eventually will guarantee survival, since butanol at increasing concentrations of up to 2% also exerts a toxic effect. Solventogenesis thus only serves as a means to provide the cell with enough time to complete endospore production (for reviews see Jones and Woods, 1986; Dürre and Bahl, 1996; Dürre, 1998).

Biological butanol synthesis served as a major feedstock of this solvent for industrial purposes until about 1950. Then, increasing substrate prices and cheap raw oil availability gave the petrochemical industry a massive economic advantage which led to the decline of the fermentation process. The oil crisis in the 1970s provoked a renewed interest in the underlying biological processes in order to eventually establish a competitive fermentation procedure. However, only at the end of the 1980s, the first molecular biological investigations on *C. acetobutylicum* were reported (for a review see Dürre, 1998). Since then, most of the genes required for solventogenesis have been cloned and sequenced. These data were the starting point for elucidation of signals and regulatory mechanisms, which is pursued by *in vitro* and *in vivo* studies, RNA analyses, reporter gene constructions, and gene knock outs. It is the aim of this mini review to present the current status of the picture thus emerging.

Cardinal Operons for Solvent Formation

Essential genes required for solventogenesis are arranged in the *sol* and *adc* operons. They were found to be located on an app. 210-kbp megaplasmid (Cornillot and Soucaille, 1996; Cornillot *et al.*, 1997). Meanwhile, sequencing revealed this plasmid to have a size of 192 kbp (see last paragraph). Both operons represent a contiguous stretch of DNA, but with convergent direction of transcription (Figure 2).

1. *sol* Operon

This transcription unit comprises genes encoding a small peptide (OrfL), a bifunctional butyraldehyde/butanol dehydrogenase (AdhE), and the two subunits of acetoacetyl-CoA: acetate/butyrate-coenzyme A transferase (CtfA/B) (Fischer *et al.*, 1993). Northern experiments revealed that this operon is one of the first being induced at the onset of solventogenesis (Sauer and Dürre, 1995). The essential role of the *adhE* gene (or *aad*, as it was designated by an American group) in butanol synthesis was also confirmed by successful complementation of a solvent-negative mutant (Nair and Papoutsakis, 1994) and the significant decrease of butanol production upon its insertional inactivation (Green and Bennett, 1996). In contrast to the CoA transferase (Wiesenborn *et al.*, 1989), the AdhE enzyme has not been characterized. Using the his-tag technology, the protein could be purified, but did no longer show enzymatic activity. Two translation products are obtained from the *adhE* gene in both *C. acetobutylicum* and *Escherichia coli:* the bifunctional enzyme and separately the butanol dehydrogenase

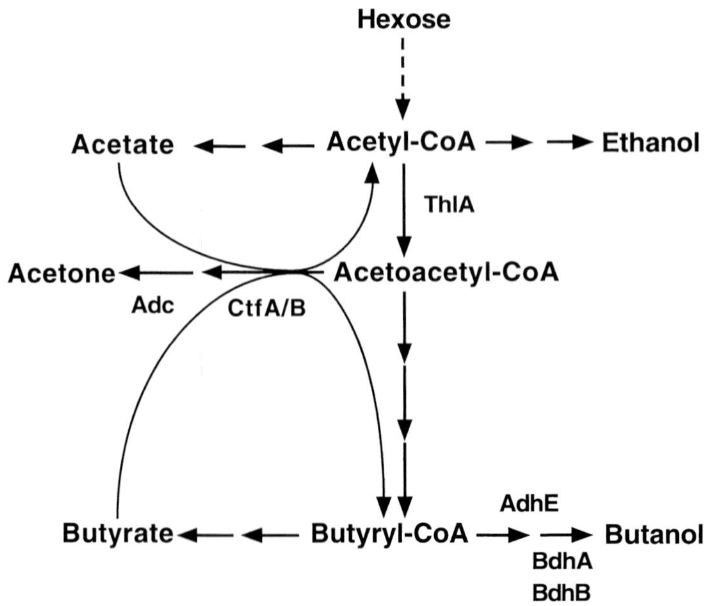

Figure 1. Catabolic pathways of acid and solvent formation in *Clostridium acetobutylicum.* Gene products discussed in the text are listed above, aside, and below the arrows indicating the catalyzed reaction(s). Adc, acetoacetate decarboxylase; AdhE, bifunctional butyraldehyde/butanol dehydrogenase; BdhA, butanol dehydrogenase; BdhB, butanol dehydrogenase; CtfA/B, acetoacetyl-CoA: acetate/butyrate-coenzyme A transferase; ThlA, thiolase.

domain (Thormann *et al.*, 2002). The physiological relevance of this phenomenon is still unknown. Based on primer extension experiments, the *sol* operon was proposed to be controlled by two promoters, the distal P_1 (or S_2) and the proximal P_2 (or S_1) (Fischer *et al.*, 1993; Nair *et al.*, 1994). Signal intensities indicated that most transcripts originated from P_2 (Fischer *et al.*, 1993). However, this putative promoter structure showed 5 mismatches to the consensus in the −10 and −35 regions as well as an unusual spacing of these two boxes (Fischer *et al.*, 1993; Nair *et al.*, 1994; Dürre *et al.*, 1995). For further investigations, a reporter gene technology has been developed (Dürre *et al.*, 1995; Tummala *et al.*, 1999) that is based on the *lacZ* gene from *Thermoanaerobacterium thermosulfurigenes* (Burchhardt and Bahl, 1991). Recent experiments revealed that indeed only one promoter exists, which is P_1. The second primer extension signal (P_2) is not due to a promoter structure, but rather represents an mRNA processing site (Thormann *et al.*, 2002).

DNA topology was found to play an important role in induction of the *sol* operon. Relaxation (i.e. less negative supercoiling), induced by inhibition of the DNA gyrase, caused an immediate increase in transcription (Ullmann *et al.*, 1996). This is in perfect accordance with the finding that DNA from *C. acetobutylicum* becomes less negatively supercoiled at the onset of solventogenesis (Wong and Bennett, 1996). Since the environmental conditions at this time point (still excess of substrate, pH below 4.5, high concentrations of acetate and butyrate) are all known to affect DNA topology directly, it is assumed that the change in DNA supercoiling serves as a signal to induce solventogenesis (Dürre *et al.*, 1995; Dürre, 1998; Ullmann and Dürre, 1998).

Recently, it has been proposed that the *sol* operon is regulated by a transcriptional repressor (Orf5), which is encoded by the monocistronically organized gene directly upstream of *adhE* (Figure 2) (Nair *et al.*, 1999). The conclusion of the authors was based on experiments indicating that i) insertional inactivation of the *orf5* gene (then designated *solR*) yielded strains with significantly increased solvent production, ii) overexpression of this gene yielded strains with a solvent-negative phenotype, and iii) a potential helix-turn-helix (HTH) DNA-binding motif is present in the *orf5* gene product (Nair *et al.*, 1999). However, this suggestion was disproved by data showing that Orf5 i) is localized at the extracellular side of the cytoplasmic membrane, ii) is involved in glycosylation/deglycosylation reactions, and iii) contains a tetratrico peptide repeat motif for protein-protein interactions rather than a HTH-motif (Thormann and Dürre, 2001). The dominant glycosylated protein detected in the course of these experiments represents most probably flagellin, which recently was found to be post-translationally modified by a terminal sialyl residue (Lyristis *et al.*, 2000). Most strikingly, the claim that overexpression of *orf5* leads to a solvent-negative phenotype could not be reproduced. Respective constructions even produced 15% more butanol than the wild type (Thormann and Dürre, 2001). Further investigations revealed that the results obtained by Nair *et al.,* stemmed from erroneous subcloning of a part of the regulatory region of the *sol* operon together with the *orf5* gene. This fragment, when subcloned, yielded a solvent-negative phenotype (thus explaining the result obtained by Nair *et al.*) and was found to be the control region of the P_1 promoter of the *sol* operon (Thormann *et al.*, 2002). It contained an imperfect three

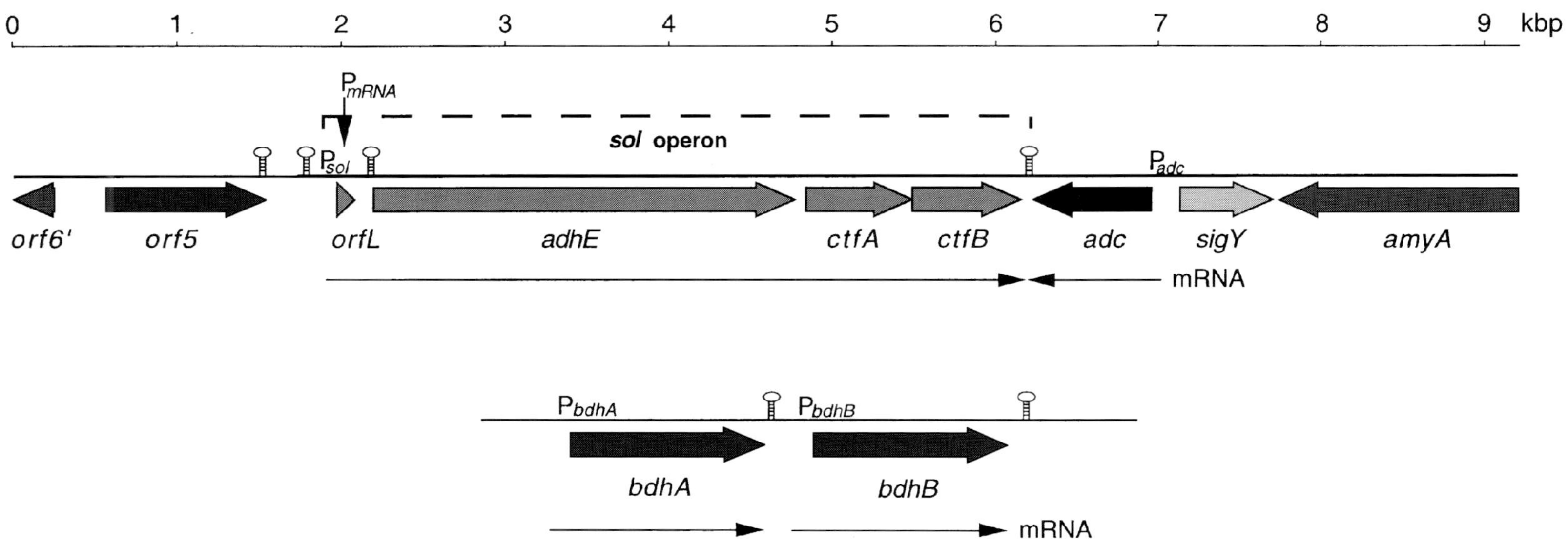

Figure 2. Schematic map of *C. acetobutylicum* DNA regions, encoding essential enzymes for solventogenesis. Arrows and arrowheads represent lengths, locations, and orientation of genes; lines with arrows indicate primary mRNA transcripts. A prime at the end of a gene indicates truncation of the respective open reading frame. Promoter positions of operons directly involved in solventogenesis are marked by P_{sol}, P_{adc}, P_{bdhA}, P_{bdhB}. P_{mRNA} indicates a putative mRNA processing site; possible stem-loop structures are indicated by hairpin symbols. Functionally characterized proteins from indicated genes: *adhE*, bifunctional butyraldehyde/butanol dehydrogenase; *ctfA/B*, acetoacetyl-CoA: acetate/butyrate-coenzyme A transferase; *adc*, acetoacetate decarboxylase; *sigY*, alternative sigma factor; *amyA*, α-amylase; *bdhA*, butanol dehydrogenase A; *bdhB*, butanol dehydrogenase B.

times repeat and a potential binding site for the transcription factor Spo0A. Targeted mutagenesis of this box revealed that Spo0A is indeed involved in the control of the *sol* operon (Thormann *et al.*, 2002). This is in perfect agreement with a recent report that the other cardinal operon of solventogenesis (*adc*) is also regulated by Spo0A (Ravagnani *et al.*, 2000). Spo0A is not only a key regulator of sporulation, but directly or indirectly controls competence, motility, and production of peptide antibiotics and extracellular hydrolases in *Bacillus* (Strauch and Hoch, 1993). The findings that in *C. acetobutylicum* this transcription factor is involved in regulation of *sol*, *adc*, and *ptb-buk* operons (the latter encoding enzymes for butyrate formation) (Ravagnani *et al.*, 2000; Thormann *et al.*, 2002) provides the long sought-for link between solventogenesis and sporulation. It must be added, however, that the reported data do not rule out the participation of additional proteins in the transcriptional regulation.

The genome sequencing project (see last paragraph) revealed that a second *adhE* gene is present on the megaplasmid. It is monocistronically organized and does not play a role in the usual acetone/butanol formation. Induction takes place under conditions that lead to a so-called 'alcohologenic' fermentation, when only butanol and ethanol, but no acetone, are formed (Girbal and Soucaille, 1998). This happens during simultaneous feeding of glucose and glycerol (Vasconcelos *et al.*, 1994). Artificial induction can be achieved by adding methyl viologen to an actively growing culture (Rao and Mutharasan, 1986, 1987). RNA analyses had shown that none of the at that time known genes of solventogenic enzymes was involved in this type of fermentation. A weak hybridization obtained with an *adhE* probe had led to the suggestion that a homologous gene might be present (Dürre *et al.*, 1995). This prediction has now been substantiated by the sequencing project and analysis of the *adhE2* gene (Fontaine *et al.*, 2002). The monocistronic operon encodes another butyraldehyde/butanol dehydrogenase. As in case of the *sol* operon, two transcription start points were detected upstream of *adhE2*. A putative FNR-binding motif is located 34 bp upstream from the start codon (Fontaine *et al.*, 2002), whereas no obvious Spo0A-binding site could be found.

2. *adc* Operon

The *adc* operon encoding acetoacetate decarboxylase, which catalyzes the last step of acetone formation, was the first cloned gene involved in solventogenesis (Gerischer and Dürre, 1990; Petersen and Bennett, 1990). It is located on the megaplasmid and shares the transcription terminator (which is working in both directions) with the *sol* operon (Figure 2). These two operons are the first to be induced at the onset of solventogenesis (Sauer and Dürre, 1995). The *adc* gene is controlled by a single promoter (Gerischer and Dürre, 1992), which recently was shown to be induced by Spo0A (Ravagnani *et al.*, 2000). However, destruction of the Spo0A binding sites did lower, but not completely abolish the promoter activity (determined by the *gusA* reporter gene system) (Ravagnani *et al.*, 2000), which might indicate the involvement of additional transcription factors. Two-dimensional protein gels loaded with extracts from acidogenic and solventogenic cells, respectively, and subsequent N-terminal amino acid sequencing revealed two different spots for acetoacetate decarboxylase (Schaffer *et al.*, 2002). Since according to the genome sequence only one *adc* gene is present, this finding indicates a post-translational modification of Adc. Both, nature and physiological relevance of this modification are still unknown.

As for the *sol* operon, relaxation of DNA was found to have an inducing effect on the *adc* gene (Ullmann *et al.*, 1996). Adjacent to *adc*, but with divergent direction of transcription, is the gene for an alternative sigma factor (*sigY*) located, whose target operons are so far unknown (Behrens *et al.*, 2000). Interestingly, at the late solventogenic phase a protein was induced that was found to bind to the *sigY* promoter region by gel retardation studies (Nakotte, 1998). It is thus tempting to speculate that globally changed DNA supercoiling in combination with Spo0A (and possibly other transcription factors) leads to the induction of *adc*, which, however, is only transient (Sauer and Dürre, 1995). Termination could then be achieved by starting transcription at the *sigY* promoter leading to a deactivation of the adjacent *adc* promoter by local reverse DNA topology changes according to the 'twin transcriptional-loop model' (Liu and Wang, 1987). According to this model, transcription from a divergent promoter (in this case P_{sigY}) causes an increased negative supercoiling in an area of app. 250 bp behind the RNA polymerase (i.e. at the *adc* promoter). This would exactly match the regulatory pattern of the *adc* promoter, i.e. inactive at a high degree of negative supercoiling, induction after DNA relaxation due to global DNA topology changes by altered environmental conditions, shut-down late in the solventogenic phase due to a locally increased degree of negative supercoiling. Future experiments will have to prove or disprove this hypothesis.

At first glance, it might appear strange that the genes encoding the acetone-forming enzymes (Ctf and Adc) are arranged in different operons. However, at the onset of solventogenesis a high concentration of acids is converted by the CoA transferase to mainly butyryl-CoA, which is in turn transformed into butanol by AdhE. The respective genes are arranged in the *sol* operon. Adc is only needed to remove the end product of the CoA transferase reaction (acetoacetate), since this conversion is thermodynamically less favorable. Thus, Adc is not required if, for example, plenty of reducing equivalents are available (Dürre *et al.*, 1995). Indeed, depending on the substrates used, the ratio of butanol to acetone was found to increase up to 50fold (Bahl *et al.*, 1986).

Other Genes Related to Solventogenesis

1. *bdhA* and *bdhB* Operons

These two operons are consecutively located on the chromosome and are arranged in monocistronic transcription units each (Walter *et al.*, 1992). RNA analyses indicated that *bdhB* plays a major role in butanol formation and is induced later than the *adc* and *sol* operons. On the other hand, the function of the *bdhA* gene product seems to be a sink for excess reducing equivalents by low level butanol formation (Sauer and Dürre, 1995; Dürre *et al.*, 1995). Upstream of both promoters, one and two Spo0A-binding sites, respectively, can be deduced from the sequence, suggesting an involvement of this transcription factor in *bdhB* and *bdhA* induction as well (Ravagnani *et al.*, 2000).

2. *thlA* Operon

Two thiolase-encoding genes were identified in *C. acetobutylicum* (Stim-Herndon *et al.*, 1995; Winzer *et al.*, 2000). *thlB* forms an operon together with *thlR* (a putative regulator gene) and *thlC* (no significant homology to known genes or their deduced products). The physiological function of ThlB has yet to be elucidated (Winzer *et al.*, 2000). *thlA* is monocistronically organized and codes for the essential metabolic enzyme involved in both, acido- and solventogenesis. Interestingly, the gene is actively transcribed during the exponential, acidogenic growth, repressed in the transition phase, and induced again together with the *sol* and *adc* operons at the onset of solventogenesis (Winzer *et al.*, 2000). No other known gene of *C. acetobutylicum* shows such a regulatory pattern. Since no putative Spo0A-binding boxes could be found upstream of the *thlA* promoter, its control mechanism still needs to be resolved.

3. *ser* Operon

In the proteome analysis approach mentioned above, several proteins were significantly induced at the onset of solventogenesis (among them AdcI and AdcII). N-terminal amino acid sequencing revealed that three spots represented enzymes with high homology to serine aminotransferase (SerC), 3-phosphoglycerate dehydrogenase (SerA), and a seryl-tRNA synthetase (SerS) (Schaffer *et al.*, 2002). The respective genes form a common operon together with a gene encoding a protein of unknown function (*serX*) in the sequence *serC-serA-serX-serS*, whose transcription is induced concomitantly with the *sol* and *adc* operons (Schaffer *et al.*, 2002). A promoter structure could be deduced from primer extension experiments (Schaffer *et al.*, 2002), but no Spo0A boxes were found in its vicinity. The gene products catalyze the biosynthesis of serine and its incorporation into proteins (via the tRNA synthetase). The only enzyme missing is the 3-phosphoserine phosphatase. SerX does not show homology to such an enzyme, but might catalyze such an activity. The reason, why *C. acetobutylicum* enhances serine biosynthesis and incorporates this amino acid into proteins at the onset of solventogenesis has yet to be elucidated. There are no known proteins with an unusually high serine content that play a role in solventogenesis or sporulation.

4. *gap* Operon

The proteome analysis also showed that the protein level of glyceraldehyde-3-phosphate dehydrogenase (Gap) was app. twofold higher during solventogenesis (Schaffer *et al.*, 2002). Cloning and sequencing of the respective gene region revealed a common operon consisting of *gap* – *pgk* (encoding phosphoglycerate kinase) – *tpi* (encoding triosephosphate isomerase). Adjacently located is the *pgm*(i) gene coding for a 2,3-bisphosphoglycerate-independent phosphoglycerate mutase (Schreiber and Dürre, 1999). Transcription analysis indicated promoters upstream of the *gap* and *pgm*(i) operons, as well as an internal start point within the *gap* operon upstream of *tpi* (Schreiber and Dürre, 2000). The *gap* promoter proved to be constitutively active (although the resulting transcript was processed in different patterns according to the growth conditions (Schreiber and Dürre, 2000)), which is in accordance with the finding that although the protein level was higher in solventogenic cells, its synthesis rate determined by incorporation of labelled methionine remained unchanged (Schaffer *et al.*, 2002). On the other hand, the *tpi* promoter showed an enhanced activity during solventogenesis if compared to the *gap* control unit (Schreiber and Dürre, 2000). The physiological relevance of this phenomenon is not yet understood.

Perspective and Future Directions

Recombinant DNA methods are meanwhile well developed for *C. acetobutylicum* (e.g., Nakotte *et al.*, 1998; Young *et al.*, 1999). Thus, use of reporter genes and knock-out mutations has contributed enormously to our understanding of the regulatory mechanisms in this bacterium (and will continue to do so at an even increasing rate). Most importantly, at the end of the 1990s, the type strain has been selected for a genome sequencing project, which was performed by Genome Therapeutics Corp. (USA) as a component of the US Department of Energy (DOE) Microbial Genome Project. Sequencing is meanwhile complete and the data are available in the internet (http://www.ncbi.nlm.nih.gov and http://www.genomecorp.com/programs/sequence_data_clost.shtml) (Nölling *et al.*, 2001). This will enable a functional analysis by transcriptional profiling, supported by the proteome approach already mentioned, and renders other technologies such as e.g., differential RNA display obsolete. The primary goal will be comprehension of the coherent regulatory networks of solventogenesis and sporulation, thus forming the basis for new industrial and medical applications in solvent formation and cancer therapy by recombinant clostridial spores (Bahl and Dürre, 2001).

Acknowledgements

This project was supported by grants from the Deutsche Forschungsgemeinschaft within the priority program "Molekulare Analyse von Regulationsnetzwerken in Bakterien".

References

Bahl, H., Gottwald, M., Kuhn, A., Rale, V., Andersch, W., and Gottschalk, G. 1986. Nutritional factors affecting the ratio of solvents produced by *Clostridium acetobutylicum*. Appl. Environ. Microbiol. 52: 169–172.

Bahl, H., and Dürre, P. (eds.), 2001. Clostridia: Biotechnology and Medical Applications. Wiley-VCH, Weinheim, Germany.

Behrens, S., Meyer, U., Schankin, H., Lonetto, M., Fischer, R.-J., and Bahl, H. 2000. Identification of two genes encoding putative new members of the ECF subfamily of eubacterial RNA polymerase sigma factors in *Clostridium acetobutylicum*. J. Mol. Microbiol. Biotechnol. 2: 265–269.

Burchhardt, G., and Bahl, H. 1991. Cloning and analysis of the β-galactosidase-encoding gene from *Clostridium thermosulfurogenes* EM1. Gene 106: 13–19.

Cornillot, E., and Soucaille, P. 1996. Solvent-forming genes in clostridia. Nature 380: 489.

Cornillot, E., Nair, R., Papoutsakis, E.T., and Soucaille, P. 1997. The genes for butanol and acetone formation in *Clostridium acetobutylicum* ATCC 824 reside on a large plasmid whose loss leads to degeneration of the strain. J. Bacteriol. 179: 5442–5447.

Dürre, P., Fischer, R.-J., Kuhn, A., Lorenz, K., Schreiber, W., Stürzenhofecker, B., Ullmann. S., Winzer, K., and Sauer, U. 1995. Solventogenic enzymes of *Clostridium acetobutylicum*: catalytic properties, genetic organization, and transcriptional regulation. FEMS Microbiol. Rev. 17: 251–262.

Dürre, P., and Bahl, H. 1996. Microbial production of acetone/butanol/isopropanol. In: Biotechnology, vol. 6, 2nd ed. M. Roehr, (ed.), VCH Verlagsgesellschaft mbH, Weinheim, Germany. p. 229–268.

Dürre, P. 1998. New insights and novel developments in clostridial acetone/butanol/isopropanol fermentation. Appl. Microbiol. Biotechnol. 49: 639–648.

Fischer, R.J., Helms, J., and Dürre, P. 1993. Cloning, sequencing, and molecular analysis of the *sol* operon of *Clostridium acetobutylicum*, a chromosomal locus involved in solventogenesis. J. Bacteriol. 175: 6959–6969.

Fontaine, L., Meynial-Salles, I., Girbal, L., Yang, X., Croux, C., and Soucaille, P. 2002. Molecular characterization and transcriptional analysis of *adhE2*, the gene encoding the NADH-dependent aldehyde/alcohol dehydrogenase responsible for butanol production in alcohologenic cultures of *Clostridium acetobutylicum* ATCC 824, J. Bacteriol. 184: 821–830.

Gerischer, U., and Dürre, P. 1990. Cloning, sequencing, and molecular analysis of the acetoacetate decarboxylase gene region from *Clostridium acetobutylicum*. J. Bacteriol. 172: 6907–6918.

Gerischer, U., and Dürre, P. 1992. mRNA analysis of the *adc* gene region of *Clostridium acetobutylicum* during the shift to solventogenesis. J. Bacteriol. 174: 426–433.

Girbal, L., and Soucaille, P. 1998. Regulation of solvent production in *Clostridium acetobutylicum*. Trends Biotechnol. 16: 11–16.

Green, E.M, and Bennett, G.N. 1996. Inactivation of an aldehyde/alcohol dehydrogenase gene from *Clostridium acetobutylicum* ATCC 824. Appl. Biochem. Biotechnol. 57/58: 213–221.

Jones, D.T., and Woods, D.R. 1986. Acetone-butanol fermentation revisited. Microbiol. Rev. 50: 484–524.

Liu, L.F., and Wang, J.C. 1987. Supercoiling of the DNA template during transcription. Proc. Natl. Acad. Sci. USA 84: 7024–7027.

Lyristis, M., Boynton, Z.L., Petersen, D., Kan, Z., Bennett, G.N., and Rudolph, F.B. 2000. Cloning, sequencing, and characterization of the gene encoding flagellin, *flaC*, and the post-translational modification of flagellin, FlaC, from *Clostridium acetobutylicum* ATCC824. Anaerobe 6: 69–79.

Nair, R.V., Bennett, G.N., and Papoutsakis, E.T. 1994. Molecular characterization of an aldehyde/alcohol dehydrogenase gene from *Clostridium acetobutylicum* ATCC 824. J. Bacteriol. 176: 871–885.

Nair, R.V., and Papoutsakis, E.T. 1994. Expression of plasmid-encoded *aad* in *Clostridium acetobutylicum* M5 restores vigorous butanol production. J. Bacteriol. 176: 5843–5846.

Nair, R.V., Green, E.M., Watson, D.E., Bennett, G.N., and Papoutsakis, E.T. 1999. Regulation of the *sol* locus genes for butanol and acetone formation in *Clostridium acetobutylicum* ATCC 824 by a putative transcriptional repressor. J. Bacteriol. 181: 319–330.

Nakotte, S. 1998. Funktionelle Analyse von Regulationsstrukturen essentieller Gene für die Lösungsmittelbildung in *Clostridium acetobutylicum*. Ph.D. thesis, University of Ulm.

Nakotte, S., Schaffer, S., Böhringer, M., and Dürre, P. 1998. Electroporation of, plasmid isolation from and plasmid conservation in *Clostridium acetobutylicum* DSM 792. Appl. Microbiol. Biotechnol. 50: 564–567.

Nölling, J., Breton, G., Omelchenko, M.V., Makarova, K.S., Zeng, Q., Gibson, R., Lee, H.M., Dubois, J., Qiu, D., Hitti, J., GTC Sequencing Center Production, Finishing, and Bioinformatics Teams, Wolf, Y.I., Tatusov, R.L., Sabathe, F., Doucette-Stamm, L., Soucaille, P., Daly, M.J., Bennett, G.N., Koonin, E.V., and Smith, D.R. 2001. Genome sequence and comparative analysis of the solvent-producing bacterium *Clostridium acetobutylicum*. J. Bacteriol. 183: 4823–4838.

Petersen, D.J., and Bennett, G.N. 1990. Purification of acetoacetate decarboxylase from *Clostridium acetobutylicum* ATCC 824 and cloning of the acetoacetate decarboxylase gene in *Escherichia coli*. Appl. Environ. Microbiol. 56: 3491–3498.

Rao, G., and Mutharasan, R. 1986. Alcohol production by *Clostridium acetobutylicum* induced by methyl viologen. Biotechnol. Lett. 8: 893–896.

Rao, G., and Mutharasan, R. 1987. Altered electron flow in continuous cultures of *Clostridium acetobutylicum* induced by viologen dyes. Appl. Environ. Microbiol. 53: 1232–1235.

Ravagnani, A., Jennert, K.C. B., Steiner, E., Grünberg, R., Jefferies, J.R., Wilkinson, S.R., Young, D.I., Tidswell, E.C., Brown, D.P., Youngman, P., Morris, J.G., and Young, M. 2000. Spo0A directly controls the switch from acid to solvent production in solvent-forming clostridia. Mol. Microbiol. 37: 1172–1185.

Sauer, U., and Dürre, P. 1995. Differential induction of genes related to solvent formation during the shift from acidogenesis to solventogenesis in continuous culture of *Clostridium acetobutylicum*. FEMS Microbiol. Lett. 125: 115–120.

Schaffer, S., Isci, N., Zickner, B., and Dürre, P. 2002. Changes in protein synthesis and identification of proteins specifically induced during solventogenesis in *Clostridium acetobutylicum*. Electrophoresis 23: 110–121.

Schreiber, W., and Dürre, P. 1999. The glyceraldehyde-3-phosphate dehydrogenase of *Clostridium acetobutylicum*: isolation and purification of the enzyme, and sequencing and localization of the *gap* gene within a cluster of other glycolytic genes. Microbiol. UK 145: 1839–1847.

Schreiber, W., and Dürre, P. 2000. Differential expression of genes within the *gap* operon of *Clostridium acetobutylicum*. Anaerobe 6: 291–297.

Stim-Herndon, K.P., Petersen, D.J., and Bennett, G.N. 1995. Characterization of an acetyl-CoA acetyltransferase (thiolase) gene from *Clostridium acetobutylicum* ATCC 824. Gene 154: 81–85.

Thormann, K., and Dürre, P. 2001. Orf5/SolR: a transcriptional repressor of the *sol* operon of *Clostridium acetobutylicum*? J. Ind. Microbiol. Biotechnol., 27: 307–313.

Thormann, K., Feustel, L., Lorenz, K., Nakotte, S., and Dürre, P. 2002. Control of butanol formation in *Clostridium acetobutylicum* by transcriptional activation. J. Bacteriol. 184: 1966–1973.

Tummala, S.B., Welker, N.E., and Papoutsakis, E.T. 1999. Development and characterization of a gene expression reporter system for *Clostridium acetobutylicum* ATCC 824. Appl. Environ. Microbiol. 65: 3793–3799.

Ullmann, S., Kuhn, A., and Dürre, P. 1996. DNA topology and gene expression in *Clostridium acetobutylicum*: implications for the regulation of solventogenesis. Biotechnol. Lett. 18: 1413–1418.

Ullmann, S., and Dürre, P. 1998. Changes in DNA topology are involved in triggering the onset of solventogenesis in *Clostridium acetobutylicum*. Recent Res. Dev. Microbiol. 2: 281–294.

Vasconcelos, I., Girbal, L., and Soucaille, P. 1994. Regulation of carbon and electron flow in *Clostridium acetobutylicum* grown in chemostat culture at neutral pH on mixtures of glucose and glycerol. J. Bacteriol. 176: 1443–1450.

Walter, K.A., Bennett, G.N., and Papoutsakis, E.T. 1992. Molecular characterization of two *Clostridium acetobutylicum* ATCC 824 butanol dehydrogenase isozyme genes. J. Bacteriol. 174: 7149–7158.

Wiesenborn, D.P., Rudolph, F.B., and Papoutsakis, E.T. 1989. Coenzyme A transferase from *Clostridium acetobutylicum* ATCC 824 and its role in the uptake of acids. Appl. Environ. Microbiol. 55: 323–329.

Winzer, K., Lorenz, K., Zickner, B., and Dürre, P. 2000. Differential regulation of two thiolase genes from *Clostridium acetobutylicum* DSM 792. J. Mol. Microbiol. Biotechnol. 2: 531–541.

Wong, J., and Bennett, G.N. 1996. The effect of novobiocin on solvent production by *Clostridium acetobutylicum*. J. Ind. Microbiol. 16: 354–359.

Young, D.I., Evans, V.J., Jefferies, J.R., Jennert, K.C.B., Phillips, Z.E.V., Ravagnani, A., and Young, M. 1999. Genetic methods in clostridia. Methods Microbiol. 29: 191–207.

Chapter 21

Network Regulation of the *Escherichia coli* Maltose System

Anja Schlegel, Alex Böhm, Sung-Jae Lee, Ralf Peist, Katja Decker and Winfried Boos

Abstract

The genes of the *Escherichia coli* maltose regulon are controlled by MalT, the specific transcriptional activator which, together with the inducer maltotriose and ATP, is essential for *mal* gene transcription. Network regulation in this system affects the function of MalT and occurs on two levels. The first concerns the expression of *malT*. It has long been known that *malT* is under catabolite repression and thus under the control of the cAMP/CAP complex. We found that, in addition, the global regulator Mlc is a repressor for *malT* transcription. The repressor activity of Mlc is controlled by the transport status of the glucose-specific enzyme EIICB of the PTS that causes sequestration (and inactivation as a repressor) of Mlc when glucose is transported. The second level of MalT regulation affects its activity. MalT is activated by maltotriose which is not only formed when the cells are growing on any maltodextrin but also, in low amounts, endogenously when the cells grow on non-maltodextrin carbon sources. Thus, cellular metabolism, for instance degradation of galactose or trehalose, can cause *mal* gene induction. It was found that unphosphorylated internal glucose takes part in endogenous maltodextrin biosynthesis and is therefore a key element in endogenous *mal* gene expression. In addition to the maltotriose-dependent activation, MalT can interact with three different enzymes that lead to its inactivation as a transcriptional activator. The first is MalK, the energy transducing ABC subunit of the maltodextrin transport system. Transport controls the interaction of MalK and MalT thus affecting gene expression. The second enzyme is MalY, a pyridoxal phosphate containing enzyme exhibiting cystathionase activity. The crystal structure of MalY was established and mutations in MalY that reduce *mal* gene repression map in a hydrophobic MalT interaction patch on the surface of the enzyme. The last enzyme is a soluble esterase of as yet unknown function. When overproduced, this enzyme specifically reduces *mal* gene expression and affects the activity of MalT in an *in vitro* transcription assay.

Introduction

The *E. coli* maltose regulon (Boos *et al.*, 1998) consists of 10 genes in five operons whose products are four enzymes, five transport proteins and one periplasmic protein, MalM (Gilson *et al.*, 1986), the function of which remains obscure. The enzymes (MalP, MalQ, MalS and MalZ) catalyze the degradation of maltose and maltodextrins to glucose and glucose-1-phosphate. The transport proteins consist of the λ receptor (maltoporin) in the outer membrane (Schirmer *et al.*, 1995) as well as a binding protein-dependent ABC transporter composed of the periplasmic maltose binding protein, MalE, the translocation complex, MalF/G, in the membrane, and the associated ATP-hydrolyzing subunit, MalK. All *mal* genes are controlled by MalT, the specific *mal* gene activator of the system which, together with the inducer maltotriose and ATP, is needed for their expression (Richet *et al.*, 1989). MalT is in an equilibrium of inactive monomers and active multimers, the latter being stabilized by maltotriose (Schreiber *et al.*, 1999). The system is induced by the presence of maltose or any maltodextrin in the medium, but only maltotriose which is always an intermediate of maltodextrin metabolism can activate MalT (Raibaud *et al.*, 1987). Mutants lacking MalT cannot grow on maltose and mutations in *malT* exist that exhibit constitutive *mal* gene expression (*malT*^{c} mutants). Figure 1 is an overview of the genetic organization of the maltose regulon. For reviews on the maltose system, see (Schwartz, 1987) (Boos *et al.*, 1998).

Several curious observations concerning *mal* gene expression have been noticed in the past but have not obtained sufficient attention at the time. They were the start point for the studies on network regulation of the maltose system outlined in this review:

1. Sugars other than maltodextrins are able to induce the maltose system. For instance galactose or lactose are inducers in a mutant lacking phosphoglucomutase (Pgm) (Adhya *et al.*, 1971). Trehalose metabolism induces the system even though trehalose is neither a substrate of the transport system nor an effector of MalT (Klein *et al.*, 1993). This indicated that the inducer maltotriose can be formed endogenously in the absence of maltodextrins in the medium. In fact, it turned out that there are at least two ways by which maltotriose can be synthesized endogenously; one was by degradation of glycogen and the other seemed to involve

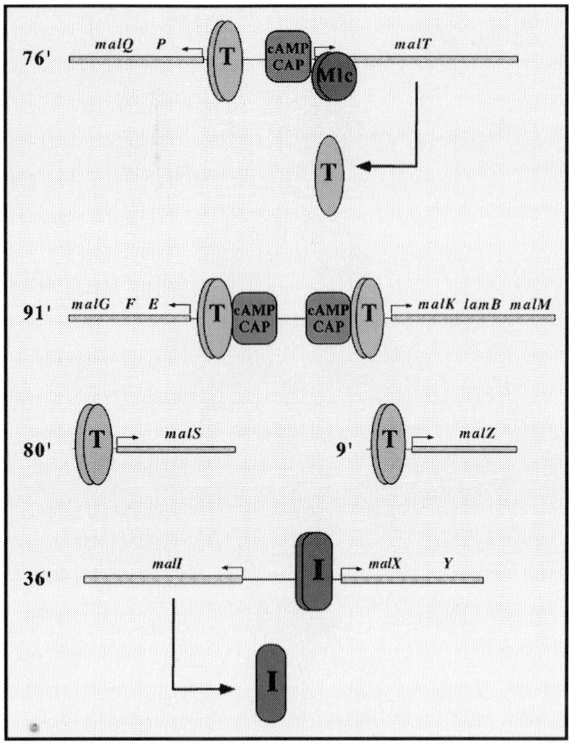

Figure 1. Genetic organization of the *mal* genes and the *malI/malXY* gene cluster. Boxes depict genes or operons; their location in min on the *E. coli* map is given on the left. Transcriptional start points and the direction of transcription are indicated by small arrows. Transcription enhancing proteins are in green, repressors in red. One-letter abbreviations stand for the corresponding *mal* genes or proteins.

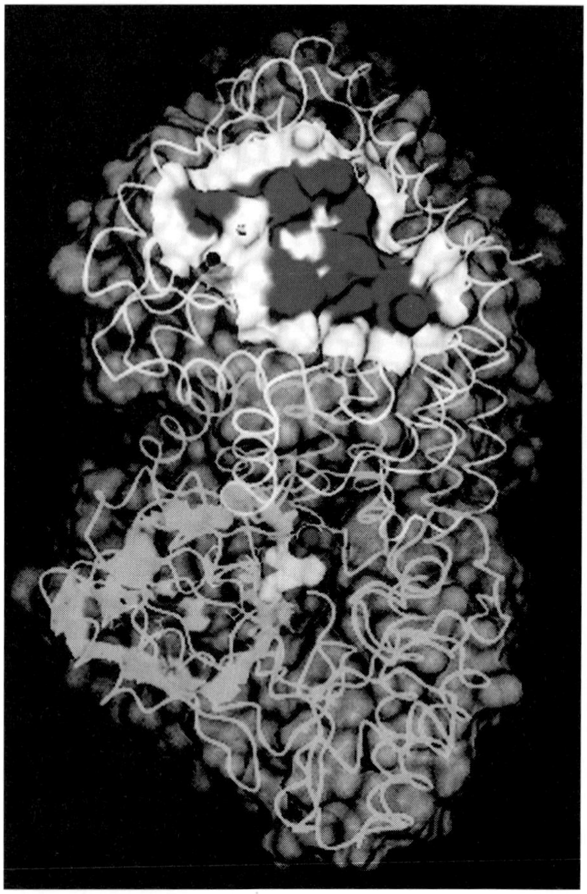

Figure 2. Representation of the crystal structure of the MalY dimer. The carbon backbone traces of the monomers, shaded green and white, are overlaid with a transparent surface. Highlighted as a white solid surface is the MalT-interaction patch that has been defined by repression negative mutations (red). The pyridoxal-5′-phosphate cofactor is depicted in the van der Waals representation. Note that the MalT binding face and the cystathionase active site are on opposite sides of the individual monomers.

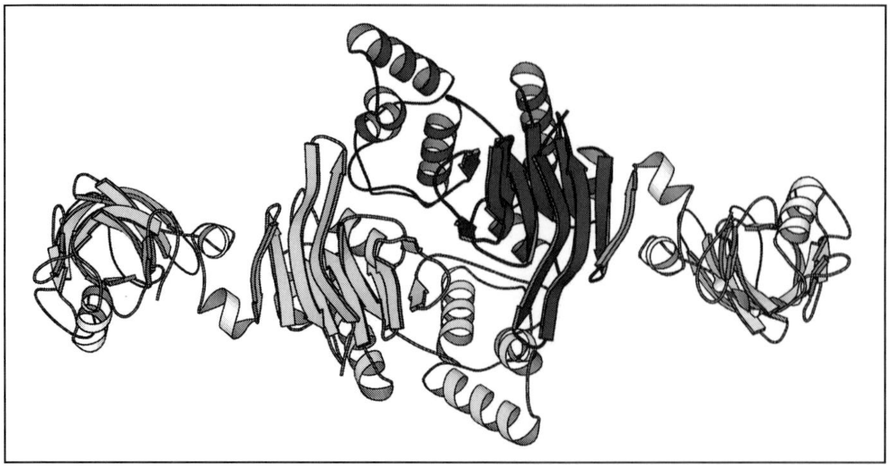

Figure 4. The crystal structure of the MalK dimer, the ABC subunit of the trehalose/maltose transporter of *Thermococcus litoralis* as a ribbon representation. The ABC domains of the individual monomers are shown in blue and yellow. The distinct regulatory barrel-like domains are shown in grey. We propose that *E. coli* MalK harbours a similar regulatory domain carrying the MalT interaction site. For details on the crystal structure see (Diederichs *et al.*, 2000).

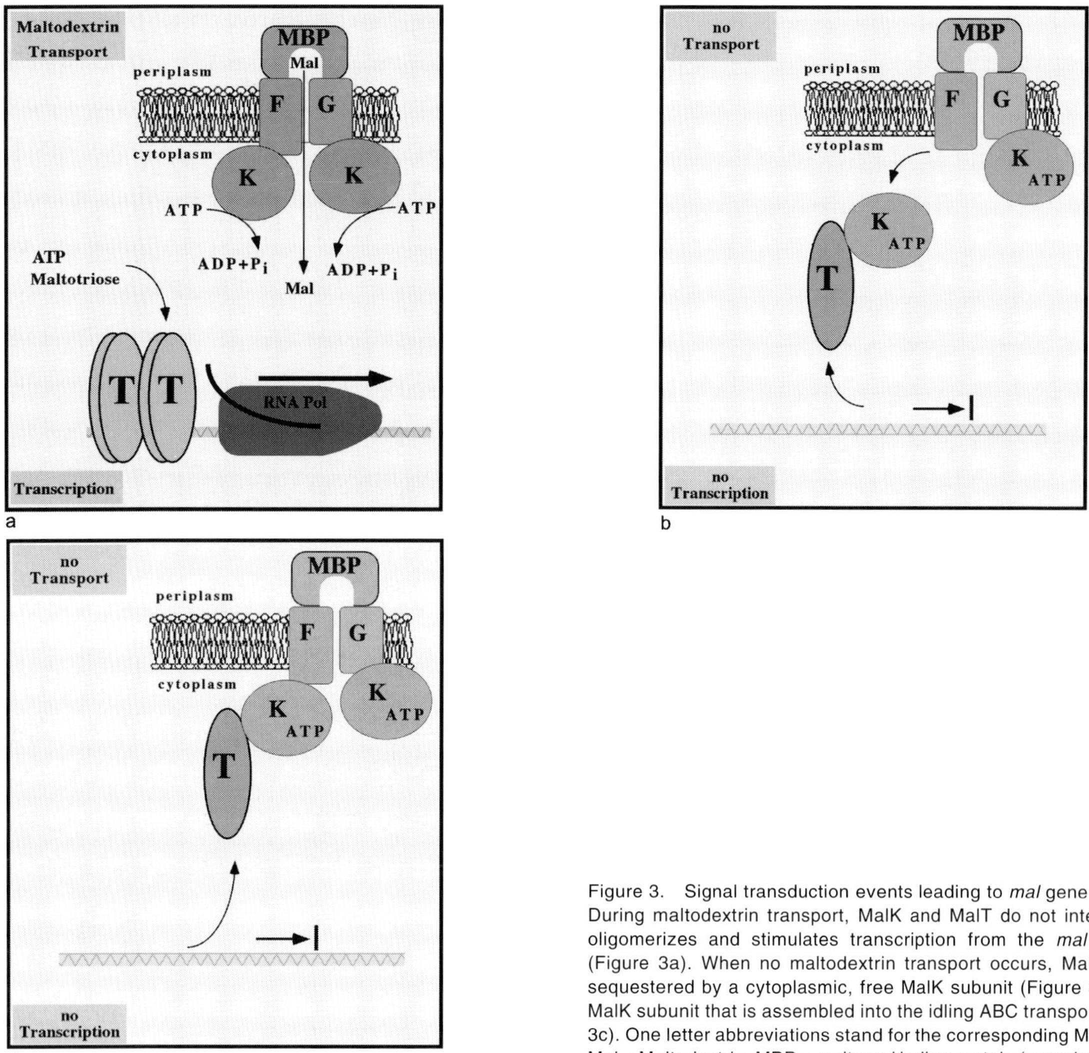

Figure 3. Signal transduction events leading to *mal* gene regulation. During maltodextrin transport, MalK and MalT do not interact; MalT oligomerizes and stimulates transcription from the *mal* promoters (Figure 3a). When no maltodextrin transport occurs, MalT is either sequestered by a cytoplasmic, free MalK subunit (Figure 3b) or by a MalK subunit that is assembled into the idling ABC transporter (Figure 3c). One letter abbreviations stand for the corresponding Mal proteins. Mal = Maltodextrin. MBP = *m*altose *b*inding *p*rotein (encoded by *malE*).

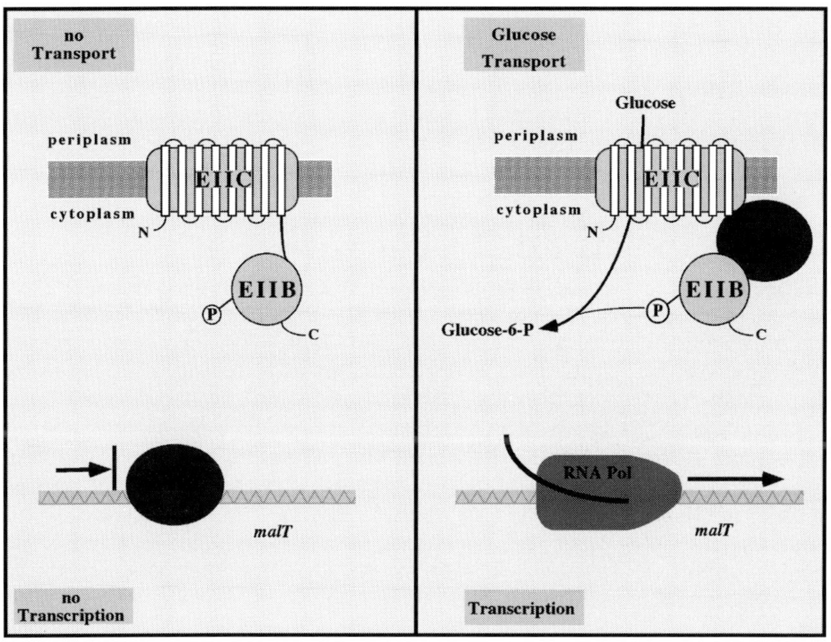

Figure 5. Glucose transport via the EIICB of the PTS leads to derepression of *malT*. When no glucose is transported by EIICB, Mlc binds to the *malT* promoter and represses transcription (left). During glucose transport, Mlc is sequestered from the *malT* promoter by unphosphorylated EIICB and transcription can occur (right). Phosphate groups are depicted as circled "P".

unphosphorylated internal glucose as well as glucose-1-phosphate (Decker *et al.*, 1993). Thus, gluconeogenesis had to play a role in the endogenous induction of the maltose system.

2. It had been noticed early on that some mutants defective in maltose transport showed elevated expression of *malQ* and *malP* genes (Hofnung *et al.*, 1974). Later on, it became clear that it was MalK that was associated with this phenomenon. *malK* mutants cause constitutive expression (Bukau *et al.*, 1986) whereas the overproduction of MalK caused repression of the system. In fact, strains harboring *malK* expression plasmids can no longer grow on maltose (Reyes *et al.*, 1988). The C-terminus of MalK was found to be responsible for repression (Kühnau *et al.*, 1991). The mechanism by which MalK acted as a repressor was unclear.
3. In an attempt to isolate mutants in the still unknown enzyme(s) involved in the biosynthesis of the internal inducer we had isolated the *malI* (for *mal*tose *i*nductor) mutation that abolished the constitutivity of *malK* mutants (Ehrmann *et al.*, 1987). The sequencing of *malI* revealed that it did not encode an enzyme in maltotriose synthesis but an as yet unknown repressor (Reidl *et al.*, 1989). Subsequently, it became clear that MalI was the repressor of an adjacent operon consisting of *malX* and *malY*. *malX* encodes an enzyme IIBC of the phosphotransferase system (PTS), *malY* a pyridoxal phosphate-containing enzyme. It was the derepressed synthesis of MalY in the *malI* mutant that caused the repression of the *mal* genes (Reidl *et al.*, 1991). How this enzyme interfered with *mal* gene expression was unclear.

Conclusions

The Role of Glucokinase

We had suspected that one pathway of endogenous maltotriose biosynthesis involved unphosphorylated internal glucose. Evidence came from a triple mutant that was unable to phosphorylate glucose due to mutations in *ptsG*, *ptsM* (encoding EII's of the PTS specific for glucose) and a mutation in *glk* (encoding glucokinase). Glucose can enter this strain only via the galactose permease (without phosphorylation) and the addition of glucose to the growth medium causes induction of the maltose system (Decker *et al.*, 1993). The glucose-dependent induction in this strain can be abolished by overexpression of plasmid encoded glucokinase. Overexpression of glucokinase also abolishes the constitutivity of *malK* mutants by removing internal free glucose via phosphorylation. Overexpression of glucokinase has no effect on the constitutivity of mutants in *malQ*, the gene coding for amylomaltase, which is due to the formation of maltotriose from glycogen where glucose is not an intermediate. Thus, these experiments strongly support the idea that *E. coli* does contain free internal glucose and that this internal glucose plays a decisive role in the glycogen-independent synthesis of endogenous maltotriose (Meyer *et al.*, 1997). At present it is not clear which reaction leads to glucose-dependent maltotriose synthesis. One possibility, supported by the phenotype of *pgm* mutants, was that the critical enzyme is a maltose/maltotriose phosphorylase that would produce maltose from glucose and glucose-1-phosphate and maltotriose from maltose and a second glucose-1-phosphate. Yet, no maltose phosphorylase with the expected specificity could be found to date in cellular extracts. Another possibility, suggested by the kinetics of maltotriose formation (a very early reaction product) from ^{14}C glucose is that an unknown maltodextrin precursor, possibly protein bound, would transfer maltosyl residues onto the non reducing end of ^{14}C glucose. The origin of free unphosphorylated glucose in the cell is still an enigma.

The MalY-MalT Intermezzo

The overproduction of MalY in *malI* mutants (or when expressed from plasmids) abolishes *mal* gene expression (Reidl, 1992). The identification of MalY as a βC-S lyase (cystathionase) did not give any clue about its function in *mal* gene regulation. The enzyme when induced can complement a *metC* mutation that is normally auxotrophic for methionine since *metC* encodes a cystathionase involved in the biosynthesis of methionine. Exchanging the lysine residue that is critical for pyridoxal phosphate binding yielded a mutant MalY protein that had lost cystathionase activity but was still functioning as a *mal* gene repressor. On the other hand, mutations in MalY were isolated that exhibit reduced *mal* gene repression but are still fully active in cystathionase activity. These regulation minus mutations cluster on three different locations along the polypeptide chain. Taken together, these studies showed that enzymatic activity and repressor function are separable entities. How MalY represses *mal* gene expression became clear when it was demonstrated that purified MalY binds to monomeric MalT and inactivates it in an *in vitro* transcription assay using purified proteins. The interaction of MalY and MalT as well as the strength of inhibition is counteracted by the addition of maltotriose (Schreiber *et al.*, 2000). Thus, the overproduction of MalY leads to sequestration and a reversible inactivation of MalT. MalY was crystallized and its three-dimensional structure determined (Figure 2). It is a homodimer; each monomer is composed of two clearly separable domains. The regulation minus mutations in MalY cluster in a hydrophobic patch on the surface of the protein that is surrounded by a circle of hydrophilic residues. This patch must represent the site where MalY interacts with MalT (Clausen *et al.*, 2000). Conversely, it is the N-terminal portion of MalT (AA 1-250) that is the minimal target for MalY (A. Schlegel, unpublished observations). At present, it is unknown what the natural substrate of MalY is. The clustering of *malY* with *malX* in an operon and the common regulation by MalI (whose inducer is also unknown) indicates that an unknown sugar, possibly a cysteinyl derivative, might be the substrate of these gene products. One could imagine that the carbon flux through this system exerts a specific repression on the maltose system. A *malI/malX/malY* region with high sequence identity (67%) exists in *Vibrio furnissii* (Bouma *et al.*, 1996). In this organism, the MalX homologue encodes the only glucose transporter.

The Esterase Connection

In search for the gene encoding a maltose phosphorylase which we expected to be involved in the synthesis of endogenous maltotriose, we screened a plasmid bank of *E. coli* genes (under their own promoter) to confer a Mal^+ phenotype to a *malQ* mutant. A *malQ* mutant is unable to grow on maltose (due to the loss of amylomaltase) and is sensitive to maltose. We found a clone that conferred a resistant phenotype on McConkey maltose indicator plates. However, the plasmid did not encode a maltose phosphorylase nor did it allow growth on maltose. Instead, it encodes a protein that represses the maltose system. This became clear when the plasmid was tested in a *malK-lacZ* strain. Analysis of the plasmid-encoded protein revealed that it is a cytoplasmic enzyme exhibiting homology to human lipases including a soluble conserved signature sequence. The purified protein (now called Aes for **A**cetyl **es**terase) is able to hydrolyze para-nitrophenyl acetate but is not a lipase. So far nothing has been learned about the control of monocistronic *aes*. Activity of *aes-lacZ* fusions does not vary significantly under different growth conditions (Peist *et al.*, 1997) (Kanaya *et al.*, 1998). Like MalY, Aes interferes with transcription initiation in an *in vitro* system using purified components and inhibits maltotriose binding by MalT *in vitro* (E. Richet, unpublished) which suggests that Aes too directly interacts with MalT.

The MalK (or the MalT) Cycle

Whereas the physiological relevance of the control exerted by MalY and Aes on MalT activity remains unclear, the role of MalK in controlling MalT activity seems to be logical. Obviously, it makes sense to repress a metabolic system when its substrate is not transported and to relieve repression when the transport machinery is active. Transport of any maltodextrin by the maltose/maltodextrin ABC transporter will free MalT from its inactive stage and may be regarded as a prelude for the final induction by increasing concentrations of internal maltotriose.

The decisive insight into the repressor[1] function of MalK came through the demonstration that MalT, biotinylated at its C-terminus and bound to Avidin coated beads, specifically retained MalK (Panagiotidis *et al.*, 1998). The interaction of MalK with MalT indicated that MalK, like MalY and Aes, is able to inactivate MalT as a transcriptional activator. Indeed, preliminary experiments with purified MalK in the MalT-dependent transcription assay support this notion. Obviously, soluble cytoplasmic MalK can exert repression since the plasmid-encoded protein abolishes any MalT-dependent transcription in a *malEFG* background (Decker, 1998). But also when in complex with the other transport components MalK participates in MalT regulation and its ATPase activity must play a decisive role in controlling its regulatory activity. This was concluded from the effects of mutations in *malF* (*malF* 500) causing a MalE-independent transport activity. This mutant transport complex is uncoupled for its ATPase activity and causes *mal* gene transcription to become partially constitutive, despite being wild type in MalK (Panagiotidis *et al.*, 1998). More evidence comes from the G137A mutation within the ATP binding motif of MalK that leads to loss of ATP hydrolysis but not ATP binding. This MalK mutant is an even stronger repressor ("super-repressor") than the wild type protein (Panagiotidis *et al.*, 1998). Thus, in the absence of transport, i.e., in the absence of ATP hydrolysis, MalK has an increased affinity for MalT sequestrating it from its transcriptional activity and leading to *mal* gene repression.

It is less clear how this regulation is realized on the molecular level. One could imagine that upon binding to MalT, MalK dissociates from the membrane and keeps MalT in an inactive (monomeric) state (Figure 3b). The presence of substrate-loaded binding protein on the periplasmic side of the membrane would increase the affinity of the complex for MalK, initiate transport and set MalT free for transcriptional activity (Figure 3a). The binding affinity of isolated MalK for MalT is in favor of such a scenario. Alternatively, the close connection of transport and regulation is in strong favor of a model where MalT is sequestrated by the idling ABC transporter (Figure 3c) (Boos and Böhm, 2000). The recent structure determination of *Thermococcus litoralis* MalK, a functional homologue of the *E. coli* protein, has revealed the C-terminus as an independent domain, forming a β-barrel (Diederichs *et al.*, 2000). Even though nothing is known about a possible homologue of MalT in *T. litoralis*, it is suggested from the sequence homology as well as from the structural prediction of *E. coli* MalK that the C-termini in both proteins have similar functions and might constitute a new type of regulation module (Figure 4). MalK is also known to interact with $EIIA^{Glc}$, a component of the phosphotransferase system (PTS) that is mediating catabolite repression and inducer exclusion in non PTS systems (van der Vlag *et al.*, 1994). From the analysis of mutations in MalK rendering maltose transport insensitive to inducer exclusion it is clear that the binding sites in MalK for $EIIA^{Glc}$ and MalT are close but not overlapping (Kühnau *et al.*, 1991; Böhm *et al.*, J. Biol. Chem. 2002, in press).

[1]We used the term repressor for the function of MalK, MalY and Aes when inhibiting the transcriptional activity of MalT, even though these proteins are not repressors in the sense of DNA binding proteins. But since they reduce gene transcription by interaction with a gene activator the term repressor is kept in lack of a better term.

Sequestration of a Global Repressor as a Means of *mal* Gene Regulation

In the attempt to identify proteins needed for transmission of the repressor function of MalK we searched for mutants that would abolish the repressing effect of overproduced MalK on *mal* gene expression. Such a mutant was found. It had an insertion in *mlc*, a gene involved in the regulation of glucose utilization (Hosono *et al.*, 1995). But instead of mediating the MalK-MalT interaction Mlc was shown to be a protein with repressor function; it binds to the regulatory region of *malT* at a sequence with palindromic structure, the Mlc-box. Thus, Mlc did not interfere at the level of MalT activity but *malT* expression. Increasing the level of MalT was successful in over-riding the inhibiting activity of MalK (Decker *et al.*, 1998). In search of a potential inducer of Mlc, glucose-6-phosphate or a metabolic product were suspected, but their involvement could not be proven. This prompted us to probe for another possibility. Since it was known that transport of glucose via enzyme IICB for glucose (PtsG) of the PTS was needed for derepression of MalT it seemed possible that Mlc was bound by transporting PtsG and would thus be sequestrated from the DNA. Indeed, we found that Mlc was bound to PtsG-containing membranes and that phosphorylation of PtsG controlled Mlc binding. Elements necessary for Mlc binding were the B domain extending into the cytoplasm and the hinge region connecting the B domain and the membrane-bound C domain. In addition, the *in vitro* activity of PtsG in phosphorylation of α–methyl glucoside, a substrate analogue of glucose, was inhibited by increasing concentrations of Mlc (Lee *et al.*, 2000) (Figure 5). The effect of Mlc on the maltose system appears paradoxical. On one hand, in catabolite repression, transport of glucose reduces the expression of *malT* by lowering the cAMP concentration, on the other hand, transport of glucose *via* PtsG leads to sequestration of Mlc, a repressor for *malT*, thus dampening the effect of catabolite repression. The teleology behind this arrangement is at present unclear.

Acknowledgment

This work was only possible due to the generous collaboration of several laboratories: Howard Shuman and his crew at Columbia University in New York (MalK), Tim Clausen and collaborators at the Max Planck Institut in München (crystallization of MalY), Evelyne Richet and her coworkers at the Institut Pasteur in Paris (MalT) and Jackie Plumbridge at the Institut de Biologie Physico-chimique in Paris (Mlc).

References

Adhya, S., and Schwartz, M. 1971. Phosphoglucomutase mutants of *Escherichia coli* K12. J. Bacteriol. 108: 621–626.

Boos, W., and Böhm, A. 2000. Learning new tricks from an old dog. Trends Genetics 16: 404–409.

Boos, W., and Shuman, H.A. 1998. The maltose/maltodextrin system of *Escherichia coli*; transport, metabolism and regulation. Microbiol. Mol. Biol. Rev. 62: 204–229.

Bouma, C.L., and Roseman, S. 1996. Sugar transport by the marine chitinolytic bacterium Vibrio furnissii. Molecular cloning and analysis of the glucose and N-acetylglucosamine permeases. J. Biol. Chem. 271: 33457–33467.

Bukau, B., Ehrmann, M., and Boos, W. 1986. Osmoregulation of the maltose regulon in *Escherichia coli*. J. Bacteriol. 166: 884–891.

Clausen, T., Schlegel, A., Peist, R., Schneider, E., Steegborn, C., Chang, Y.-S., Haase, A., Bourenkov, G.P., Bartunik, H.D., and Boos, W. 2000. X-ray structure of MalY from *Escherichia coli*: a pyridoxal 5′-phosphate-dependent enzyme acting as a modulator in mal gene expression. EMBO J. 19: 831–842.

Decker, K. 1998. Untersuchungen zur Regulationsfunktion von MalK, der ATP-bindenden Untereinheit des Maltosetransportsystems aus *Escherichia coli*. Ph.D. thesis, Faculty of Biology, University of Konstanz.

Decker, K., Peist, R., Reidl, J., Kossmann, M., Brand, B., and Boos, W. 1993. Maltose and maltotriose can be formed endogenously in *Escherichia coli* from glucose and glucose-1-phosphate independently of enzymes of the maltose system. J. Bacteriol. 175: 5655–5665.

Decker, K., Plumbridge, J., and Boos, W. 1998. Negative transcriptional regulation of a positive regulator: the expression of malT, encoding the transcriptional activator of the maltose regulon of *Escherichia coli*, is negatively controlled by Mlc. Mol. Microbiol. 27: 381–390.

Diederichs, K., Diez, J., Greller, G., Müller, C., Breed, J., Schnell, C., Vonrhein, C., Boos, W., and Welte, W. 2000. Crystal structure of MalK, the ATPase subunit of the trehalose/maltose ABC transporter of the archaeon Thermococcus litoralis. EMBO J. 19: 5951–5961.

Ehrmann, M., and Boos, W. 1987. Identification of endogenous inducers of the mal system in *Escherichia coli*. J. Bacteriol. 169: 3539–3545.

Gilson, E., Rousset, J.-P., Charbit, A., Perrin, D., and Hofnung, M. 1986. malM, a new gene of the maltose regulon in *Escherichia coli* K12. I. malM is the last gene of the malK-lamB operon and encodes a periplasmic protein. J. Mol. Biol. 191: 303–311.

Hofnung, M., Hatfield, D., and Schwartz, M. 1974. malB region in *Escherichia coli* K-12: characterization of new mutations. J. Bacteriol. 117: 40–47.

Hosono, K., Kakuda, H., and Ichihara, S. 1995. Decreasing accumulation of acetate in a rich medium by *Escherichia coli* on introduction of genes on a multicopy plasmid. Biosci. Biotech. Biochem. 59: 256–261.

Kanaya, S., Koyanagi, T., and Kanaya, E. 1998. An esterase from *Escherichia coli* with a sequence similarity to hormone-sensitive lipase. Biochem. J. 332: 75–80.

Klein, W., and Boos, W. 1993. Induction of the λ receptor is essential for the effective uptake of trehalose in *Escherichia coli*. J. Bacteriol. 175: 1682–1686.

Kühnau, S., Reyes, M., Sievertsen, A., Shuman, H.A., and Boos, W. 1991. The activities of the *Escherichia coli* MalK protein in maltose transport, regulation and inducer exclusion can be separated by mutations. J. Bacteriol. 173: 2180–2186.

Lee, S.-J., Boos, W., Bouché, J.-P., and Plumbridge, J. 2000. Signal transduction between a membrane bound transporter, PtsG, and a soluble transcription factor, Mlc, of *Escherichia coli*. EMBO J.: in press.

Meyer, D., Schneider-Fresenius, C., Horlacher, R., Peist, R., and Boos, W. 1997. Molecular characterization of glucokinase from *Escherichia coli* K-12. J. Bacteriol. 179: 1298–1306.

Panagiotidis, C.H., Boos, W., and Shuman, H.A. 1998. The ATP-binding cassette subunit of the maltose transporter MalK antagonizes MalT, the activator of the *Escherichia coli* mal regulon. Mol. Microbiol. 30: 535–546.

Peist, R., Koch, A., Bolek, P., Sewitz, S., Kolbus, T., and Boos, W. 1997. Characterization of the aes gene of *Escherichia coli* encoding an enzyme with esterase activity. J. Bacteriol. 179: 7679–7686.

Raibaud, O., and Richet, E. 1987. Maltotriose is the inducer of the maltose regulon. J. Bacteriol. 169: 3059–3061.

Reidl, J. 1992. Identifizierung der malY-abhängigen Regulation des Maltose-Systems in *Escherichia coli*. Ph.D. thesis. University of Konstanz, Konstanz, Germany.

Reidl, J., and Boos, W. 1991. The malX malY operon of *Escherichia coli* encodes a novel enzyme II of the phosphotransferase system recognizing glucose and maltose and an enzyme abolishing the endogenous induction of the maltose system. J. Bacteriol. 173: 4862–4876.

Reidl, J., Römisch, K., Ehrmann, M., and Boos, W. 1989. MalI, a novel protein involved in regulation of the maltose system of *Escherichia coli*, is highly homologous to the repressor proteins GalR, CytR, and LacI. J. Bacteriol. 171: 4888–4899.

Reyes, M., and Shuman, H.A. 1988. Overproduction of MalK protein prevents expression of the *Escherichia coli* mal regulon. J. Bacteriol. 170: 4598–4602.

Richet, E., and Raibaud, O. 1989. MalT, the regulatory protein of the *Escherichia coli* maltose system, is an ATP-dependent transcriptional activator. EMBO J. 8: 981–987.

Schirmer, T., Keller, T.A., Wang, Y.F., and Rosenbusch, J.P. 1995. Structural basis for sugar translocation through maltoporin channels at 3.1 Å resolution. Science 267: 512–514.

Schreiber, V., and Richet, E. 1999. Self-association of the *Escherichia coli* transcription activator MalT in the presence of maltotriose and ATP. J. Biol. Chem. 274: 33220–33226.

Schreiber, V., Steegborn, C., Clausen, T., Boos, W., and Richet, E. 2000. A new mechanism for the control of a prokaryotic transcriptional regulator: antagonistic binding of positive and negative effectors. Mol. Microbiol. 35: 765–776.

Schwartz, M. 1987. The maltose regulon. *Escherichia coli* and Salmonella typhimurium: cellular and molecular biology. F. C. Neidhardt. Washington D.C., American Society of Microbiology. 2: 1482–1502.

van der Vlag, J., van Dam, K., and Postma, P.W. 1994. Quantification of the regulation of glycerol and maltose metabolism by IIA^{Glc} of the phosphoenolpyruvate-dependent glucose phosphotransferase system in Salmonella typhimurium. J. Bacteriol. 176: 3518–3526.

Chapter 22

Catabolite Control Protein CcpA-Dependent Glucose Repression in *Staphylococcus xylosus*: Efficient Activation of CcpA by Glucose Transported Independently from the Phosphotransferase System

Ivana Jankovic, Jens Meyer, and Reinhold Brückner

Abstract

Carbon catabolic repression (CR) by the catabolite control protein CcpA has been analyzed in *Staphylococcus xylosus*. Genes encoding components needed to utilize lactose, sucrose, and maltose were found to be repressed by CcpA. In addition, the *ccpA* gene is under negative autogenous control. Among several tested sugars, glucose caused strongest CcpA-dependent repression. Glucose can enter *S. xylosus* in nonphosphorylated form via the glucose uptake protein GlcU. Internal glucose is then phosphorylated by the glucose kinase GlkA. Alternatively, glucose can be transported and concomitantly phosphorylated by glucose-specific permease(s) of the phosphotransferase system (PTS). *S. xylosus* mutant strains deficient in GlcU or GlkA showed partial relief of glucose-specific, CcpA-dependent repression. Likewise, blocking PTS activity completely by inactivation of the gene encoding the general PTS protein enzyme I resulted in diminished glucose-mediated repression. Thus, both glucose entry routes contribute to glucose-specific CR in *S. xylosus*. The sugar transport activity of the PTS is not required to trigger glucose-specific repression. The phosphocarrier protein HPr however, is absolutely essential for CcpA activity. Inactivation of the HPr gene led to a complete loss of CR. Repression is also abolished upon inactivation of the HPr kinase gene or by replacing serine at position 46 of HPr by alanine. These results clearly show that HPr kinase provides the signal, seryl-phosphorylated HPr, to activate CcpA in *S. xylosus*.

Introduction

Carbon catabolite repression (CR) is a regulatory process in microorganisms, in which the presence of a rapidly metabolizable carbon source inhibits expression of alternate catabolic functions. Although the final outcome of CR is uniform, reduced expression of certain genes and operons, the mechanisms leading to repression may be quite diverse (Saier, 1991). The presence of a repressing carbon source can result in lower concentrations of inducers specific for alternate routes of catabolism (Saier *et al.*, 1996; Saier and Crasnier, 1996), in altered activities of specific regulators (Stülke *et al.*, 1998), or in the activation of global control proteins, such as the cAMP receptor protein in enteric bacteria (Saier *et al.*, 1995) or the catabolite control protein CcpA in low-GC Gram-positive bacteria (Henkin, 1996). In many cases, genes are subject to multiple levels of control. It is therefore important to distinguish gene- or operon-specific regulatory processes from global control to be able to dissect the underlying mechanisms.

In low-GC Gram-positive bacteria, one consequence of the availability of a rapidly metabolizable carbon source is the regulation of gene expression by CcpA, the central transcriptional regulator of CR (Henkin, 1996). CcpA, a member of the LacI/GalR family of transcription factors (Nguyen and Saier, 1995), shows a relatively weak affinity for its cognate operator sites, termed catabolite responsive elements (*cre*) (Hueck *et al.*, 1994). Therefore, CcpA must be activated in order to bind efficiently to *cre*. While conflicting results have been reported with regard to the effectors enabling CcpA to bind to *cre in vitro* (Fujita *et al.*, 1995; Galinier *et al.*, 1999; Gösseringer *et al.*, 1997; Jones *et al.*, 1997; Kim *et al.*, 1998; Miwa *et al.*, 1997; Ramseier *et al.*, 1995), a unifying concept emerges regarding the *in vivo* situation. In *Bacillus subtilis*, *Staphylococcus xylosus*, and *Lactobacillus casei*, inactivation of the gene encoding HPr kinase led to a complete relief of CR (Dossonnet *et al.*, 2000; Galinier *et al.*, 1998; Huynh *et al.*, 2000; Reizer *et al.*, 1998). HPr kinase phosphorylates serine at position 46 in HPr, the general phosphocarrier protein of the phosphoenolpyruvate:carbohydrate phosphotransferase system (PTS) (Postma *et al.*, 1993), or Crh, an HPr-like protein detected in *B. subtilis* (Galinier *et al.*, 1997). Accordingly, replacement of serine-46 by alanine in HPr abolished CR in *L. casei* and *Lactococcus lactis* (Monedero *et al.*, 2001; Viana *et al.*, 2000), whereas concomitant exchange in HPr and Crh was required to block CR in *B. subtilis* (Deutscher *et al.*, 1994; Galinier *et al.*, 1999; Martin-Verstraete *et al.*, 1999; Zalieckas *et al.*, 1999). These results clearly show that HPr kinase-mediated phosphorylation of HPr (Crh) is the key event triggering CR in Gram-positive bacteria.

Besides the regulatory role of HPr in CR, the PTS constitutes the main sugar uptake system in many bacteria. Sugars transported by the PTS, especially glucose, are generally regarded as preferred carbohydrates causing strongest CR. However, glucose was shown to be transported by both, PTS-dependent and PTS-independent uptake systems, in several Gram-positive bacteria (Cvitkovitch *et al.*, 1995; Fiegler *et al.*, 1999; Paulsen *et al.*, 1998; Reizer *et al.*, 1989; Russell, 1990; Tangney *et al.*, 1993; Wagner *et al.*, 2000). Therefore, the observed dominance of glucose may sometimes be a consequence of multiple uptake routes. In addition, it has recently been demonstrated that in *Streptococcus thermophilus* the non-PTS sugar lactose activated CcpA-dependent regulation most efficiently (van Den Bogaard *et al.*, 2000). Although common principles in signal transduction leading to CR exist in AT-rich Gram-positive bacteria, it appears that specific details have to be worked out for each bacterial species individually.

In the last couple of years, we have analyzed sugar utilization and CR in *S. xylosus*, a nonpathogenic staphylococcal species that is applied in food fermentations. In the course of this study, we could demonstrate that *S. xylosus* possesses an efficient PTS-independent glucose uptake system that participates in CR (Fiegler *et al.*, 1999; Wagner *et al.*, 1995). In this review, we will present recent data on the participation of the PTS in CR and we will discuss our current view of CR in this organism. Especially, we will focus our attention to glucose-mediated signal transduction leading to CcpA-dependent repression.

Review

Genes that are Subject to CcpA-Mediated Repression

In *S. xylosus* molecular data concerning the utilization of xylose (Sizemore *et al.*, 1991; Sizemore *et al.*, 1992), sucrose (Brückner *et al.*, 1993; Gering and Brückner, 1996; Wagner *et al.*, 1995), lactose (Bassias and Brückner, 1998), and maltose (Egeter and Brückner, 1995) are available. For sucrose and lactose, genes encoding degradative enzymes and the major transporters have been cloned, whereas information on xylose or maltose transporters is still missing. Expression of the xylose, sucrose, and lactose utilization genes is induced in the presence of the cognate sugar. In addition, the genes are subject to global CcpA-mediated repression. In contrast, expression of α-glucosidase, needed to cleave maltose, is not specifically induced by maltose, but is under CcpA control. In the promoter regions of these genes, *cre* sites serving as operators for CcpA are found. Two of them, in the *xyl* and *mal* system, were shown to be of functional significance by deletion analysis (Egeter and Brückner, 1996; Hueck *et al.*, 1994). The remaining putative *cre* sites conferred CcpA-dependent repression upon a constitutive promoter proving their potential to direct CcpA regulation (Jankovic *et al.*, 2001). Besides the genes for sugar utilization, the *ccpA* gene itself was found to be regulated by CcpA (Egeter and Brückner, 1996). Activation by CcpA, well established in *B. subtilis* and *L. lactis* (Grundy *et al.*, 1993; Luesink *et al.*, 1998; Moreno *et al.*, 2001; Tobisch *et al.*, 1999), has not yet been detected in *S. xylosus.*

Since CR of the sugar utilization systems mentioned above did not only rely on CcpA (Bassias and Brückner, 1998), promoters were constructed that were exclusively regulated by CcpA (Jankovic *et al.*, 2001). This was achieved by insertion of *cre* sites into a constitutively expressed promoter downstream of the transcriptional start site at positions +2 or +3. The synthetic promoters were placed in the genome of *S. xylosus* in front of a promoterless β-galactosidase gene. With these promoters and the promoter probe system, we were able to analyze CcpA-mediated regulation and the signal transduction leading to its activation (Jankovic *et al.*, 2001).

The Catabolite Control Protein CcpA

S. xylosus was the first Gram-positive AT-rich bacterium besides *B. subtilis* and *Bacillus megaterium* (Henkin *et al.*, 1991; Hueck *et al.*, 1995), where the role of CcpA in CR has clearly been demonstrated (Egeter and Brückner, 1996), and it was *S. xylosus* in which negative autoregulation of the *ccpA* gene was first described. Transcription of one of the two *ccpA* promoters is repressed in the presence of glucose and production of the CcpA protein is reduced about twofold under these conditions (Egeter and Brückner, 1996). Measuring β-galactosidase expression directed by the two *ccpA* promoters confirmed the twofold repression in the presence of glucose (Jankovic *et al.*, 2001). In addition, *ccpA* promoter activity was about threefold higher in the absence of a functional CcpA indicating that the *ccpA* gene of *S. xylosus* is repressed even in the absence of sugar. Most recently, genes that are regulated in such a way in *B. subtilis*, CcpA-repressed with or without glucose, but stronger with glucose, were designated class 2 genes, with respect to the mode of CcpA regulation (Moreno *et al.*, 2001). Interestingly, negative autoregulation of *ccpA* is not generally observed in Gram-positive bacteria. While *ccpA* expression appeared to be constitutive in the bacilli, *L. casei* and *L. lactis* (Hueck *et al.*, 1995; Luesink *et al.*, 1998; Miwa *et al.*, 1994; Monedero *et al.*, 1997), it is negatively autoregulated in *Lactobacillus pentosus* and *Lactobacillus delbrueckii* (Mahr *et al.*, 2000; Morel *et al.*, 2001). Autorepression of *ccpA* in the presence of carbon sources apparently reflects the need to balance CcpA production, when preformed CcpA is activated to carry out CR.

CcpA-Mediated Repression in Mutants Deficient in PTS-Independent Glucose Utilization

Transposon mutagenesis using the expression of β-galactosidase as a screening system, led to the discovery of a PTS-independent glucose uptake and utilization system in *S. xylosus.* The system is comprised of GlcU, a membrane protein responsible for glucose uptake (Fiegler *et al.*, 1999), and the glucose kinase GlkA, needed to phosphorylate internalized unmodified glucose (Wagner *et al.*, 1995). Inactivation of the respective genes, *glcU* and *glkA*, partially relieved several catabolic enzymes from glucose repression. The mutant strains were still able to utilize glucose as carbon source. Subsequent analyses, using CcpA-controlled promoters in the β-galactosidase based promoter probe system, showed that glucose-dependent CcpA-mediated repression was about half as strong as in the wild type (Jankovic *et al.*, 2001). Hence, PTS-independent glucose utilization significantly contributes to CR in *S. xylosus.*

A second glucose kinase activity had been detected in GlkA-deficient strains leading to the assumption that GlkA plays a special regulatory role (Wagner *et al.*, 1995). Subsequent analysis revealed that the glucose kinase assays were compromised by a rather strong glucose dehydrogenase activity, whose gene, *gdh,* is located downstream of and cotranscribed with *glcU* (Fiegler *et al.*, 1999). However, examination of glucose kinase activity in an *S. xylosus* strain with deletions in *gdh* and *glkA* indeed yielded a second glucose kinase activity, albeit weaker as originally determined (I. Jankovic and R. Brückner, unpublished). Further investigations identified the gene responsible for this minor activity as the fructokinase gene *scrK* (nucleotide accession number X67744) located downstream of the sucrase gene *scrB*. Inactivation of *scrK* abolished fructokinase activity as well as the minor glucose kinase activity. The gene is cotranscribed with *scrB* and is therefore part of the sucrose regulon controlled by the repressor ScrR (Gering and Brückner, 1996). In wild type cells, ScrK contributes only marginally to glucose phosphorylation, since expression of the *scrBK* operon is repressed by glucose. Interestingly, in the absence of a functional GlkA, glucose repression is partially relieved and *scrK* expression is slightly enhanced. As *scrK* requires sucrose in the medium to be fully expressed, ScrK levels remain low, but detectable, in glucose-grown cells. Therefore, GlkA is the only physiologically relevant glucose kinase in *S. xylosus.*

CcpA-Mediated Repression in PTS Mutant Strains

In the course of the transposon mutagenesis to isolate mutants resistant to CR, PTS genes, anticipated to be involved in CR, were not detected. Since the PTS constitutes an important sugar uptake system, the genes *ptsH* and *ptsI,* encoding the general PTS proteins HPr and enzyme I, were cloned by a PCR-based approach. As found in many other bacteria, *ptsH* and *ptsI* form an operon transcribed from a promoter in front of *ptsH.* As we were interested to determine the consequences of *pts* mutations on CR, *ptsH* and *ptsI* deletion mutants were constructed. Inactivation of *ptsI* resulted in a partial relief of glucose-specific CcpA-dependent CR (I. Jankovic and R. Brückner, unpublished). Repression was even slightly stronger than in the GlcU mutant strain. As expected, sugars exclusively transported by the PTS, e.g., sucrose and fructose, did no longer cause CR in the *ptsI* mutant strain. In contrast to the *ptsI* mutation, deletion of *ptsH* abolished CR by all sugars. Although PTS sugar transport activity is not required to trigger CR, when glucose enters independently from PTS, the general PTS protein HPr is essential for CR. Inactivation of the HPr kinase gene *hprK* also led to a complete loss of CR (Huynh *et al.*, 2000). Therefore, HPr phosphorylated at serine-46 (P-Ser-HPr) seems to be the signal triggering CR in *S. xylosus.* Consequently, exchange of serine at position 46 by alanine, preventing phosphorylation at this position, blocked CR irrespective of the sugar. Thus, unlike *B. subtilis, S. xylosus* does not appear to possess an HPr-like protein able to replace HPr in regulation.

Since deletion of *ptsI* resulted in partially diminished glucose-specific CR, at least some glucose must enter *S. xylosus* by PTS-mediated transport. Screening a genomic *S. xylosus* library in *E. coli* for complementation of glucose transport deficiency, yielded three different genetic loci (R. Brückner, unpublished), one of which contained a gene, *ptsG*, encoding a protein with high identity to the glucose-specific PTS permeases GlcA and GlcB of *Staphylococcus carnosus* (Christiansen and Hengstenberg, 1999). Deduced from similarities of neighboring genes, PTS systems specific for β-glucosides and N-acetyl-glucosamine are located at the other two loci. Inactivation of *ptsG* in a strain already deficient for GlcU resulted in an *S. xylosus* mutant which was still able to utilize glucose. A *ptsI-glcU* double mutant however, lost this ability. These results strongly suggest that *S. xylosus* transports glucose by more than one PTS permease.

Inspection of the presumptive promoter region of the *ptsG* gene identified a putative *cre* site and a region of dyad symmetry resembling antitermination regions of catabolic genes (Rutberg, 1997). Therefore, *ptsG* could be under dual control, negatively by CcpA and positively by an antiterminator protein (Stülke *et al.*, 1998). Indeed, analysis of PTS-mediated glucose transport indicated that uptake is repressed, when CcpA is present, while inducibility is apparent in a *ccpA* mutant strain (Huynh *et al.*, 2000) and, interestingly, in a GlcU-deficient mutant (Fiegler *et al.*, 1999). Therefore, glucose internalized independently from PTS is able to repress PTS-dependent glucose uptake via CcpA, a situation so far unique to *S. xylosus.*

Glucose sensitivity of *S. xylosus* upon inactivation of the HPr kinase gene

From the studies of CcpA repression in PTS mutant strains mentioned above, it follows that an *S. xylosus* strain devoid of HPr kinase activity should be deficient in CcpA-mediated regulation, which was indeed the case (Huynh *et al.*, 2000; Jankovic *et al.*, 2001). In addition, CR of the lactose operon, which is only partially mediated by CcpA (Bassias and Brückner, 1998), is also lost in the mutant strain (Huynh *et al.*, 2000). This result strongly suggests that processes such as inducer exclusion or expulsion, which are the likely cause of CcpA-independent repression of the lactose utilization operon, require a functional HPr kinase.

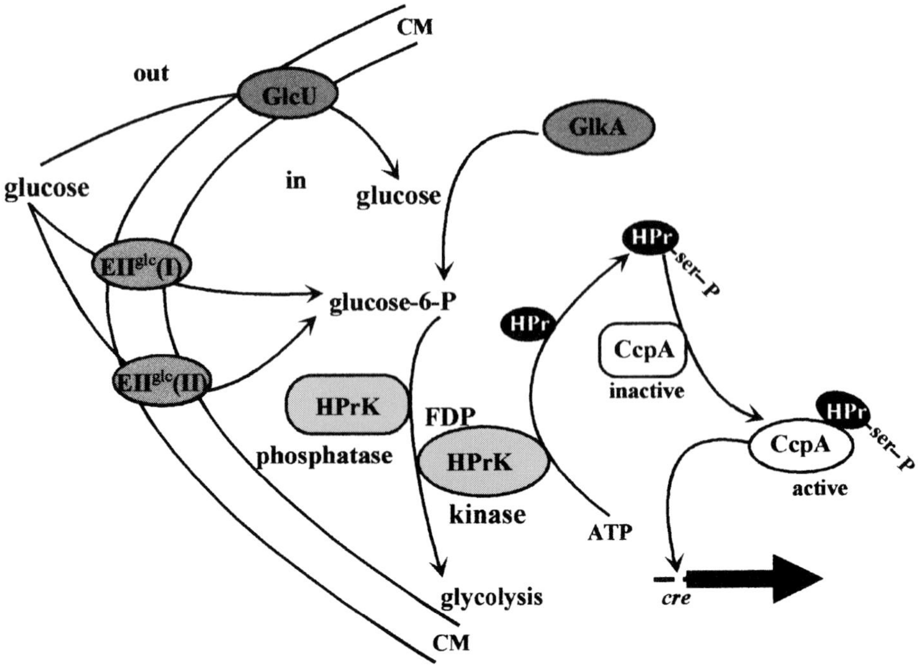

Figure 1. Glucose-mediated carbon catabolite repression by the catabolite control protein CcpA in *S. xylosus*. The components involved in regulation are: CcpA, carbon catabolite control protein; GlkA, glucose kinase; GlcU, PTS-independent glucose uptake protein; HPr, general PTS phosphocarrier protein; HPrK, HPr kinase/P-HPr phosphatase; EII^{glc} glucose-specific PTS permease. FDP: fructose-1,6-diphosphate. Except for the gene encoding the second PTS permease, all other genes are cloned and mutants are available. The general PTS protein enzyme I, essential for the transport of all PTS sugars, is omitted for clarity. The thick arrow represents a gene that is subject to carbon catabolite repression by CcpA via *cre* interaction. The position of the CcpA binding site *cre* in the promoter region is indicated. The double line marked with CM represents the cytoplasmic membrane. The fact that some proteins may only function as dimers or multimers is not depicted.

An unexpected phenotype of the HPr kinase mutant was the severe growth retardation in the presence of glucose, which was accompanied by the production of methylglyoxal (Huynh *et al.*, 2000). The amount of the produced methylglyoxal, however, was not sufficient to explain the growth defect. The adverse effect of glucose on growth was clearly dependent on the glucose concentration in the medium. Apparently, the balance of carbohydrate transport and catabolic capacities of the cell is severely disturbed in the absence of a functional HPr kinase. Since the *ccpA* mutant did not show strong growth retardation (Egeter and Brückner, 1996), glucose sensitivity cannot be due to the loss of regulation by CcpA. Interestingly, the glucose-mediated growth defect did not occur in the HPr Ser-46-Ala mutant indicating that this mutation is not absolutely equivalent to the loss of HPr kinase activity.

Suppression of glucose sensitivity in the HPr kinase mutant

A conceivable consequence of the inability of the HPr kinase mutant strain to phosphorylate serine-46 of HPr would be enhanced transport activity of the PTS. It has been shown that HPr phosphorylated at serine-46 is a poor substrate for enzyme I (Monedero *et al.*, 2001). Consequently, phosphorylation of HPr by HPr kinase limits sugar uptake by the PTS and loss of HPr kinase activity would result in unrestricted PTS glucose transport leading to deleterious levels of metabolites. Inactivation of PTS transport should therefore suppress glucose sensitivity. To test this prediction, an *S. xylosus* mutant strain was constructed devoid of HPr kinase and PTS enzyme I activity. Surprisingly, the double mutant still showed growth retardation in the presence of glucose. Likewise, inactivation of the non-PTS glucose transporter GlcU did not relieve glucose sensitivity. Both double mutants grew slightly better than the wild type, but glucose in the medium had still adverse effects on growth. *S. xylosus* without a functional HPr kinase is apparently not able do deal with high amounts of glucose, regardless of the route of glucose entry. Thus, enhanced PTS activity cannot account for the glucose phenotype of the HPr kinase mutant strain.

In an approach to isolate spontaneous suppressors enabling the HPr kinase mutant to grow on glucose, colonies of the mutant grown on complex media in the absence of additional sugar were patched onto agar plates containing 0.5 % glucose and incubated at 30°C, since previous attempts to isolate suppressors at 37°C failed. After several days growth occurred in some of the patches. Streaking these cells on the same media resulted in colonies after incubation at 30°C as well at 37°C. Inspection of the phenotype of the suppressors revealed a striking resemblance to HPr mutants. Therefore a *ptsH-hprK* mutant was constructed and tested for growth in glucose-containing media. And indeed, the double mutant was able to grow. Glucose sensitivity was completely abolished. We have no explanation for this surprising result at the moment, but if HPr is present, phosphorylation of serine-46 by HPr kinase is a prerequisite survival when *S. xylosus* encounters high concentrations of glucose.

Conclusions

Signal transduction leading to sugar-dependent CcpA activation and, consequently, to CcpA-dependent CR in *S. xylosus* has been elucidated (Fig. 1). P-Ser-HPr is absolutely required as a corepressor to enable CcpA to regulate transcription. Modulation of CcpA binding activity by secondary corepressors or modulation of P-Ser-HPr interaction with CcpA (Deutscher *et al.*, 1995) may contribute to fine regulation, but without P-Ser-HPr CcpA is virtually inactive. Consequently, the bifunctional HPr kinase/P-HPr phosphatase HPrK is the second essential component in the signaling cascade. Glycolytic intermediates like fructose-1-6-diphosphate (FDP), perhaps together with elevated ATP levels, shift HPrK activity towards kinase activity producing P-Ser-HPr. This process can be reversed by the HPrK P-Ser-HPr phosphatase activity, when glycolytic intermediates are low. Glucose, causing strongest CR in *S. xylosus*, activates phosphorylation of HPr by HPr kinase via two uptake routes. Since non-PTS glucose uptake is particularly efficient in *S. xylosus*, its influence may be more pronounced than in other bacteria possessing both, PTS-dependent and -independent glucose uptake systems.

The glucose sensitivity of the HPr kinase mutant strain clearly demonstrates that carbon catabolite repression is important even if only one carbon source is present. This form of autoregulatory limitation of sugar uptake and consumption is indispensable when carbon sources are plentiful (Brückner and Titgemeyer, 2002) and may be even more important for bacteria than the ability to utilize certain substrates preferentially.

Acknowledgments

I would like to thank E. Wagner, H. Fiegler, M. Gering, O. Egeter, and J. Bassias, who made substantial contributions at various stages of the project.

Most recent work was supported by the Deutsche Forschungsgemeinschaft (DFG) within the priority program, Molecular Analysis of Regulatory Networks in Bacteria (BR947/4-1). The DFG also supported earlier parts of the project by an individual grant (BR947/3-1).

References

Bassias, J., and Brückner, R. 1998. Regulation of lactose utilization genes in *Staphylococcus xylosus*. J. Bacteriol. 180: 2273–2279.

Brückner, R., Wagner, E., and Götz, F. 1993. Characterization of a sucrase gene from *Staphylococcus xylosus*. J. Bacteriol. 175: 851–857.

Brückner, R., and Titgemeyer, F. 2002. Carbon catabolite repression in bacteria: choice of the carbon source and autoregulatory limitation of sugar utilization. FEMS Microbiol. Lett. 209: 141–148.

Christiansen, I., and Hengstenberg, W. 1999. Staphylococcal phosphoenolpyruvate-dependent phosphotransferase system – two highly similar glucose permeases in *Staphylococcus carnosus* with different glucoside specificity: protein engineering *in vivo*? Microbiology 145: 2881–2889.

Cvitkovitch, D.G., Boyd, D.A., Thevenot, T., and Hamilton, I.R. 1995. Glucose transport by a mutant of *Streptococcus mutans* unable to accumulate sugars via the phosphoenolpyruvate phosphotransferase system. J. Bacteriol. 177: 2251–2258.

Deutscher, J., Küster, E., Bergstedt, U., Charrier, V., and Hillen, W. 1995. Protein kinase-dependent HPr/CcpA interaction links glycolytic activity to carbon catabolite repression in gram-positive bacteria. Mol. Microbiol. 15: 1049–1053.

Deutscher, J., Reizer, J., Fischer, C., Galinier, A., Saier, M.H., Jr., and Steinmetz, M. 1994. Loss of protein kinase-catalyzed phosphorylation of HPr, a phosphocarrier protein of the phosphotransferase system, by mutation of the *ptsH* gene confers catabolite repression resistance to several catabolic genes of *Bacillus subtilis*. J. Bacteriol. 176: 3336–3344.

Dossonnet, V., Monedero, V., Zagorec, M., Galinier, A., Pérez-Martínez, G., and Deutscher, J. 2000. Phosphorylation of HPr by the bifunctional HPr Kinase/P-Ser-HPr phosphatase from *Lactobacillus casei* controls catabolite repression and inducer exclusion but not inducer expulsion. J. Bacteriol. 182: 2582–2590.

Egeter, O., and Brückner, R. 1995. Characterization of a genetic locus essential for maltose-maltotriose utilization in *Staphylococcus xylosus*. J. Bacteriol. 177: 2408–2415.

Egeter, O., and Brückner, R. 1996. Catabolite repression mediated by the catabolite control protein CcpA in *Staphylococcus xylosus*. Mol. Microbiol. 21: 739–749.

Fiegler, H., Bassias, J., Jankovic, I., and Brückner, R. 1999. Identification of a gene in *Staphylococcus xylosus* encoding a novel glucose uptake protein. J. Bacteriol. 181: 4929–4936.

Fujita, Y., Miwa, Y., Galinier, A., and Deutscher, J. 1995. Specific recognition of the *Bacillus subtilis gnt cis*-acting catabolite-responsive element by a protein complex formed between CcpA and seryl-phosphorylated HPr. Mol. Microbiol. 17: 953–960.

Galinier, A., Deutscher, J., and Martin-Verstraete, I. 1999. Phosphorylation of either Crh or HPr mediates binding of CcpA to the *Bacillus subtilis xyn cre* and catabolite repression of the *xyn* operon. J. Mol. Biol. 286: 307–314.

Galinier, A., Haiech, J., Kilhoffer, M.C., Jaquinod, M., Stülke, J., Deutscher, J., and Martin-Verstraete, I. 1997. The *Bacillus subtilis crh* gene encodes a HPr-like protein involved in carbon catabolite repression. Proc. Natl. Acad. Sci. USA 94: 8439–8444.

Galinier, A., Kravanja, M., Engelmann, R., Hengstenberg, W., Kilhoffer, M.C., Deutscher, J., and Haiech, J. 1998. New protein kinase and protein phosphatase families mediate signal transduction in bacterial catabolite repression. Proc. Natl. Acad. Sci. USA 95: 1823–1828.

Gering, M., and Brückner, R. 1996. Transcriptional regulation of the sucrase gene of *Staphylococcus xylosus* by the repressor ScrR. J. Bacteriol. 178: 462–469.

Gösseringer, R., Küster, E., Galinier, A., Deutscher, J., and Hillen, W. 1997. Cooperative and non-cooperative DNA binding modes of catabolite control protein CcpA from *Bacillus megaterium* result from sensing two different signals. J. Mol. Biol. 266: 665–676.

Grundy, F.J., Waters, D.A., Allen, S.H., and Henkin, T.M. 1993. Regulation of the *Bacillus subtilis* acetate kinase gene by CcpA. J. Bacteriol. 175: 7348–7355.

Henkin, T.M. 1996. The role of CcpA transcriptional regulator in carbon metabolism in *Bacillus subtilis*. FEMS Microbiol. Lett. 135: 9–15.

Henkin, T.M., Grundy, F.J., Nicholson, W.L., and Chambliss, G.H. 1991. Catabolite repression of alpha-amylase gene expression in *Bacillus subtilis* involves a trans-acting gene product homologous to the Escherichia coli lacI and galR repressors. Mol. Microbiol. 5: 575–584.

Hueck, C.J., Hillen, W., and Saier, M.H., Jr. 1994. Analysis of a *cis*-active sequence mediating catabolite repression in gram-positive bacteria. Res. Microbiol. 145: 503–518.

Hueck, C.J., Kraus, A., Schmiedel, D., and Hillen, W. 1995. Cloning, expression and functional analyses of the catabolite control protein CcpA from *Bacillus megaterium*. Mol. Microbiol. 16: 855–864.

Huynh, P.L., Jankovic, I., Schnell, N.F., and Brückner, R. 2000. Characterization of an HPr kinase mutant of *Staphylococcus xylosus*. J. Bacteriol. 182: 1895–1902.

Jankovic, I., Egeter, O., and Brückner, R. 2001. Analysis of catabolite control protein A-dependent repression in *Staphylococcus xylosus* by a genomic reporter gene system. J. Bacteriol. 183: 580–586.

Jones, B.E., Dossonnet, V., Küster, E., Hillen, W., Deutscher, J., and Klevit, R.E. 1997. Binding of the catabolite repressor protein CcpA to its DNA target is regulated by phosphorylation of its corepressor HPr. J. Biol. Chem. 272: 26530–26535.

Kim, J.H., Voskuil, M.I., and Chambliss, G.H. 1998. NADP, corepressor for the *Bacillus*catabolite control protein CcpA. Proc. Natl. Acad. Sci. USA 95: 9590–9595.

Luesink, J.L., van Herpen, R.E.M.A., Grossiord, B.P., Kuipers, O.P., and M., d.V.W. 1998. Transcriptional activation of the glycolytic *las* operon and catabolite repression of the *gal* operon in *Lactococcus lactis* are mediated by the catabolite control protein CcpA. Mol. Microbiol. 30: 789–798.

Mahr, K., Hillen, W., and Titgemeyer, F. 2000. Carbon catabolite repression in *Lactobacillus pentosus*: Analysis of the *ccpA* region. Appl. Enveron. Microbiol. 66: 277–283.

Martin-Verstraete, I., Deutscher, J., and Galinier, A. 1999. Phosphorylation of HPr and Crh by HprK, early steps in the catabolite repression signalling pathway for the *Bacillus subtilis* levanase operon. J. Bacteriol. 181: 2966–2969.

Miwa, Y., Nagura, K., Eguchi, S., Fukuda, H., Deutscher, J., and Fujita, Y. 1997. Catabolite repression of the *Bacillus subtilis gnt* operon exerted by two catabolite-responsive elements. Mol. Microbiol. 23: 1203–1213.

Miwa, Y., Saikawa, M., and Fujita, Y. 1994. Possible function and some properties of the CcpA protein of *Bacillus subtilis*. Microbiology 140: 2567–2575.

Monedero, V., Gosalbes, M.J., and Pérez-Martínez, G. 1997. Catabolite repression in *Lactobacillus casei* ATCC 393 is mediated by CcpA. J. Bacteriol. 179: 6657–6664.

Monedero, V., Kuipers, O.P., Jamet, E., and Deutscher, J. 2001. Regulatory Functions of Serine-46-Phosphorylated HPr in *Lactococcus lactis*. J. Bacteriol. 183: 3391–3398.

Monedero, V., Poncet, S., Mijakovic, I., Fieulaine, S., Dossonnet, V., Martin-Verstraete, I., Nessler, S. and Deutscher, J. 2001. Mutations lowering the phosphatase activity of HPr kinase/phosphatase switch off carbon metabolism. Embo J. 20: 3928–3937.

Morel, F., Lamarque, M., Bissardon, I., Atlan, D., and Galinier, A. 2001. Autoregulation of the biosynthesis of the CcpA-like protein, PepR1, in *Lactobacillus delbrueckii* subsp *bulgaricus*. J. Mol. Microbiol. Biotechnol. 3: 63–66.

Moreno, M.S., Schneider, B.L., Maile, R.R., Weyler, W., and Saier, M.H. 2001. Catabolite repression mediated by the CcpA protein in *Bacillus subtilis:* novel modes of regulation revealed by whole-genome analyses. Mol. Microbiol. 39: 1366–1381.

Nguyen, C.C., and Saier, M.H., Jr. 1995. Phylogenetic, structural and functional analyses of the LacI-GalR family of bacterial transcription factors. FEBS Lett 377: 98–102.

Paulsen, I.T., Chauvaux, S., Choi, P., and Saier, M.H., Jr. 1998. Characterization of glucose-specific catabolite-resistant mutants of *Bacillus subtilis*: Identification of a novel hexose: H^+ symporter. J. Bacteriol. 180: 498–504.

Postma, P.W., Lengeler, J.W., and Jacobson, G.R. 1993. Phosphoenolpyruvate: carbohydrate phosphotransferase systems of bacteria. Microbiol. Rev. 6: 543–594.

Ramseier, T.M., Reizer, J., Küster, E., Hillen, W., and Saier, M.H. 1995. *In vitro* binding of the CcpA protein of *Bacillus megaterium* to *cis-* acting catabolite responsive elements (CREs) of gram-positive bacteria. FEMS Microbiol. Lett. 129: 207–213.

Reizer, J., Hoischen, C., Titgemeyer, F., Rivolta, C., Rabus, R., Stülke, J., Karamata, D., Saier, M.H.J., and Hillen, W. 1998. A novel protein kinase that controls carbon catabolite repression in bacteria. Mol. Microbiol. 27: 1157–1169.

Reizer, J., Sutrina, S.L., Saier, M.H., Stewart, G.C., Peterkofsky, A., and Reddy, P. 1989. Mechanistic and physiological consequences of HPr(ser) phosphorylation on the activities of the phosphoenolpyruvate: sugar phosphotransferase system in gram-positive bacteria: studies with site-specific mutants of HPr. Embo J. 8: 2111–2120.

Russell, J.B. 1990. Low-Affinity, high-capacity system of glucose transport in the ruminal bacterium *Streptococcus bovis:* Evidence for a mechanism of facilitated diffusion. Appl. and Environm. Microbiol. 56: 3304–3307.

Rutberg, B. 1997. Antitermination of transcription of catabolic operons. Mol. Microbiol. 23: 413–241.

Saier, M.H., Jr. 1991. A multiplicity of potential carbon catabolite repression mechanisms in prokaryotic and eukaryotic microorganisms. New Biol. 3: 1137–1147.

Saier, M.H., Jr., Chauvaux, S., Cook, G.M., Deutscher, J., Paulsen, I.T., Reizer, J., and Ye, J.J. 1996. Catabolite repression and inducer control in Gram-positive bacteria. Microbiology 142: 217-230.

Saier, M.H., Jr., Chauvaux, S., Deutscher, J., Reizer, J., and Ye, J.J. 1995. Protein phosphorylation and regulation of carbon metabolism in gram-negative versus gram-positive bacteria. Trends Biochem. Sci. 20: 267–271.

Saier, M.H., Jr., and Crasnier, M. 1996. Inducer exclusion and the regulation of sugar transport. Res. Microbiol. 147: 482–489.

Sizemore, C., Buchner, E., Rygus, T., Witke, C., Götz, F., and Hillen, W. 1991. Organization, promoter analysis and transcriptional regulation of the *Staphylococcus xylosus* xylose utilization operon. Mol. Gen. Genet. 227: 377–384.

Sizemore, C., Wieland, B., Götz, F., and Hillen, W. 1992. Regulation of *Staphylococcus xylosus* xylose utilization genes at the molecular level. J. Bacteriol. 174: 3042–3048.

Stülke, J., Arnaud, M., Rapoport, G., and Martin-Verstrate, I. 1998. PRD – a protein domain involved in PTS-dependent induction and carbon catabolite repression of catabolic operons in bacteria. Mol. Microbiol. 28: 865–874.

Tangney, M., Priest, F.G., and Mitchel, W.J. 1993. Two glucose transport systems in *B. licheniformis*. J. Bacteriol. 175: 2137–2142.

Tobisch, S., Stülke, J., and Hecker, M. 1999. Regulation of the *lic* operon of *Bacillus subtilis* and characterization of potential phosphorylation sites of the LicR regulator protein by site-directed mutagenesis. J. Bacteriol. 181: 4995–5003.

van Den Bogaard, P.T., Kleerebezem, M., Kuipers, O.P., and de Vos, W.M. 2000. Control of lactose transport, β-galactosidase activity, and glycolysis by CcpA in *Streptococcus thermophilus*: evidence for carbon catabolite repression by a non-phosphoenolpyruvate-dependent phosphotransferase system sugar. J. Bacteriol. 182: 5982–5989.

Viana, R., Monedero, V., Dossonnet, V., Vadeboncoeur, C., Perez-Martinez, G., and Deutscher, J. 2000. Enzyme I and HPr from *Lactobacillus casei*: their role in sugar transport, carbon catabolite repression and inducer exclusion. Mol. Microbiol. 36: 570–584.

Wagner, A., Küster-Schöck, E., and Hillen, W. 2000. Sugar uptake and carbon catabolite repression in *Bacillus megaterium* strains with inactivated *ptsHI*. J. Mol. Microbiol. Biotechnol. 2: 587–592.

Wagner, E., Marcandier, S., Egeter, O., Deutscher, J., Götz, F., and Brückner, R. 1995. Glucose kinase-dependent catabolite repression in *Staphylococcus xylosus*. J. Bacteriol. 177: 6144–6152.

Zalieckas, J.M., Wray, L.V., Jr., and Fisher, S.H. 1999. Trans-acting factors affecting carbon catabolite repression of the *hut* operon in *Bacillus subtilis*. J. Bacteriol. 181: 2883–2888.

Chapter 23

Carbon Catabolite Repression in *Bacillus subtilis*: Mechanisms Beside the Main Regulator CcpA

Michael K. Dahl

Abstract

The past decade has witnessed an exiting unveiling of numerous molecular mechanisms that characterize signal transduction by protein-protein interaction. The recent findings encouraged an increasing effort to understand the sequential metabolism of different sugars available as energy sources at the same time. It seems probable that at least three principle mechanisms which act together or separately, mediate carbon catabolite repression (CCR) depending on the system which is under metabolic control: i) by the main signal transducing chain via the ATP-dependent HPr-kinase, HPr(Ser46~P) or alternatively Crh via the central component CcpA and its interaction with *cre*, ii) by signals sensed from the specific regulators directly or via phosphorylation by HPr, iii) by inducer exclusion based on the concurrence of the enzyme IIA^{Glc} domain of the glucose permease, and other PTS-dependent permeases composed only of the B and C domains and lacking the enzyme IIA domain.

Introduction

Together with *Escherichia coli, Bacillus subtilis* is one of the best-understood prokaryotes. Its entire DNA sequence has been available since autumn 1997, when the genome project was completed (Kunst *et al.*, 1997; Moszer, 1998; SubtiList, 1999), yielding a wealth of insights that allow multifarious new approaches to investigating biological molecular mechanisms. *B. subtilis* has a single 4,214,814 bp length chromosome, which harbors about 4,100 open reading frames, reflecting potential protein-encoding genes (Kunst *et al.*, 1997; Moszer, 1998; SubtiList, 1999). General properties of the genes in the genome were classified according to their functional features. At present, about half of the genes can be assigned to proteins with a defined or probable function, and half of the open reading frames – about 2800 – have no clear or biochemically identified function (Kunst *et al.*, 1997; Moszer, 1998; SubtiList, 1999).

Currently, three main topics are of significant interest: i) clarifying the protein function of gene products with no clear or biochemically identified function, ii) structure-function relations of proteins with known functions in particular fields and iii) regulatory networks.

One of these numerous networks plays an important global regulatory role in the sequential metabolism of different carbon sources, such as sugars available as energy sources at the same time. In order to survive and divide, bacterial cells need to take up nutrients such as carbon sources. Bacterial transporters are generally quite specific for one or a few substrates. As a consequence, there are a great variety of transport proteins with corresponding metabolic enzymes (Saier, 2000). The expression of each of these systems underlies specific regulation by induction mediated by the presence of the corresponding substrate in the growth medium and, in many – if not all – cases by regulatory networks (Stulke and Hillen, 2000). These networks are necessary to coordinate all biochemical processes of the cell depending on its physiological state and the environmental conditions. For example, the presence of two different sugars as carbon and energy sources in the media of Bacilli results in a sequential metabolism leading to the hierarchical use of saccharides (Hueck and Hillen, 1995). The expression of genes involved in catabolism of many substrates depends on their presence (induction) and the absence of carbon sources that can be well metabolized (catabolite repression; CCR) (Saier Jr. *et al.*, 1996). This phenomenon is well-known in the Gram-negative bacteria *E. coli,* where the cytoplasmic cAMP concentration plays a crucial role, as it modulates the activity of the cAMP-binding protein CAP. This protein activates the genes and operons under general carbon regulation in its cAMP bound form (Busby and Ebright, 1999; Postma *et al.*, 1993). However, in the Gram-positive bacteria *B. subtilis* no analogous system was identified, neither the cAMP level plays an important role nor a CRP homologous protein could be found (Saier Jr. *et al.*, 1995). In contrast, CCR in *B. subtilis* is negatively regulated by a repressor protein CcpA, belonging to the LacI / GalR repressor family (Chambliss, 1993; Henkin *et al.*, 1991; Hueck and Hillen, 1995; Weickert and Adhya, 1992). In this regulatory network one transporter family, the phosphoenolpyruvate-dependent sugar phosphotransferase system (PTS), has an important regulatory function. Operationally, the multi-component PTS comprises both membrane-localized and cytoplasmic proteins that concertedly catalyze the simultaneous phosphorylation and vectorial translocation of sugar across the cytoplasmic membrane. In the presence of phosphoenolpyruvate in the cytoplasm, enzyme I is autophosphorylated on a histidine residue, and subsequently enzyme I~P transfers the phosphoryl group to a histidine residue at position 15 on HPr (Figure

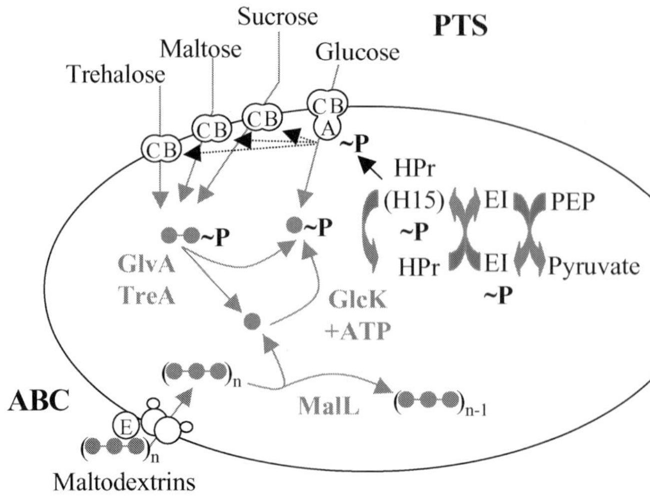

Figure 1. Appearance of glucose-6-phosphate in *B. subtilis*. The upper part of the picture presents PT systems mentioned in the text, the lower part presents the maltodextrin ABC transporter. C, B and A denote functional domains of PTS permeases as described in the text. E denotes the maltodextrin-binding protein, which is linked by a lipid anker to the cell wall. The maltodextrin system metabolizes different sizes of dextrins. This variation is indicated by n. P denotes the phosphoryl group. Doted lines indicate alternative interactions for the phosphoryl transfer from enzymeIIA^{Glc} to enzymeIIB of several PTS permeases as indicated. Grey dots denotes one glucose unit with the exception for the sucrose system for which it should be seen as a glucose and fructose subunit. For details see text.

1) (Postma *et al.*, 1993). In turn, HPr(His~P) phosphorylates the active site histidine in enzyme IIA (which in *B. subtilis* is a functional domain of one polypeptide enzyme IICBA^{Glc} glucose permease) (Gonzy-Tréboul *et al.*, 1991; Reizer *et al.*, 1999; Sutrina *et al.*, 1990; Zagorec and Postma, 1992), from where the phosphoryl group is transferred to a cysteine or histidine residue in the enzyme IIB component. Upon phosphorylation of enzyme IIB, the membrane spanning enzyme IIC domain is activated and translocation takes place, followed by carbohydrate phosphorylation (for an overview see Postma *et al.*, 1993 and Saier Jr. and Reizer, 1994). In the case of disaccharide-transporting PTS, the intracellular disaccharide phosphates must be hydrolyzed to their constituent hexose 6-phosphate and a glycone moieties, prior to catabolism via energy-yielding pathways.

Alternatively to phosphorylation of the histidine residue at position 15 of the phosphocarrier protein HPr during the phosphoryl transfer, the protein is also regulatorily phosphorylated at serine position 46 by an ATP-dependent, metabolically activated HPr kinase (Deutscher *et al.*, 1994; Reizer *et al.*, 1998). HPr kinase activity is severely inhibited by P_i and HPr (His~P) and stimulated by a catabolic intermediate, fructose 1,6-bisphosphate (FBP). The HPr kinase from *B. subtilis* forms homo-oligomeric structures, which are related to cooperative binding of FBP (Jault *et al.*, 2000). As mentioned above, in Bacilli the regulatory mechanism of CCR contains the central component CcpA (Henkin *et al.*, 1991; Hueck and Hillen, 1995; Weickert and Adhya, 1992). CcpA interacts with *cis* active DNA elements (*cre*) (Fujita *et al.*, 1995; Gösseringer *et al.*, 1997) located in the promoter regions of the genes or operons subject to CCR (Weickert and Chambliss, 1990). This CcpA-DNA interaction is triggered by the HPr protein of the phosphoenolpyruvate-dependent phosphotransferase system (PTS) when it is phosphorylated at Ser46 by the ATP dependent HPr kinase (Deutscher *et al.*, 1994; Reizer *et al.*, 1998).

A gene encoding an HPr-like protein Crh (catabolite repression HPr) composed of 85 amino acids residues has been discovered in *B. subtilis*. It contains 45% identical amino acids compared to HPr (Galinier *et al.*, 1997). However, Crh is not phosphorylated by PEP and Enzyme I, because the active site His-15 of HPr is replaced by glutamine. Nevertheless, the protein is phosphorylated at Ser-46 by an ATP-dependent HPr-kinase (Galinier *et al.*, 1998). Results obtained by mutational analysis and combinatory gene disruption of both genes suggested that both HPr and Crh mediate CCR in gram-positive bacteria.

However, on one hand inactivation of the central regulatory component CcpA leads to the loss of CCR of many systems under carbon mediated regulation which fits in with the model of the concerted action of HPr-kinase, HPr and CcpA-*cre* interaction. On the other hand, several systems or monitored enzymatic activities still underlie glucose or fructose mediated CCR, even in genetic backgrounds leading to the loss of CcpA or carrying the no longer regulatory phosphorylatable variant HPr(S46A) (*ptsH1* mutant)(Deutscher *et al.*, 1994; Helfert *et al.*, 1995). Glycerol kinase and maltose inducible α-glucosidase expression which remain sensitive to CCR both in a *ptsH1* strain and a CcpA minus strain are examples for this phenomenon (Deutscher *et al.*, 1994; Schönert *et al.*, 1998). Another interesting observation has been made studying α-amylase expression, which also underlies CcpA-dependent CCR (Henkin *et al.*, 1991; Nicholson *et al.*, 1987). CcpA and *cre* are essential for glucose repression, but the *ptsH1* mutation did not prevent glucose repression of α-amylase production (Voskuil and Chambliss, 1996). Furthermore, as we have reported previously in context with an analysis of the trehalose PTS of *B. subtilis*, CcpA inactivation leads to the loss of fructose repression of phospho-α-(1,1)-glucosidase TreA synthesis (see below), but does not affect glucose repression (Helfert *et al.*, 1995). In addition, the *ptsH1* mutation abolishes neither glucose nor fructose repression (Helfert *et al.*, 1995). Therefore, although central components mainly mediate CCR, there are indications that additional mechanisms might exist for different carbon metabolizing systems (Bürklen *et al.*, 1998; Dahl, 1997; Dahl and Hillen, 1995; Dahl *et al.*, 1995; Helfert *et al.*, 1995; Rosana-Ani *et al.*, 1999). The following chapters will mainly focus on such alternative signaling pathways suggested and known so far which are involved in CcpA-independent CCR in the Gram-positive organism *B. subtilis*.

The Trehalose PTS

While a great number of organisms synthesize the non-reducing sugar trehalose and use it as an osmoprotectant, *B. subtilis* itself does not (Arguelles, 2000; DeSmet *et al.*, 2000; Thevelein, 1984). On the contrary, when trehalose is present in the growth medium, *B. subtilis* degrades the sugar and uses it as its sole carbon and energy source (Helfert *et al.*, 1995; Kennett and Sueoka, 1971). This trehalose metabolizing system belongs to the PTS family and harbors a specific permease enzyme II, composed of the two functional domains B and C (Figure 1). It is therefore grouped to the sucrose permease family (Reizer *et al.*, 1999). After all, the enzymeIIBC^{Tre} HPr-(His15)-P-dependently phosphorylates the disaccharide during translocation over the membrane to trehalose-6-phosphate presumably in cooperation with the enzymeIIA^{Glc} domain (see below (Dahl, 1997). Besides the specific enzyme IIBC^{Tre}, also named TreP, *B. subtilis* contains a cytoplasmic phospho-α-1,1-glucosidase (phosphotrehalase TreA) able to hydrolyze trehalose-6-phosphate into the two monosaccharide subunits glucose-6-phosphate and glucose (Gotsche and Dahl, 1995; Helfert *et al.*, 1995). The phosphorylated form of the glucose can directly enter glycolysis, whereas the resulting non-phosphorylated glucose requires further ATP-dependent phosphorylation by the glucose kinase (Figure 1) (Gotsche and Dahl, 1995; Rosana-Ani *et al.*, 1999; Skarlatos and Dahl, 1998). Interestingly, TreA is also able to hydrolyze the substrate analogous *para*-nitophenyl-glucopyranoside (PNPG), making the gene suitable as a reporter gene (Schöck *et al.*, 1996; Spiegelhalter and Bremer, 1998).

DNA sequence and Northern-blot analysis clearly demonstrated that the genes *treP* (coding for the trehalose permease) and *treA* (coding for the phosph-α-1,1-glucosidase) are co-transcribed (*tre*-operon) (Schöck and Dahl, 1996a; Schöck and Dahl, 1996b). Another open reading frame could be identified downstream of *treA*. It codes for the specific trehalose repressor, that negatively regulates the *tre*-operon as demonstrated by a *treR* deletion, leading to constitutive expression of the *tre* operon and *in vitro* by protein DNA interaction (see below) (Bürklen *et al.*, 1998; Schöck and Dahl, 1996b). The transcription of *treR* remains unclear up to now. Two possible scenarios were discussed: i) a weak promoter might be located upstream of *treR* leading to some copies of TreR, or ii) some of transcriptional termination takes not place after *treA* and lead to a read through and in turn again to some copies of TreR (Bürklen *et al.*, 1998; Schöck and Dahl, 1996b). Both models would ultimately yield sufficient copies of TreR to regulate the *tre*-operon.

TreR has been heterologously overproduced using an overexpression system in *Bacillus megaterium* (Rygus and Hillen, 1991) and purified to homogeneity for *in vitro* characterizations (Bürklen *et al.*, 1998). It can be concluded from the results of this study, that the trehalose repressor exclusively mediates induction of the *tre* operon. Trehalose-6-phosphate was identified as the molecular inducer which inhibits TreR *tre* operator interaction in dependency of the concentration, and in turn leads to transcription of the *tre* operon (Bürklen *et al.*, 1998).

In this study, we have also provided evidence that TreR acts as a dimer (Bürklen *et al.*, 1998). Furthermore, *in vitro* titration experiments of the wild type operator containing DNA fragment revealed two different complex formations of lower and higher mobility in DNA-retardation experiments (Bürklen *et al.*, 1998). Therefore, we assumed that one TreR dimer binds each of the two discovered operators *O1* and *O2*, both of which are located upstream of *treP* (Bürklen *et al.*, 1998; Schöck and Dahl, 1996b). Additional results of DNA retardation experiments obtained with DNA fragments containing either one of the two operators support this observation, since only one complex formation was observed even at high protein concentrations (Bürklen *et al.*, 1998). However, operators *O1* and *O2* are bound by TreR with different efficiencies, as indicated by the about 25-fold higher protein: DNA ratio necessary for half-maximal DNA binding for operator *O1* compared to *O2*. Furthermore, the presence of both operators in the same DNA fragment reveals a 2.5-fold higher efficiency over *O2* and a 63-fold higher efficiency over *O1* binding. These data strongly suggest cooperative binding of TreR to the *tre* operators. The trehalose repressor presumably binds each operator as a dimer. Thus, two TreR dimers would achieve full occupation of the wild type operator. Whether TreR creates a tetramer complex in the presence of wild type operator DNA remains to be analyzed. But it should be noted that both operators are 32 bp apart, which corresponds to three DNA helix turns. Therefore, one would assume that binding of both operators takes place at the same side, supporting an interaction of two dimers. The negative influence of the molecular inducer indicates an inactivation of TreR, which results in an inability to bind *tre* operators. This inactivation could be due to the loss of the ability to create oligomers correlated with the loss of the capability to bind *tre* operators (Bürklen *et al.*, 1998).

Glucose-6-Phosphate Acts as an Anti-Inducer

An interesting perspective is opened by the fact that trehalose belongs to the PTS sugars and that after trehalose hydrolysis products occur which, when present in the growth medium, lead to repression of the trehalose system. The product of the glucose kinase-catalyzed reaction (see below) as well as from the glucose translocation over the membrane via the specific glucose PTS is glucose-6-phosphate, which has been shown to act as an anti-inducer in the TreR regulated expression of the *tre*-operon (Bürklen *et al.*, 1998). The presence of glucose-6-phosphate clearly inhibits the interaction between trehalose-6-phosphate and TreR and therefore interferes with the molecular inducer. These observations strongly indicate that different effectors regulate TreR *tre* operator interaction, which could be involved in mediating signals causing induction and CCR of the trehalose system. Such an activity of glucose-6-phosphate could explain one alternative mechanism triggering CCR on the transcriptional level.

Another known case, where the repressor is presumably involved in CCR of its specifically regulated operon, are the xylose systems of *B. subtilis* and *B. megaterium.* (Dahl *et al.*, 1995; Kraus *et al.*, 1994). Besides xylose functioning as an inducer interacting with the xylose repressor, glucose-6-phosphate was identified as an anti-inducer *in vitro* and *in vivo* analysis (Dahl and Hillen, 1995).

Inducer Exclusion in *B. subtilis*

Undoubtedly, HPr-kinase, HPr, CcpA and *cre* are the central components of carbon catabolite repression in Bacilli (Hueck and Hillen, 1995). One alternative and CcpA independent glucose triggered CCR has been discussed above showing that glucose-6-phosphate directly interacts with specific regulators acting as an anti-inducer (Bürklen *et al.*, 1998; Dahl and Hillen, 1995; Dahl *et al.*, 1995). However, it is not clear to which extent this alternative CCR contributes to the complex regulatory machinery in wild type cells. Another mechanism for CcpA-independent CCR has been postulated as inducer exclusion involving the PTS glucose permease enzyme $IIBCA^{Glc}$. Several PTS permeases are composed of the enzyme IIB and C domains in one polypeptide which is not fused to the A domain, necessary for phosphoryl transfer (Postma *et al.*, 1993; Reizer *et al.*, 1999). Growth analysis at either glucose or trehalose of a *B. subtilis* strain missing enzyme $IIBCA^{Glc}$, but still expressing *ptsH* and *ptsI* (encoding for HPr and enzyme I, respectively) results in a 2.3 to 2.6-fold increased generation time in both cases (Dahl, 1997). The interpretation of these results was that enzyme $IIBCA^{Glc}$ contributes to trehalose metabolism and might be necessary for the phosphoryl group transfer (Figure 1) (Dahl, 1997). Biochemical evidence of a kind of cross talk between domains enzyme IIA^{Glc} and enzyme IIB^{Scr} in *B. subtilis* has been described by Sutrina *et al. in vitro* (Sutrina *et al.*, 1990). They were able to demonstrate the appearance of sucrose~P by using inside out membrane vesicles containing enzyme $IIBC^{Scr}$ and harboring sucrose inside when the purified truncated enzyme IIA^{Glc} domain is present outside. Together with *in vivo* data this observation allows us to postulate that glucose permease of *B. subtilis* also energizes the sucrose, trehalose and presumably maltose permeases (see below, Figure 1), although direct interaction of enzyme IIA^{Glc} with these three permeases has not been detected so far. Nevertheless, these sets of data consequently led to the conclusion that inducer exclusion is indeed one of the additional regulatory mechanisms mediating CCR of some catabolically regulated systems. In this type of mechanism, several PTS sugar permeases interfere with the phosphoryl groups of the energized enzyme IIA^{Glc} domain. In the presence of glucose the phosphoryl group is translocated to the B domain in the same polypeptide, supporting glucose transport and phosphorylation. In turn, no phosphoryl group is available for other PTS permeases lacking an A domain. As a consequence the corresponding sugar is not transported into the cytoplasm, and, therefore, no inducer is present interacting with the specific regulators.

The Maltose PTS

Another disaccharide besides sucrose and trehalose that serves as a carbon source is maltose. Maltose transport in Gram-positive organisms has been reported to occur both via a PTS mechanism, such as in a number of streptococci (Martin and Russell, 1987) and non-PTS systems, such as in *Bacillus licheniformis* and *Bacillus popillae* (Tangney *et al.*, 1992; Tangney *et al.*, 1993; Taylor and Costilow, 1977). On one hand, previous studies on the maltose metabolism in *B. subtilis* regarding the effect of uncouplers on maltose uptake suggested that this maltose utilization system belongs to proton motif driven transport systems or ABC transporters (Tangney *et al.*, 1992). Tangney and coworkers also showed that extracts prepared from cultures of *B. subtilis* grown on maltose lacked maltose phosphotransferase system activity (Tangney *et al.*, 1992). This study also yielded evidence for the presence of maltose phosphorylase activity and a glucose kinase (which both indeed exist, see below), from which it was interpreted that accumulated maltose is degraded by maltose phosphorylase in the cytoplasm. Undoubtedly, maltose is accumulated by *B. subtilis* cells; however, recent studies clearly show that in contrast to the well-known maltose ABC-transporter system in *E. coli*, in *B. subtilis* this disaccharide is presumably transported via a maltose specific PTS-dependent system (Reizer *et al.*, 1999; Thompson *et al.*, 1998; Schönert and Dahl, in preparation). In 1996, as participants in the *Bacillus* genome project (Kunst *et al.*, 1997), Sekiguchi and co-workers determined the nucleotide sequence of a DNA-fragment near the 76° region of the *B. subtilis* genome. The sequence analysis and search for the protein data bases revealed that the deduced amino acid sequences of open reading frames *glvA* (formerly *glv*-1) and *glvC* (formerly *glv*-2) exhibited similarity to 6-phospho-β-glucosidase and enzyme IICB domains of PTS-dependent permeases. Due to the domain order, the GlvC permease (also named MalP) has been grouped, unlike SacP (sucrose) and TreP (trehalose), to the glucose permease family (Reizer *et al.*, 1999). A third open reading frame, *glvR*, which gene product acts as regulator for the maltose PTS (Yamamoto *et al.*, 2001), is located between the genes *glvA* and *glvC* (Kunst *et al.*, 1997; SubtiList, 1999). *In silico* and co-expression analysis of GlvA and GlvC indicates that these three genes might constitute an operon (Kunst *et al.*, 1997; Reizer *et al.*, 1999; Thompson *et al.*, 1998; Schönert and Dahl, in preparation). GlvC, the putative PTS dependent maltose permease, was first proposed to encode a β-glucoside-specific enzyme II (Yamamoto *et al.*, 1996). Analyzing the doubling time on maltose growth medium of a strain in which *glvC* was inactivated compared to the wild type revealed a 7.4-fold increase of the generation time (Reizer *et al.*, 1999). In this strain, maltose uptake is no longer detectable (Schönert and Dahl, in preparation). GlvA was overproduced and purified, and its enzymatic characteristics were described (Thompson *et al.*, 1998). Thompson and co-workers were able to demonstrate that GlvA is a novel 6-phospho-α-glucosidase and uses maltose 6-phosphate as natural substrate. In contrast to other phosphoglycosylhydrolases, GlvA from *B. subtilis* exhibits specific requirements for both divalent metal (Mn^{2+}, Fe^{2+}, Co^{2+}, or Ni^{2+}) and NAD(H) for activity (Thompson *et al.*, 1998). Taken together, these data provide strong evidence that in contrast to *E. coli* in *B. subtilis* maltose uptake belongs to the PTS.

Maltose in the growth medium can induce this presumed maltose PTS, which, in analogy to the related sucrose and trehalose systems, underlies CCR (Tangney *et al.*, 1992; Schönert and Dahl, in preparation). As observed by Tangney and coworkers, the glucose PTS is functionally expressed in the presence of maltose but maltose uptake is subject to glucose triggered CCR (Tangney *et al.*, 1992). Furthermore, in this study, maltose transport is almost completely abolished in an enzyme I defective strain, which fully corresponds with the observation that maltose transport mainly occurs via the PTS. One interpretation of the co-expression of enzyme $IICB^{Mal}$ and enzyme $IICBA^{Glc}$ in the presence of maltose is again the need of the A domain from the glucose permease for the phosphoryl transfer to the B domain of the maltose permease in analogy to the trehalose and the sucrose systems described above (Dahl, 1997; Sutrina *et al.*, 1990).

Interestingly, *B. subtilis* contains an additional maltose inducible α-glucosidase (sucrase, maltase, isomaltase), which is part of a nine-gene spanning operon encoding for a specific maltodextrin utilization system (see below) (Schönert *et al.*, 1998; Schönert *et al.*, 1999). Therefore, it seems that the maltose and malotdextrin systems constitute a maltodextrin regulon.

The Presumed Maltodextrin Network

As mentioned above a maltose inducible α-glucosidase is still glucose repressed in strains lacking CcpA or carrying the HPr(S46A) mutation. However, this α-glucosidase is not localized in the presumed maltose-PTS operon (Schönert *et al.*, 1998; SubtiList, 1999). The corresponding gene (*malL*) of the maltose-inducible α-glucosidase has been identified in a nine-gene spanning gene cluster which is presumably organized as an operon (Rosana-Mesak and Dahl, 2000; Schönert *et al.*, 1998; Schönert and Dahl, in preparation). The disruption of the *malL* gene leads to a complete loss of maltose inducible α-glucosidase activity, indicating the existence of one exclusively maltose inducible α-glucosidase (Schönert *et al.*, 1998). MalL activity is only observable in those cells, which were grown in maltose containing medium (Schönert *et al.*, 1998) and also in the presence of either amylose, starch or glycogen (Schönert *et al.*, 1999). This induction process probably occurs due to cleavage products derived from amylose, starch or glycogen hydrolysis, leading to smaller sugar subunits or precursors which, when taken up, directly or indirectly trigger MalL expression. This interpretation is supported by the results observed in an α-amylase minus strain in which the induction of MalL synthesis by amylose or starch is nearly completely abolished (Schönert *et al.*, 1999). This is partially also the case for induction by glycogen. Surprisingly, in these investigations we observed a temporal expression of MalL, independent of the use of inducing substrates. These data present evidence that MalL expression principally underlies two distinct temporary regulations, which resembles those observed for α-amylase activity (Weickert *et al.*, 1990). Transcription of *amyE* and concomitant amylase production is temporally activated at the onset of stationary phase (T_0) and temporarily turned off at T_2 (Weickert *et al.*, 1990). Several studies show that the ABC transporter to which MalL belongs is presumably exclusively involved in the metabolism of maltodextrins, with longer chains than maltotriose (Schönert *et al.*, 1999; Schönert and Dahl, in preparation, Figure 1). This makes it interesting to study the regulation of this ABC transporter coding operon, which is inducible by the presence of the PTS-sugar maltose in the growth medium. Both systems, maltose PTS and maltodextrin ABC-transporter are co-expressed and potentially induced by the same inducer.

The Glucose Kinase as a Component of Disaccharide and Dextrin Metabolism and its Potential Contribution on CCR

Another challenging question beside the physiological function of GlcK is its potential regulatory function. A protein family was characterized designated ROK (repressor, ORF, kinase) (Titgemeyer *et al.*, 1994) containing seven Xyl repressors from Gram-positive bacteria, the *N*-acetylglucosamine repressor (NagC) from *Escherichia coli* and the three sugar kinases: glucokinase from *Streptomyces coelicolor*, fructokinase of *Zymomonas mobilis* and fructokinase encoded within the sucrose regulon of *Streptococcus mutans*. These proteins are grouped in one superfamily according to the results of multiple alignment. The N-terminal 80 residue regions thereby exhibited low average similarities since the kinases lack the helix-turn-helix motif. The adjacent regions (alignment positions 140–420 in (Titgemeyer *et al.*, 1994)) exhibit striking sequence similarity showing six peaks of sequence conservation. These sequence analyses allow the conclusion that catalytic residues involved in ATP-binding and phosphoryl transfer might be expected to be conserved only in the kinases, while sugar-binding residues and structural residues in maintaining the three-dimensional structures involved in sugar binding are more likely to be conserved in the repressors as well as in the kinases. Discovering the glucose kinase of *B. subtilis,* which also belongs to the ROK family, it must be postulated that unphosphorylated glucose is present as a substrate for the enzyme (Skarlatos and Dahl, 1998). One mechanism leading to glucose in the cytoplasm is the hydrolysis of trehalose 6-phosphate or maltose 6-phosphate into glucose and glucose-6-phosphate, as well as the hydrolysis of maltodextrins resulting in unphosphorylated glucose (Figure 1). In *Staphylococcus xylosus* and *Streptomyces coelicolor*, glucose kinases are involved in CCR (Angell *et al.*, 1992; Wagner *et al.*, 1995) and have also been discussed for *B. megaterium* (Späth *et al.*, 1997). In contrast to *B. subtilis*, no PTS for glucose exists in *S. coelicolor.* Since glucose utilization in *B. subtilis* is dependent on sugar uptake by the glucose PTS correlated with sugar phosphorylation to glucose-6-phosphate, one would expect that in *B. subtilis* the glucose kinase is only involved in the metabolism of disaccharides and dextrins. However, it cannot be excluded that the glucose kinase acts as part of an additional glucose metabolizing system which does not belong to the PTS. Glucose PTS minus strains are still able to grow on glucose as a sole carbon source, although, with reduced generation times, which can be explained by the presence of a hexose:H^+ symporter (Gonzy-Tréboul *et al.*, 1991; Paulsen *et al.*, 1998). In a strain lacking a functional glucose kinase we observed the efflux of unphosphorylated glucose derived from disaccharide hydrolysis, which can in turn re-enters the cell via the specific glucose PTS and subsequently triggers glucose repression (Rosana-Ani *et al.*, 1999). Recently we could identify a membrane protein named GluP, encoded by a gene located in the same operon as that coding for the glucose kinase (Mesak and Dahl, in preparation). Inactivation of *gluP* leads to the loss of glucose export and surprisingly to a filamentous phenotype (Mesak and Dahl, in preparation). We could also detect glucose transport mediated by GluP, however, significant uptake is only observable when *B. subtilis* cells were grown under extremely high glucose concentrations. Nevertheless, these observed effects presumably do not represent a CcpA-independent CCR. Therefore, the direct or indirect contribution of the glucose kinase in CCR of *B. subtilis* still remains unclear.

Acknowledgements

I thank U. Ehmann for her interest in my work and her never-ending support. K. Oliva and S. Schönert are acknowledged for critical reading of the manuscript. W. Hillen is greatly appreciated for hospitality in his department of microbiology and many helpful discussions. The work in my laboratory has received generous financial support by different grants from the Deutsche Forschungsgemeinschaft DFG Bonn and the Fonds der Chemischen Industrie.

References

Angell, S., Schwarz, E., and Bibb, M.J. 1992. The glucose kinase of *Streptomyces coelicolor* A3(2): its nucleotide sequence, transcriptional analysis and role in glucose repression. Mol. Microbiol. 6: 2833–2844.

Arguelles, J.C. 2000. Physiological roles of trehalose in bacteria and yeasts: a comparative analysis. Arch. Microbiol. 174: 217–224.

Bürklen, L., Schöck, F., and Dahl, M.K. 1998. Molecular analysis of the *Bacillus subtilis* trehalose repressor TreR *tre* operator interaction. Mol. Gen. Genet. 260: 48–55.

Busby, S., and Ebright, R.H. 1999. Transcription activation by catabolite activator protein (CAP). J. Mol. Biol. 293: 199–213.

Chambliss, G.H. 1993. In A.L. Sonenshein, J.A. Hoch, and R. Losick, (eds.), Carbon source-mediated catabolite repression. American Society for Microbiology, Washington, D.C., pp. 213–218.

Dahl, M.K. 1997. Enzyme II^{Glc} contributes to trehalose metabolism in *Bacillus subtilis.* FEMS Microbiol. Lett. 148: 233–238.

Dahl, M.K., and Hillen, W. 1995. Contributions of XylR, CcpA and HPr to catabolite repression of the *xyl* operon in *Bacillus subtilis.* FEMS Microbiol. Lett. 132: 79–83.

Dahl, M.K., Schmiedel, D., and Hillen, W. 1995. Glucose and glucose-6-phosphate interaction with Xyl repressor proteins from *Bacillus* spp. may contribute to regulation of xylose utilization. J. Bacteriol. 177: 5467–5472.

DeSmet, K.A.L., Weston, A., Brown, I.N., Young, D.B., and Robertson, B.D. 2000. Three pathways for trehalose biosynthesis in mycobacteria. Microbiol. 146: 199–208.

Deutscher, J., Reizer, J., Fischer, C., Galinier, A., Saier M.H. Jr., and Steinmetz, M. 1994. Loss of protein kinase-catalyzed phosphorylation of HPr, a phosphocarrier protein of the phosphotransferase system, by mutation of the *ptsH* gene confers catabolite repression resistance to several catabolic genes of *Bacillus subtilis.* J. Bacteriol. 176: 3336–3344.

Fujita, Y., Miwa, Y., Galinier, A., and Deutscher, J. 1995. Specific recognition of the *Bacillus subtilis gnt cis*-acting catabolite-responsive element by a protein complex formed between CcpA and seryl-phosphorylated HPr. Mol. Microbiol. 17: 953–960.

Galinier, A., Haiech, J., Kilhoffer, M.C., Jaquinod, M., Stulke, J., Deutscher, J., and Martin-Verstraete, I. 1997. The *Bacillus subtilis* crh gene encodes a HPr-like protein involved in carbon catabolite repression. Proc. Natl. Acad. Sci. USA 94: 8439–8444.

Galinier, A., Kravanja, M., Engelmann, R., Hengstenberg, W., Kilhoffer, M.-C., Deutscher, J., and Haiech, J. 1998. New protein kinase and protein phosphatase families mediate signal transduction in bacterial catabolite repression. Proc. Natl. Acad. Sci. USA 95: 1823–1828.

Gonzy-Tréboul, G., de Waard, J.H., Zagorec, M., and Postma, P.W. 1991. The glucose permease of the phosphotransferase system of *Bacillus subtilis*: evidence for II^{Glc} and III^{Glc} domains. Mol. Microbiol. 5: 1241–1249.

Gösseringer, R., Küster, E., Galinier, A., Deutscher, J., and Hillen, W. 1997. Cooperative and non-cooperative DNA binding modes of catabolite control protein CcpA from *Bacillus megaterium* result from sensing two different signals. J. Mol. Biol. 266: 665–676.

Gotsche, S., and Dahl, M.K. 1995. Purification and characterization of the phospho-α-1,1-glucosidase (TreA) of *Bacillus subtilis.* J. Bacteriol. 177: 2721–2726.

Helfert, C., Gotsche, S., and Dahl, M.K. 1995. Cleavage of trehalose-phosphate in *Bacillus subtilis* is catalyzed by a phospho-α-(1,1)-glucosidase encoded by the *treA* gene. Mol. Microbiol. 16: 111–120.

Henkin, T.M., Grundy, F.J., Nicholson, W.L., and Chambliss, G.H. 1991. Catabolite repression of α-amylase gene expression in

Bacillus subtilis involves a *trans*-acting gene product homologous to the *Escherichia coli lacI* and *galR* repressors. Mol. Microbiol. 5: 575–584.

Hueck, C.J., and Hillen, W. 1995. Catabolite repression in *Bacillus subtilis*: a global regulatory mechanism for the Gram-positive bacteria? Mol. Microbiol. 15: 395–401.

Jault, J.M., Fieulaine, S., Nessler, S., Gonzalo, P., Di Pietro, A., Deutscher, J., and Galinier, A. 2000. The HPr kinase from *Bacillus subtilis* is a homo-oligomeric enzyme which exhibits strong positive cooperativity for nucleotide and fructose 1,6-bisphosphate binding. J. Biol. Chem. 275: 1773–1780.

Kennett, R.H., and Sueoka, N. 1971. Gene expression during outgrowth of *Bacillus subtilis* spores. J. Mol. Biol. 60: 31–44.

Kraus, A., Hueck, C., Gartner, D., and Hillen, W. 1994. Catabolite repression of the *Bacillus subtilis xyl* operon involves a *cis* element functional in the context of an unrelated sequence, and glucose exerts additional *xylR*-dependent repression. J. Bacteriol. 176: 1738–1745.

Kunst, F., Ogasawara, N., Moszer, I., Albertini, A.M., Alloni, G., Azevedo, V., Bertero, M.G., Bessieres, P., Bolotin, A., Borchert, S., Borriss, R., Boursier, L., Brans, A., Braun, M., Brignell, S.C., Bron, S., Brouillet, S., Bruschi, C.V., Caldwell, B., Capuano, V., Carter, N.M., Choi, S.K., Codani, J.J., Connerton, I.F., Danchin, A., *et al.* 1997. The complete genome sequence of the gram-positive bacterium *Bacillus subtilis*. Nature (London) 390: 249–256.

Martin, S.A., and Russell, J.B. 1987. Transport and phosphorylation of disaccharides by the ruminal bacterium *Streptococcus bovis*. Appl. Environ. Microbiol. 53: 2388–2393.

Moszer, I. 1998. The complete genome of *Bacillus subtilis*: from sequence annotation to data managment and analysis. FEBS Lett. 430: 28–36.

Nicholson, W.L., Park, Y.-K., Henkin, T.M., Won, M., Weickert, M.J., Gaskell, J.A., and Chambliss, G.H. 1987. Catabolite repression-resistant mutations of the *Bacillus subtilis* alpha-amylase promoter affect transcription levels and are in an operator-like sequence. J. Mol. Biol. 198: 609–618.

Paulsen, I.T., Chauvaux, S., Choi, P., and Saier, M.H., Jr. 1998. Characterization of glucose-specific catabolite repression-resistant mutants of *Bacillus subtilis*: identification of a novel hexose:H+ symporter. J. Bacteriol. 180: 498–504.

Postma, P.W., Lengeler, J.W., and Jacobson, G.R. 1993. Phosphoenolpyruvate:carbohydrate phosphotransferase systems of bacteria. Microbiol. Rev. 57: 543–594.

Reizer, J., Bachem, S., Reizer, A., Arnaud, M., Saier, M.H., Jr., and Stülke, J. 1999. Novel phosphotransferase system genes revealed by genome analysis–the complete complement of PTS proteins encoded within the genome of *Bacillus subtilis*. Microbiology 145: 3419–3429.

Reizer, J., Hoischen, C., Titgemeyer, F., Rivolta, C., Rabus, R., Stülke, J., Karamata, D., Saier, M.H., Jr., and Hillen, W. 1998. A novel protein kinase that controls carbon catabolite repression in bacteria. Mol. Microbiol. 27: 1157–1170.

Rosana-Ani, L., Skarlatos, P., and Dahl, M.K. 1999. Putative contribution of glucose kinase from *Bacillus subtilis* to carbon catabolite repression (CCR): a link between enzymatic regulation and CCR? FEMS Microbiol. Lett. 171: 89–96.

Rosana-Mesak, L., and Dahl, M.K. 2000. Purification and enzymatic characterization of PgcM, a β-phosphoglucomutase and glucose-1-phosphosphate phosphodismutase of *Bacillus subtilis*. Arch. Microbiol. 174: 256–264.

Rygus, T., and Hillen, W. 1991. Inducible high-level expression of heterologous genes in *Bacillus megaterium* using the regulatory elements of the xylose-utilization operon. Appl. Microbiol. Biotechnol. 35: 594–599.

Saier, M.H., Jr., Chauvaux, S., Cook, G.M., Deutscher, J., Paulsen, T., Reizer, J., and Ye, J.-J. 1996. Catabolite repression and inducer control in Gram-positive bacteria. 1996 142: 217–230.

Saier, M.H., Jr., Chauvaux, S., Deutscher, J., Reizer, J., and Ye, J.-J. 1995. Protein phosphorylation and regulation of carbon catabolism in Gram-negative versus Gram-positive bacteria. Trends Biochem. Sci. 20: 267–271.

Saier, M.H., Jr. 2000. Vectorial metabolism and the evolution of transport systems. J Bacteriol 182: 5029–5035.

Schöck, F., and Dahl, M.K. 1996a. Analysis of DNA flanking the *treA* gene of *Bacillus subtilis* reveals genes encoding a putative specific enzyme II^{Tre} and a potential regulator of the trehalose operon. Gene 175: 59–63.

Schöck, F., and Dahl, M.K. 1996b. Expression of the *tre* operon of *Bacillus subtilis* 168 is regulated by the repressor TreR. J. Bacteriol. 178: 4576–4581.

Schöck, F., Gotsche, S., and Dahl, M.K. 1996. Vectors using the phospho-α-(1,1)-glucosidase-encoding gene *treA* of *Bacillus subtilis* as a reporter. Gene 170: 77–80.

Schönert, S., Buder, T., and Dahl, M.K. 1998. Identification and enzymatic characterization of the maltose inducible α-glucosidase MalL (sucrase-isomaltase-maltase) of *Bacillus subtilis*. J. Bacteriol. 180: 2574–2578.

Schönert, S., Buder, T., and Dahl, M.K. 1999. Properties of maltose inducible α-glucosidase in *Bacillus subtilis:* evidence for its contribution in maltodextrin utilization. Res. Microbiol. 150: 167–177.

Skarlatos, P., and Dahl, M.K. 1998. The glucose kinase of *Bacillus subtilis*. J. Bacteriol. 180: 3222–3226.

Späth, C., Kraus, A., and Hillen, W. 1997. Contribution of glucose kinase to glucose repression of xylose utilization in *Bacillus megaterium*. J. Bacteriol. 179: 7603–7605.

Spiegelhalter, F., and Bremer, E. 1998. Osmoregulation of the *opuE* proline transport gene from *Bacillus subtilis*: contributions of the sigma A- and sigma B-dependent stress-responsive promoters. Mol. Microbiol. 29: 285–296.

Stulke, J., and Hillen, W. 2000. Regulation of carbon catabolism in *Bacillus* species. Annu. Rev. Microbiol. 54: 849–880.

SubtiList 1999. In I. Mozer, and A. Danchin (eds.), Data Release R15.1. Institut Pasteur, Paris, France, http://genolist.pasteur.fr/SubtiList/genome.

Sutrina, S.L., Reddy, P., Saier, M.H., Jr., and Reizer, J. 1990. The glucose permease of *Bacillus subtilis* is a single polypeptide chain that functions to energize the sucrose permease. J. Biol. Chem. 265: 18581–18589.

Tangney, M., Buchanan, C.J., Priest, F.G., and Mitchell, W.J. 1992. Maltose uptake and its regulation in *Bacillus subtilis*. FEMS Microbiol. Lett. 97: 191–196.

Tangney, M., Priest, F.G., and Mitchell, W.J. 1993. Two glucose transport systems in *Bacillus licheniformis*. J. Bacteriol. 175: 2137–2142.

Taylor, D.C., and Costilow, R.N. 1977. Uptake of glucose and maltose by *Bacillus popillae*. Appl. Environ. Microbiol. 34: 102–104.

Thevelein, J.M. 1984. Regulation of trehalose mobilization in fungi. Microbiol. Rev. 48: 42–59.

Thompson, J., Pikis, A., Ruvinov, S.B., Henrissat, B., Yamamoto, H., and Sekiguchi, J. 1998. The gene *glvA* of *Bacillus subtilis* 168 encodes a metal-requiring, NAD(H)-dependent 6-phospho-α-glucosidase. J. Biol. Chem. 273: 27347–27356.

Titgemeyer, F., Reizer, J., Reizer, A., and Saier, M.H., Jr. 1994. Evolutionary relationships between sugar kinases and transcriptional repressors in bacteria. Microbiology 140: 2349–2354.

Voskuil, M.I., and Chambliss, G.H. 1996. Significance of HPr in catabolite repression of α-amylase. J. Bacteriol. 178: 7014–7015.

Wagner, E., Marcandier, S., Egeter, O., Deutscher, J., Götz, F., and Brückner, R. 1995. Glucose kinase-dependent catabolite repression in *Staphylococcus xylosus*. J. Bacteriol. 177: 6144–6152.

Weickert, M.J., and Adhya, S. 1992. A family of bacterial regulators homologous to Gal and Lac Repressors. J. Biol. Chem. 267: 15869–15874.

Weickert, M.J., and Chambliss, G.H. 1990. Site-directed mutagenesis of a catabolite repression operator sequence in *Bacillus subtilis*. Proc. Natl. Acad. Sci. USA 87: 6238–6242.

Weickert, M.J., Larson, L., Nicholson, W.L., and Chambliss, G.H. 1990. In M.M. Zukowski, A.T. Ganesan, and J.A. Hoch (eds.), Negative control of amylase synthesis: mutations which eliminate catabolite repression or temporal turn off. Academic Press, San Diego, Calif., Vol. 3, pp. 237–244.

Yamamoto, H., Serizawa, M., Thompson, J., and Sekiguchi, J. 2001. Regulation of the *glv* operon in *Bacillus subtilis*: YfiA (GlvR) Is a positive regulator of the operon that is repressed through CcpA and *cre*. J. Bacteriol. 183: 5110–5121.

Yamamoto, H., Uchiyama, S., and Sekiguchi, J. 1996. Cloning and sequencing of a 40.6 kb segment in the 73 degrees-76 degrees region of the *Bacillus subtilis* chromosome containing genes for trehalose metabolism and acetoin utilization. Microbiology 142: 3057–3065.

Zagorec, M., and Postma, P.W. 1992. Cloning and nucleotide sequence of the *ptsG* gene of *Bacillus subtilis*. Mol. Gen. Genet. 234: 325–328.

Chapter 24

Structure-Function Relationship and Regulation of Two *Bacillus subtilis* DNA-Binding Proteins, HBsu and AbrB

Wolfgang Klein, and Mohamed A. Marahiel

Abstract

Microorganisms use a number of small basic proteins for organization and compaction of their DNA. By their interaction with the genome, these proteins do have a profound effect on gene expression, growth behavior, and viability. It has to be distinguished between indirect effects as a consequence of the state of chromosome condensation and relaxation that influence the rate of RNA polymerase action as represented by the histone-like proteins, and direct effects by specific binding of proteins to defined DNA segments predominantly located around promoter sequences. This latter class is represented by the transition-state regulators that are involved in integrating various global stimuli and orchestrating expression of the genes under their regulation for a better adaptation to changes in growth rate. In this article we will focus on two different but abundant DNA binding proteins of the gram-positive model organism *Bacillus subtilis*, the histone-like HBsu as a member of the unspecific and the transition state regulator AbrB as a member of specific classes of DNA binding proteins.

The HBsu Protein of *Bacillus subtilis* is a Member of the Histone-like Protein Superfamily

More than 30 members of the family of histone-like proteins, all with a size of about 90 amino acids and an overall basic net charge, have been identified so far in organisms of virtually every branch of the eubacterial kingdom, archaea, cyanobacteria, plant chloroplasts, and bacteriophages (Oberto *et al.*, 1994). The *B. subtilis* genome encodes for one histone-like protein by the *hbs* gene (Kunst *et al.*, 1997). This correlates with findings that the *hbs* gene is essential, and attempts to construct a knock out mutation have failed so far (Micka and Marahiel, 1992). The HBsu protein shows a high level of similarity to the *E. coli* heterodimeric HU protein, the so far best studied histone-like protein. It shows 57% and 51% identical amino acids with the HU-subunits HU-2 and HU-1, respectively. From the occurrence of a single gene in *B. subtilis* genome it is concluded that HBsu acts as a homodimer (see Figure 1A), and our data on the DNA-binding activity of HBsu mutant proteins support this idea. When subsaturating protein concentrations of one protein were used, they exhibited full DNA binding activity by the addition of subsaturating amounts of another mutant variant (Köhler and Marahiel, 1998), indicating the formation of a functional complex of both mutant proteins. The construction of a HBsu-GFP fusion protein allowed us to visualize for the first time the cellular localization of this protein as *in vitro* analysis has shown its ability to bind DNA. Using fluorescence microscopy, HBsu-GFP fluorescence was observed inside the cell at exactly the same location as the DAPI fluorescence that is localizing the nucleoid. Furthermore, additional expression of wild-type HBsu caused the formation of a more condensed nucleoid (Köhler and Marahiel, 1997). These results and investigations using HBsu proteins with reduced DNA binding activity emphasize its principal role in DNA packaging in *B. subtilis*.

The archetype of histone-like proteins is H-NS of *Escherichia coli* that has been identified several times independently according to the researchers focus on functions like *in vitro* transcription of phage DNA, phage transposition, and lifestyle (for a review, see Nash, 1996). *In vivo* localization data of this heterodimeric protein yielded controversial results by showing correlation with metabolically active DNA in immunocytochemical studies and even distribution throughout the entire nucleoid using fluorescein labeled HU, a H-NS homologous protein (Shellman and Pettijohn, 1991). A recent model of nucleoid organization in *E. coli* suggested that HU and H-NS are located in different domains where an internal domain, thought to contain rarely transcribed DNA, is associated with H-NS while frequently transcribed sequences complexed with HU protein form the coralline appearing outer sphere of the nucleoid (McGovern *et al.*, 1994). Although these two histone-like proteins appear to play special roles in cellular physiology, *E. coli* strains lacking either HU or H-NS are viable whereas multiple mutations severely impair cellular viability (Yasuzawa *et al.*, 1992).

Structural Features of HBsu

An understanding of the molecular structure of histone-like proteins is mainly based on X-ray crystallographic analysis of the homodimeric HBst protein from *B. sterothermophilus* (Tanaka *et al.*, 1984); see Figure 1A). As HBsu of the mesophilic *B. subtilis* and HBst display 87% identity and 94% similarity, a nearly identical 3D-structure for both can be assumed. The amino terminal half of the HBst protein is composed of two α-helices connected by a turn, followed by three β-sheets. Whereas the first sheet is involved in forming the body of the protein structure, the latter two form a β-ribbon extension called the "arm" that was shown to be involved in DNA binding (Köhler and Marahiel, 1998; see Figure 1A). The interface of the dimer is largely dependent upon interactions between several hydrophobic residues scattered along the primary sequence, an attribute that is conserved in all members of this family of proteins.

DNA-Binding by HBsu

DNA binding by HBsu is independent of cofactors or additional proteins. It has been shown to bind DNA unspecifically but with a preference for curved fragments (Köhler and Marahiel, 1998). Furthermore, HBsu enables β-recombinase-mediated recombination by stabilizing a special DNA structure (Alonso *et al.*, 1995). Even though experimental data are lacking, the preference for curved DNA could be an indication that HBsu might perturb the overall structure of duplex DNA, a fact that has been seen for the *E. coli* HU protein (Hodges-Garcia *et al.*, 1989).

In a model based on the HBst structure, each monomer of the homodimer contributes two DNA binding surfaces – the β-ribbon "arm" and a segment on the flank of the body - both of which contact the minor grove of the DNA. Based on a model that has been deduced from DNA-protein photo-crosslinking experiments with a similar DNA-binding protein (Lee *et al.*, 1992; Yang and Nash, 1994), we have performed the first site-directed mutational analysis of a histone-like protein. Our results strengthened the significance of several basic residues in DNA binding as the mutated proteins showed a dramatically reduced DNA binding activity (see Figure 1C). Furthermore, this *in vitro* activity is corroborated by *in vivo* data showing a reduced growth rate, a reduced sporulation efficiency, and a less condensed nucleoid for strains carrying only a mutant *hbs* allel (Köhler and Marahiel, 1998).

Regulation of *hbs* Expression

There are only few data on the regulation of histone-like proteins that may explain the abundance of HU protein in *E. coli* in response to different growth conditions. Values of 2 to 5 ng of HU per μg total protein might represent an average that varies little, if at all, in the life cycle (Bonnefoy *et al.*, 1994). This might reflect the operation of an autoregulatory circuit, yielding an appropriate amount of HU protein and resulting in a suitably condensed nucleoid (Kohno *et al.*, 1990). Our own data on the regulation of the *B. subtilis hbs* gene are based on two lines of evidence: two promoter structures in front of the *hbs* gene were mapped. P1 is 181 nucleotides upstream of the translation start point, shows similarities to σ^A dependent promoters, and the mRNA derived from this promoter was detected mainly in exponentially growing, non-sporulating cells. P2 localizes only 21 basepairs upstream of the ATG codon, and its σ^H promoter similarity is corroborated by determining this type of mRNA exclusively in sporulating cells (Micka *et al.*, 1991). Furthermore, different promoter fragments have been cloned in front of the β-galactosidase gene and integrated as single copy into the chromosome for monitoring *hbs*::*lacZ* expression (Köhler and Marahiel, 1997). The P1 and P2 comprising fragments triggered a high β-galactosidase activity with a slight maximum around one hour before the transition into stationary phase. Studies on different P1 promoter fragments revealed two facts: a medium level of *hbs*::*lacZ* expression during exponential growth but low expression after the transition phase. Deletion of a short DNA segment located between both promoter structures, P1 and P2, selectively reduced expression by promoter1 under all conditions tested. We therefore attributed a P1-specific enhancer function to this DNA segment. In addition, P1 seems to be the target of the HBsu autoregulatory loop as conditional overexpression of wildtype HBsu selectively reduced *hbs*:*lacZ* expression directed by P1 but not by P2. Fragments of promoter2 showed only a low activity that is about 12 fold induced around one hour after entry into the sporulation phase. This induction is dependent on sporulation promoting growth conditions and strictly dependent on σ^H as tested in *spo0H* mutants (Köhler and Marahiel, 1997). Based on these data, we propose that HBsu is regulated in exponentially growing cells by P1 and an autoregulatory loop. After the transition to stationary phase either expression is promoted by the σ^H dependent P2 if sporulation is initiated, or expression of *hbs* is reduced under non-sporulating conditions and HBsu might be in part replaced by proteins like the minor small, acid-soluble spore proteins (Ross and Setlow, 2000).

AbrB of *Bacillus subtilis* is the Major Transition State Regulator

At the end of exponential growth in a bacterial culture when conditions for growth become sub-optimal, many genes for alternative catabolic and anabolic pathways to produce substances such as antibiotics, toxins, and polymer degrading enzymes become derepressed. Therefore, the transition state between exponential growth and stationary phase can be looked at as a crossroad where cells still express some growth-related functions but begin to express new gene products that are necessary for survival in a nutrient-deprived, hostile environment. In *Bacillus* spp, typical functions of the transition state encompass the development of genetic competence for the uptake of DNA, production of antibiotics and extracellular enzymes, and the synthesis of flagella. Genes encoding such functions are regulated, at least in part, by small, global negative transcriptional factors that have been termed "transition state regulators" as they silence the expression of these functions at inappropriate times (Strauch, 1993). None of the known mutations in these regulators leads to significant sporulation defects, but they are involved in coordinating and transmitting

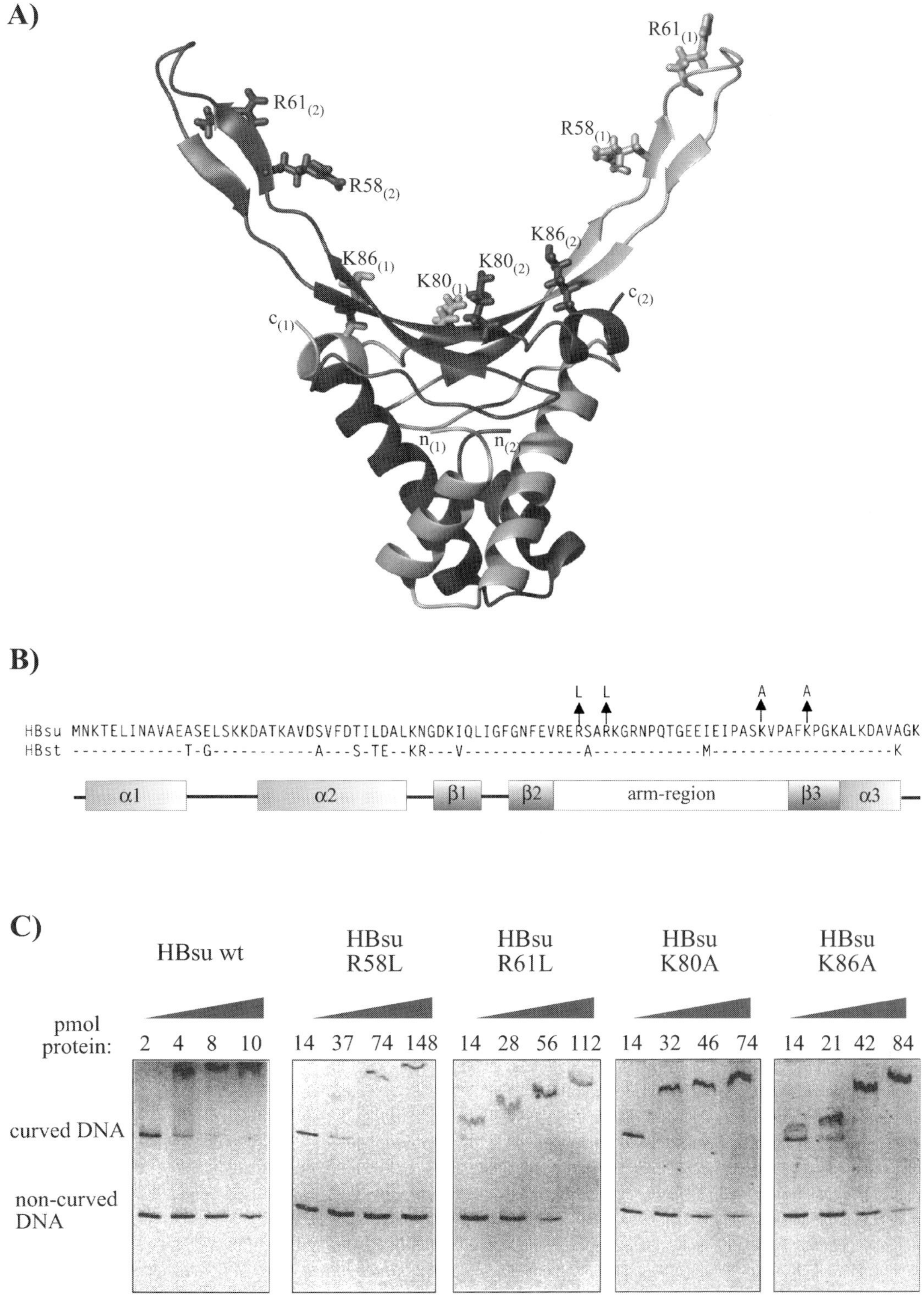

Figure 1. Structural and functional features of HBsu as a member of the histone-like protein family. A) The three-dimensional structure of the homodimer HBst protein of *B. stearothermophilus* was used as a model for site-directed mutagenesis in HBsu of *B. subtilis*. The basic residues selected for mutagenesis are shown with their side chains and labeled with single-letter amino acid code in both proteins. N- and C-termini are indicated by n and c, respectively, numbers refer to the two subunits. Modelling of the structure was done according to Konradi *et al.*, 1996. B) Comparison of the amino acid sequences of HBsu and HBst which are 86% identical. Mutationally altered amino acid residues are indicated above the sequence, secondary structure elements of HBst are shown schematically below the sequences. C) Gel retardation analysis of HBsu and mutant proteins with curved and non-curved DNA.

signals received by the sporulation-sensing network and orchestrating the transition state functions. One of these regulators is the product of the *abrB* gene that controls several transition state genes, some of which are necessary for sporulation (Perego *et al.*, 1988). In this manner, AbrB coordinates the sporulation response with other transition state processes. Furthermore, AbrB is implicated in the regulation of other transition state regulators such as Hpr (originally identified by mutations causing overexpression of extracellular proteases (Perego and Hoch, 1988)) and Sin (inhibitor of sporulation and protease production when overexpressed (Gaur *et al.*, 1988; see Figure 2)). AbrB is looked at as the central transition state regulator in *Bacillus subtilis*.

The *abrB* gene was isolated as pseudorevertants in a *spo0A* mutant background (*spo0A* mutants show impaired antibiotic production, lack of motility, inability to aquire genetic competence and to initiate sporulation) by restoring antibiotic production and motility but not sporulation (Guespin-Michel, 1971). Cloning of this locus revealed a single gene encoding for a 94 amino acid protein (10.5 kDa), and the pseudorevertants isolated resulted in loss of function (Perego *et al.*, 1988) with one class lowering the expression of *abrB*, while the other class impaired the reading frame (Zuber and Losick, 1987). The *abrB* gene is transcribed by two promoters, whose physiological significance is not yet fully understood.

DNA Binding of AbrB

Attempts to define common binding sites for the genes regulated by AbrB have failed so far. The location of AbrB-binding sites relative to the promoters shows some flexibility but suggests that AbrB might interfere or interact with the RNA polymerase (Strauch, 1995a). Footprinting analysis showed variation in size from target to target, with protected regions ranging from 30 to 120 base pairs (Strauch, 1995b). However, from these sequences no common motif can be deduced as an AbrB consensus sequence. Using an *in vitro* selection of random oligo nucleotides, seemingly optimal AbrB binding sites corresponding to a regularly spaced motif were selected (Xu and Strauch, 1996). This optimally spaced motif, however, is rare within the *B. subtilis* genome and does not correlate with the footprinting results. Therefore it is not sufficient to explain the selectivity of the AbrB-DNA-interaction. The mapped sequences show a correlation to AT-rich and bent DNA regions, but bending alone is not sufficient for promoting AbrB binding as shown by mobility shift assays using bent DNA (Klein, unpublished data). Therefore, it has been hypothesized that AbrB may actually recognize a type of DNA-three-dimensional structure that can be assumed by a finite subset of differing base sequences (Xu *et al.*, 1996).

The 10.5 kDa AbrB protein has been shown to be a hexamer in solution, and that this hexameric state is required for its DNA-binding activity (Strauch *et al.*, 1989). A sequence with slight similarity to the helix-turn-helix motif of DNA-binding proteins that was identified within the AbrB sequence was investigated by mutational analysis and failed to prove this relationship (Fürbaß and Marahiel, 1991). There is no evidence for the presence of any other known DNA-binding motif within the AbrB protein. Therefore at present no explanation can be provided for the specific and co-operative binding behavior of AbrB (Fürbaß *et al.*, 1991; Robertson *et al.*, 1989).

To understand the structural basis of AbrB specific DNA-interaction, several attempts to unravel the three dimensional structure of this proteins have been undertaken. These studies were designed to define the residues involved in DNA binding and to shed light on the oligomerization behavior. Full-length recombinant AbrB protein from *B. subtilis* was purified to homogeneity but failed to crystallize. Recently the structure of the AbrB N-terminal DNA-binding domain was solved by NMR. It was described as a looped-hinged helix fold where two conserved arginine residues required for DNA interaction are located within an α-helix flanked by a double stranded looped-hinge region (Vaughn *et al.*, 2000). This N-terminal domain was shown to retain full sequence specificity and to form dimers that are necessary for DNA interaction, but it failed to show the hexameric state that has been described for the full-length protein. NMR measurements showed that the residues of the hinge-loop region undergo significant concerted motion upon the protein's interaction with DNA, affecting the disposition of the DNA-binding helix (Zuber, 2000). According to the recent model of gene regulation by AbrB, the ordered structure of AbrB-DNA complexes varies between different target genes, suggesting changes within AbrB's tertiary and quarternary structure to conform to the site of interaction offered by the DNA target. This might be accomplished by the multimeric state of AbrB as determined by the specific target (Vaughn *et al.*, 2000).

The validity of the proposed model can only be tested on the structure of full-length AbrB bound to different DNA-target sites. Such an investigation will also reveal all the amino acid residues mediating direct interaction with the DNA, and those needed for the multimeric organization. As a first step towards structural studies on full length AbrB protein, we isolated the *abrB* gene from the thermophilic *Bacillus stearothermophilus* (76% identity to the mesophilic *B. subtilis* protein). DNA-binding activity and oligomerization behavior of the recombinant protein were studied and displayed a similar behavior as shown for the mesophilic $AbrB_{su}$. Initial crystallization attempts yielded small crystals that show diffraction at 4 Å (Klein *et al.*, 2000).

Regulation of Gene Expression by AbrB

AbrB has been characterized to act in three different ways to regulate post-exponential expression of genes which can be classified according to their cellular function as summarized in Figure 2. It is the sole repressor for genes like *spo0E* and *tycA* as *abrB* mutations cause constitutive expression of these genes

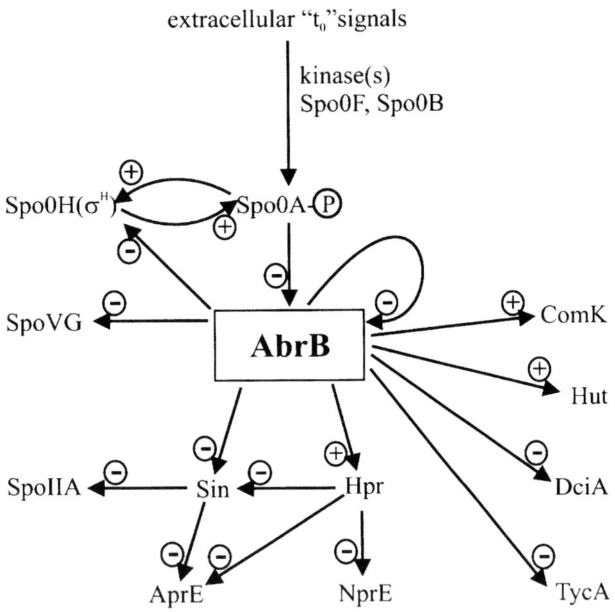

protein:	cellular function:	response:
AbrB	major transition state regulator	growth state sensing
AprE	subtilisin protease	alternative resources
ComK	transcription of competence genes	competence development
DciA	ABC transporter for dipeptides	alternative resources
Hpr	transition state regulator	growth state sensing
Hut	histidase (histidine utilization)	alternative resources
NprE	neutral protease	alternative resources
Spo0A	major response regulator for sporulation	sporulation
Spo0H	alternative σ^H for sporulation	sporulation
SpoIIA	regulation of alternative σ^F	sporulation
SpoVG	maturation of spore cortex	sporulation
TycA	tyrocidine biosynthesis	secondary metabolite formation

Figure 2. AbrB regulation of transition state gene expression. +, positive regulation; −, negative regulation. The formation of the phosphorylated, active form of Spo0A occurs at the end of exponential growth by a phosphorelay which is summarized here. t_0 means time zero according to the sporulation pathway. The components of this regulatory circuit and their cellular functions are summarized.

(Marahiel *et al.*, 1987; Perego and Hoch, 1991). For *spoVG* and *aprE*, AbrB is termed a 'preventer': while *abrB* mutations restore their expression in Spo0 mutants, the absence of AbrB does not cause constitutive expression, indicating additional temporal control acting in concert with AbrB (Ferrari *et al.*, 1988). In a third manner AbrB regulates expression as an activator. Genetic data indicate that AbrB positively affects expression of the transition state regulator Hpr (Strauch and Hoch, 1993), factors of the competence pathway (Dubnau, 1991), and enzymes involved in histidine utilization (Fisher *et al.*, 1994). Whereas in the case of repression the binding of AbrB to DNA segments around the promoter region has been verified, this has not yet been shown clearly for positively regulated genes. By the combination of several levels of macromolecular assembly with the flexibility of the looped-hinge helix DNA recognition fold to alter its orientation, AbrB can serve as specific regulatory protein for diverse promoter elements. This structure-function model sets AbrB as an archetype for a new class of regulatory proteins (Zuber, 2000).

To date, *abrB* expression is known to be controlled by both negative autoregulation and repression by phoshorylated Spo0A protein (Perego *et al.*, 1988). It is believed that the autoregulatory loop ensures an intracellular concentration of AbrB during exponential growth high enough to silence transition state functions and sensitive enough to changes in transcription and/or protein activity. The cooperative DNA-binding activity of AbrB (see above) could magnify the effects of slight decreases in the AbrB levels on its target genes. Phosphorylated Spo0A protein, being the primary sporulation signal as it is phosphorylated by a regulatory phosphorelay cascade in response to nutrient deprivation, seems to be the sole factor responsible for *abrB* repression (Strauch *et al.*, 1990). This sporulation initiating signal cascade represses transcription of the *abrB* gene and causes the AbrB level to drop below its threshold, thereby releasing the repression of transition state functions. This genetic model is supported by data showing that the cellular amount of AbrB is indeed growth phase dependent (O'Reilly and Devine, 1997). Both *abrB* mRNA and AbrB protein accumulate in the early exponential growth phase, but whereas the transcript level sharply decreases in the mid-exponential phase, the AbrB protein is gradually decreasing along the exponential growth with an abrupt reduction at the onset of stationary phase. Furthermore, analysis of *abrB* (partially deletion) and *spo0A* mutant strains confirmed the *abrB* regulation model on the level of mRNA and protein quantification (O'Reilly and Devine, 1997). Even though not related on the amino acid level, the pattern of *abrB* expression, its size, and overall net charge are similar to that of Fis from *E. coli.* But both proteins bind to highly degenerate consensus sequences with a bias for A and T residues, both are negatively autoregulated, and both regulate expression of a variety of genes of diverse functions. Interestingly, so far no Fis homolog has been found in gram-positive bacteria, and AbrB seems to be a characteristic of the genus *Bacillus.* This has been taken as an indication that these two proteins evolved independently but have similar cellular functions (O'Reilly and Devine, 1997).

Acknowledgements

We would like to thank all the members in our lab that have contributed to the research in these topics as well as all our colleagues that have sheared ideas and results. We apologize to all those whose work was not cited appropriately due to limitations in article size. Our research was supported by grants from the DFG.

References

Alonso, J.C., Gutierrez, C., and Rojo, F. 1995. The role of chromatin-associated protein Hbsu in beta-mediated DNA recombination is to facilitate the joining of distant recombination sites. Mol. Microbiol. 18: 471–478.

Bonnefoy, E., Takahashi, M., and Yaniv, J.R. 1994. DNA-binding parameters of the HU protein of *Escherichia coli* to cruciform DNA. J. Mol. Biol. 242: 116–129.

Dubnau, D. 1991. The regulation of genetic competence in *Bacillus subtilis.* Mol. Microbiol. 5: 11–18.

Ferrari, E., Henner, D.J., Perego, M., and Hoch, J.A. 1988. Transcription of *Bacillus subtilis* subtilisin and expression of subtilisin in sporulation mutants. J. Bacteriol. 170: 289–295.

Fisher, S.H., Strauch, M.A., Atkinson, M.R., and Wray, L.V., Jr. 1994. Modulation of *Bacillus subtilis* catabolite repression by transition state regulatory protein AbrB. J. Bacteriol. 176: 1903–1912.

Fürbaß, R., Gocht, M., Zuber, P., and Marahiel, M.A. 1991. Interaction of AbrB, a transcriptional regulator from *Bacillus subtilis* with the promoters of the transition state-activated genes *tycA* and *spoVG.* Mol. Gen. Genet. 225: 347–354.

Fürbaß, R., and Marahiel, M.A. 1991. Mutant analysis of interaction of the *Bacillus subtilis* transcription regulator AbrB with the antibiotic biosynthesis gene *tycA.* FEBS 287: 153–156.

Gaur, N.K., Cabane, K., and Smith, I. 1988. Structure and expression of the *Bacillus subtilis sin* operon. J. Bacteriol. 170: 860–869.

Guespin-Michel, J.F. 1971. Phenotypic reversion in some early blocked sporulation mutants of *Bacillus subtilis.* Mol. Gen. Genet. 112: 243–254.

Hodges-Garcia, Y., Hagerman, P.J., and Pettijohn, D.E. 1989. DNA ring closure mediated by protein HU. J. Biol. Chem. 264: 14621–14623.

Klein, W., Winkelmann, D., Hahn, M., Weber, T., and Marahiel, M.A. 2000. Molecular characterization of the transition state regulator AbrB from *Bacillus stearothermophilus.* Biochim. Biophys. Acta 1493: 82–90.

Köhler, P., and Marahiel, M.A. 1997. Association of the histone-like protein HBsu with the nucleoid of *Bacillus subtilis.* J. Bacteriol. 179: 2060–2064.

Köhler, P., and Marahiel, M.A. 1998. Mutational analysis of the nucleoid-associated protein HBsu of *Bacillus subtilis.* Mol. Gen. Genet. 260: 487–491.

Kohno, K., Wada, M., Kano, Y., and Imamoto, F. 1990. Promoters and autogenous control of the *Escherichia coli hupA* and *hupB* genes. J. Mol. Biol. 213: 27–36.

Konradi, R., Billeter, M., and Wüthrich, K. 1996. MOLMOL: a program for display and analysis of macromolecular structures. J. Mol. Graphics 14: 51–55.

Kunst, F., Ogasawara, N., Moszer, I., Albertini, A.M., Alloni, G., Azevedo, V., Bertero, M.G., Bessieres, P., Bolotin, A., Borchert, S., Borriss, R., Boursier, L., Brans, A., Braun, M., Brignell, S.C., Bron, S., Brouillet, S., Bruschi, C.V., Caldwell, B., Capuano, V., Carter, N.M., Choi, S.K., Codani, J.J., Connerton, I.F., and Danchin, A., *et al.* 1997. The complete genome sequence of the gram-positive bacterium *Bacillus subtilis.* Nature 390: 249–256.

Lee, E.C., Hales, L.M., Gumport, R.I., and Gardner, J.F. 1992. The isolation and characterization of mutants of the integration host factor (IHF) of *Escherichia coli* with altered, expanded DNA-binding specificities. EMBO J. 11: 305–313.

Marahiel, M.A., Zuber, P., Czekay, G., and Losick, R. 1987. Identification of the promoter for a peptide antibiotic biosynthesis gene from *Bacillus brevis* and its regulation in *Bacillus subtilis.* J. Bacteriol. 169: 2215–2222.

McGovern, V., Higgins, N.P., Chiz, R.S., and Jaworski, A. 1994. H-NS over-expression induces an artificial stationary phase by silencing global transcription. Biochimie 76: 1019–1029.

Micka, B., Groch, N., Heinemann, U., and Marahiel, M.A. 1991. Molecular cloning, nucleotide sequence, and characterization of the *Bacillus subtilis* gene encoding the DNA-binding protein HBsu. J. Bacteriol. 173: 3191–3198.

Micka, B., and Marahiel, M.A. 1992. The DNA-binding protein HBsu is essential for normal growth and development in *Bacillus subtilis.* Biochimie 74: 641–650.

Nash, H.A. 1996. The Hu and IHF proteins: accessory factors for complex protein-DNA assemblies. In: E.C.C. Lin, and A.S. Lynch (eds.), Regulation of gene expression in *Escherichia coli.* RG Landes Company, Austin, Texas, pp. 149–179.

Oberto, J., Drlica, K., and Rouviere-Yaniv, J. 1994. Histones, HMG, HU, IHF: Meme combat. Biochimie 76: 901–908.

O'Reilly, M., and Devine, K.M. 1997. Expression of AbrB, a transition state regulator from *Bacillus subtilis,* is growth phase dependent in a manner resembling that of Fis, the nucleoid binding protein from *Escherichia coli.* J. Bacteriol. 179: 522–529.

Perego, M., and Hoch, J.A. 1988. Sequence analysis and regulation of the *hpr* locus, a regulatory gene for protease production and sporulation in *Bacillus subtilis.* J. Bacteriol. 170: 2560–2567.

Perego, M., Spiegelman, G.M., and Hoch, J.A. 1988. Structure of the gene for the transition state regulator, AbrB: regulator synthesis is controlled by the *spo0A* sporulation gene in *Bacillus subtilis.* Mol. Microbiol. 2: 689–699.

Perego, M., and Hoch, J.A. 1991. Negative regulation of *Bacillus subtilis* sporulation by the *spo0E* gene product. J. Bacteriol. 173: 2514–2520.

Robertson, J.B., Gocht, M., Marahiel, M.A., and Zuber, P. 1989. AbrB, a regulator of gene expression in *Bacillus,* interacts with the transcription initiation regions of a sporulation gene and an antibiotic biosynthesis gene. Proc. Natl. Acad. Sci. USA 86: 8457–8461.

Ross, M.A., and Setlow, P. 2000. The *Bacillus subtilis* HBsu protein modifies the effects of alpha/beta-type, small acid-soluble spore proteins on DNA. J. Bacteriol. 182: 1942–1948.

Shellman, V.L., and Pettijohn, D.E. 1991. Introduction of proteins into living bacterial cells: distribution of labeled HU protein in *Escherichia coli.* J. Bacteriol. 173: 3047–3059.

Strauch, M.A., Spiegelman, G.B., Perego, M., Johnson, W.C., Burbulys, D., and Hoch, J.A. 1989. The transition state transcription regulator *abrB* of *Bacillus subtilis* is a DNA binding protein. EMBO J. 8: 1615–1621.

Strauch, M., Webb, V., Spiegelman, G., and Hoch, J.A. 1990. The SpoOA protein of *Bacillus subtilis* is a repressor of the *abrB* gene. Proc. Natl. Acad. Sci. USA 87: 1801–1805.

Strauch, M.A., and Hoch, J.A. 1993. Transition-state regulators: sentinels of *Bacillus subtilis* post-exponential gene expression. Mol. Microbiol. 7: 337–342.

Strauch, M.A. 1993. AbrB, a Transition State Regulator. In: A.L. Sonenshein, J.A. Hoch, and R. Losick (eds.), *Bacillus subtilis* and other gram-positive bacteria. Am. Soc. Microbiol., Washington DC, pp. 757–763.

Strauch, M.A. 1995a. Delineation of AbrB-binding sites on the *Bacillus subtilis spo0H*, *kinB*, *ftsAZ*, and *pbpE* promoters and use of a derived homology to identify a previously unsuspected binding site in the *bsuB1* methylase promoter. J. Bacteriol. 177: 6999–7002.

Strauch, M.A. 1995b. *In vitro* binding affinity of the *Bacillus subtilis* AbrB protein to six different DNA target regions. J. Bacteriol. 177: 4532–4536.

Tanaka, I., Appelt, K., Dijk, J., White, S.W., and Wilson, K.S. 1984. 3-A resolution structure of a protein with histone-like properties in prokaryotes. Nature 310: 376–381.

Vaughn, J.L., Feher, V., Naylor, S., Strauch, M.A., and Cavanagh, J. 2000. Novel DNA binding domain and genetic regulation model of *Bacillus subtilis* transition state regulator AbrB. Nat. Struct. Biol. 7: 1139–1146.

Xu, K., Clark, D., and Strauch, M.A. 1996. Analysis of *abrB* mutations, mutant proteins, and why *abrB* does not utilize a perfect consensus in the -35 region of its σ^A promoter. J. Biol. Chem. 271: 2621–2626.

Xu, K., and Strauch, M.A. 1996. *In vitro* selection of optimal AbrB-binding sites: comparison to known in vivo sites indicates flexibility in AbrB binding and recognition of three-dimensional DNA structures. Mol. Microbiol. 19: 145–158.

Yang, S.-W., and Nash, H.A. 1994. Specific photocrosslinking of DNA-protein complexes: identification of contacts between integration host factor and its target DNA. Proc. Natl. Acad. Sci. USA 91: 12183–12187.

Yasuzawa, K., Hayashi, N., Goshima, N., Kohno, K., Imamoto, F., and Kano, Y. 1992. Histone-like proteins are required for cell growth and constraint of supercoils in DNA. Gene 122: 9–15.

Zuber, P., and Losick, R. 1987. Role of AbrB in Spo0A- and Spo0B-dependent utilization of a sporulation promoter in *Bacillus subtilis*. J. Bacteriol. 169: 2223–2230.

Zuber, P. 2000. Specificity through flexibility. Nat. Struct. Biol. 7: 1079–1081.

Chapter 25

Regulation of Ribosomal RNA Synthesis in *E. coli*: Effects of the Global Regulator Guanosine Tetraphosphate (ppGpp)

Rolf Wagner

Abstract

The global regulatory nucleotides (p)ppGpp are major effectors for the control of ribosomal RNA in bacteria. The effector molecules accumulate to different cellular levels at amino acid deprivation or during different growth rates. They change the activity of RNA polymerase to transcribe from sensitive promoters (e.g. ribosomal RNA promoters). Sensitive promoters are characterized by a GC-rich discriminator element in addition to further structural requirements not completely understood. ppGpp must also be regarded as a mediator for growth rate control although it appears that ppGpp-independent regulatory mechanisms exist. Inhibition occurs at various steps during initiation but also during elongation where RNA polymerase pausing is observed. From the existing data a mechanistic model for the action of ppGpp is suggested considering structural details of RNA polymerase obtained at high resolution.

Introduction

The goal of this review is to serve three aspects. First it should provide the reader with some basic information on the global network of ppGpp-dependent regulation and explain how it affects ribosomal RNA (rRNA) transcription in bacteria. This part should highlight results that are undisputed but also present findings that are controversial and need further clarification. I will then summarize and review results collected over a period of several years in my own group undertaken to better understand some of the crucial steps of ppGpp-dependent regulation of rRNA transcription. These findings will be discussed with results or models obtained by other groups working on related subjects. Finally, I like to introduce some new ideas on a possible mechanism of ppGpp-dependent control. The ideas are based on the three-dimensional structure of the RNA polymerase transcription complex obtained at high resolution considering most of the available data on ppGpp-dependent regulation.

Ribosomal RNA Synthesis is Tightly Coupled to Cell Growth

In bacteria cell growth and the capacity for translation are directly linked. Changes in the translational capacity are generally reflected by changes in the number of ribosomes. This is a consequence of the fact that the rate of translation is relatively constant. Since ribosomes are extremely large ribonucleoprotein complexes they represent a very costly machinery for the cell. Their number is thus carefully adapted to the growth demands of the cell. The biosynthesis of ribosomes is determined, however, by the rate of rRNA synthesis while the synthesis of the protein components is a subordinate process. Hence, rRNA synthesis is intricately coupled to cell growth and largely determines the translational capacity of bacterial cells. It is not surprising that a process so central for the cell is tightly and efficiently regulated. In fact, the synthesis of rRNAs is controlled by a complex set of interacting regulatory networks (Wagner, 1994; Condon *et al.*, 1995; Gourse *et al.*, 1996; Wagner, 2000). Here I will focus on the global regulation governed by hormone-like effector nucleotides guanosine 3′,5′-bis(diphosphate) (ppGpp) and guanosine 3′-diphosphate 5′-triphosphate (pppGpp). These two regulatory compounds are major players in the adaptation of rRNA synthesis upon changes in the environmental conditions and during nutritional deprivation.

Despite many years of intensive research the molecular mechanisms of (p)ppGpp-dependent transcriptional regulation are still unsolved and many controversial ideas have been put forward to explain the regulatory effects. In the following section I like to focus on those (regrettably few) steps of the regulatory cascade which are more or less undisputed and for which consensus has been reached by several independent scientific approaches.

The response provoked by the effector nucleotides (p)ppGpp is rapid and can either be dramatic, for example, under conditions of amino acid starvation when rRNA synthesis is shut-off immediately. On the other hand, the response can lead to a gradual and balanced adaptation of rRNA synthesis rates to accommodate different steady state growth rates of the cell. Both reactions are linked to changes in the levels of (p)ppGpp but can be distinguished physiologically. One distinction can be made by the concentration of the mediator that is reached in the cell. The fast and dramatic response that is triggered by high (millimolar) concentrations of the effector is generally termed *stringent regulation.* In contrast, the steady state adaptation of rRNA synthesis to an altered growth rate, for which the concentrations of the effector nucleotides vary in the micromolar range, is termed *growth rate control* (Cashel *et al.*, 1996). It is one characteristic of growth rate regulation that changes in the rate of rRNA synthesis are roughly proportional to the square of the changes in growth rates. It should be noted at this point, however, that growth rate control of rRNA promoters has also been described in strains devoid of (p)ppGpp (see below). Stringent control, on the other hand, summarises a complex set of metabolic reactions triggered by amino acid deprivation. Again, the most outstanding and direct effect during the stringent control is the immediate repression of stable RNA (rRNAs and tRNAs) transcription.

In summary, the hormone-like effector molecules (p)ppGpp are able to trigger two types of transcriptional regulation, stringent and growth rate control. Both types of regulation are linked to changes in the cellular concentration of the effector nucleotides, which are either in the micro- or the millimolar range. We will see below that two different enzymatic pathways are responsible for the accumulation or maintenance of the different effector nucleotide levels.

The Chain of Events Leading to Stringent Control

The stringent response represents the consequence of amino acid deprivation in the cell. A lack in the availability of one or several amino acids causes that the corresponding tRNAs cannot be sufficiently charged at their 3′ acceptor ends. Hence, the ratio of amino acylated versus non-amino acylated tRNAs changes. The appearance of deacylated tRNAs, not free amino acids, represents the signal for the cell to change a large number of metabolic activities. The actual sensors for the stringent control are translating ribosomes (Gallant, 1979; Krohn and Wagner, 1995). When a non-amino acylated tRNA is bound in a codon-dependent way to the ribosomal A-site the ribosome-associated RelA protein (ppGpp synthetase I or PSI) catalyses the synthesis of pppGpp by transferring pyrophosphate from ATP to the 3′ OH group of GTP. The guanosine pentaphosphate is then rapidly converted to ppGpp by a pppGpp 5′ phosphohydrolase (GPPase) (see Figure 1). It is not completely clear whether the penta- or the tetraphosphate exhibit different regulatory features. Based on *in vitro* studies, however, both effector molecules appear to be indistinguishable with respect to their regulatory activity (Heinemann and Wagner, 1997).

With the rapid increase of the cellular (p)ppGpp level from micromolar to roughly millimolar concentrations pleiotropic changes in many metabolic activities can be observed. Immediate repression of stable RNA transcription is clearly the most dramatic effect. However, a number of additional cellular changes can be observed as direct or indirect consequence of the increased (p)ppGpp level. For instance, concerted inhibition is observed for the synthesis of the components, which constitute the translation apparatus (including ribosomal RNAs, ribosomal proteins translation initiation and elongation factors as well as tRNAs). In addition, the synthesis of RNA polymerase subunits is repressed while the expression of stress-specific sigma factors, like σ^S and σ^H, is increased. The transcription of amino acid biosynthetic operons is activated. Moreover, proteolysis is activated and transport, with the exception of the uptake for branched amino acids, is inhibited. DNA synthesis, cell wall synthesis and phospholipid metabolism are all inhibited. Peptidoglycan synthesis and penicillin tolerance are also changed. Furthermore, the fidelity of translation is reduced and an enhanced mutation frequency has been observed. Many of the reactions must be regarded as indirect consequences. There appears to be a common logic behind this set of pleiotropic changes. They all seem to serve one crucial purpose, namely to overcome the amino acid limitation and to adjust the protein synthesizing capacity to an altered cellular substrate pool (for further reviews see Galant, 1979; Cashel *et al.*, 1996; Wagner, 2000).

What is the Target for (p)ppGpp Action?

Within the long list of cellular responses to increasing (p)ppGpp levels transcription is clearly the most prominent step affected directly. Hence, the most likely candidate where (p)ppGpp exerts its effect is RNA polymerase. This assumption is supported already from the early analysis of mutants which show reduced sensitivity to enhanced ppGpp levels (Little *et al.*, 1983; Tedin and Bremer, 1992). Several such mutations have been mapped in the RNA polymerase β subunit (some have also been characterized within the σ^{70} subunit (Hernandez and Cashel, 1995)). This had initiated many attempts to show ppGpp-RNA polymerase interactions directly. Cross-linking experiments, CD-measurements and fluorescence labelling with ppGpp analogues indicate that the β subunit is the target for interaction. The experiments also show that a likely site of interaction is in the vicinity of the active center for nucleotide addition (Reddy *et al.*, 1995). Together the results are fairly convincing. They do not provide absolute proof, however. For instance, the direct binding of genuine ppGpp has never been achieved and a certain degree of unspecific binding of the analogues cannot be completely eliminated. Attempts to cross-link radioactive ppGpp directly by UV irradiation were not conclusive, for example, but rather indicated that contaminating GTP was cross-linked to the NTP binding site (Heinemann and Wagner, unpublished). Moreover, in a recent cross-linking study the β′ not the β subunit was identified as the major target for ppGpp binding. It was proposed therefore that part of the β together with part of the β′ subunit constitute a modular ppGpp binding site

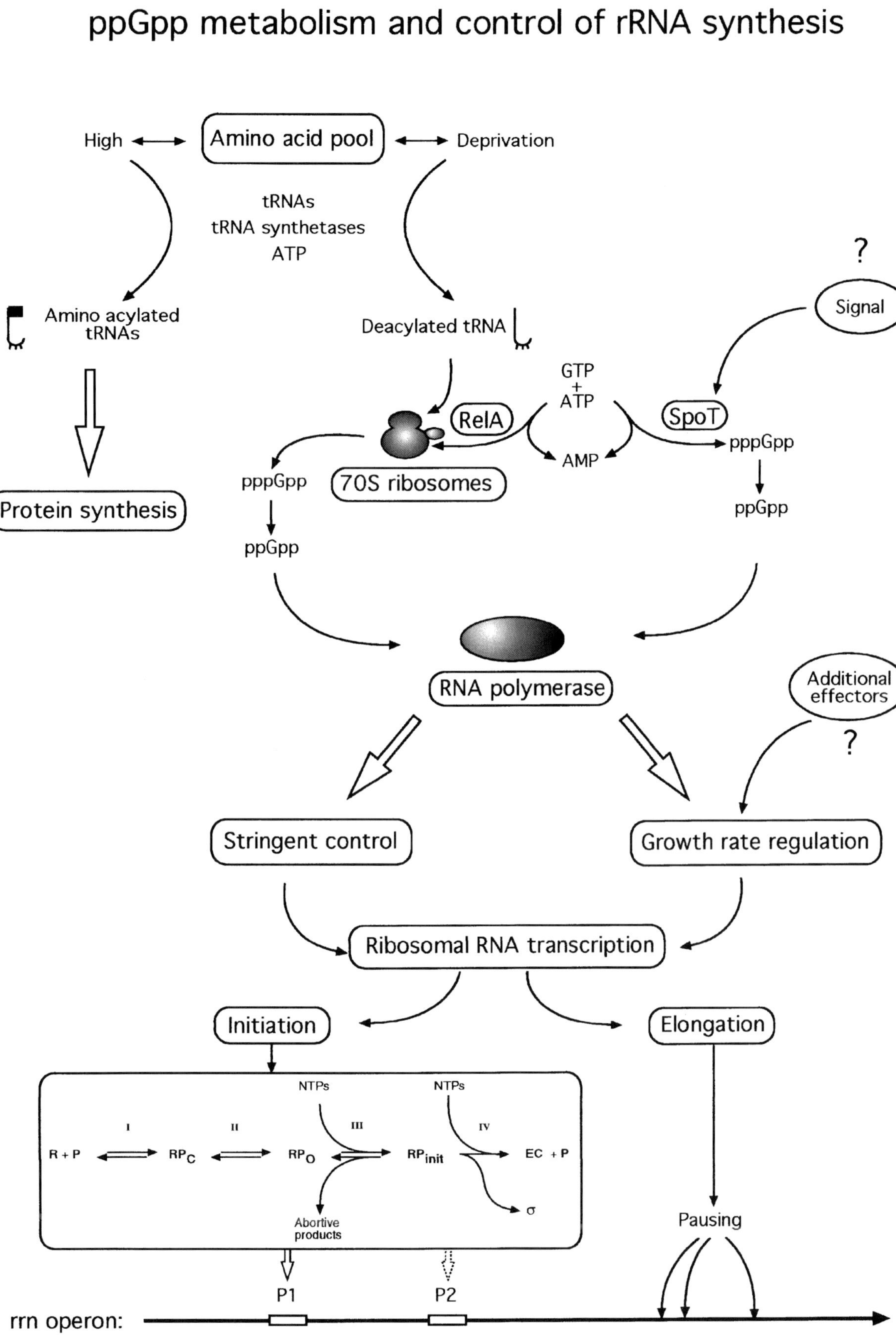

Figure 1. ppGpp metabolism and control of rRNA synthesis. Details see text.

(Toulokhonov *et al.*, 2000). There is also some evidence that the stability of σ-core interaction may be affected by (p)ppGpp (Hernandez and Cashel, 1995; Michels and Wagner, unpublished). No indication for binding of ppGpp to the σ subunit has been presented so far.

In conclusion, there is little doubt that RNA polymerase is the target for ppGpp interaction and binding very likely occurs close to the RNA polymerase active center constituted by the β and partly by the β′ subunit. Unequivocal proof where exactly the effector molecules bind is still pending, however.

The Promoter Structure Discriminates Stringently or Non-Stringently Transcribed Genes

Whether a gene is under stringent control or not is determined by the promoter structure. A conserved GC-rich consensus sequence, the GCGC discriminator motif, localized immediately downstream to the −10 promoter element has been recognized as important for negative stringent control (Travers, 1984). The involvement of the discriminator element in the stringent control as well as during growth rate regulation has been demonstrated in a number of cases (Mizushima-Sugano and Kaziro, 1985; Zacharias *et al.*, 1989; Davies and Drabble, 1996). We had shown, for example, that an unregulated promoter could be converted to stringency by changing the non-consensus discriminator into a GCGC consensus sequence. Interestingly, the same base change also resulted in growth rate sensitivity of the altered promoter. However, introduction of the GCGC sequence did not convert the synthetic and unregulated *tac* promoter to stringency (nor to growth rate dependence). In conclusion, the GCGC discriminator must be viewed as an element, *necessary* but not *sufficient* for negative stringent control (Zacharias *et al.*, 1989). Fusion of different sequence elements upstream and downstream of the discriminator site obtained from stringently and non-stringently regulated promoters revealed that not a defined sequence element but the complete promoter structure in unknown sterical consequence defines whether the resulting promoter is stringently regulated or not (Zacharias *et al.*, 1990). Whether the GCGC promoter determinants are generally identical for stringent control *and* growth rate-dependent regulation cannot be answered unambiguously today. Although the involvement of the GCGC discriminator sequence in growth rate-dependence has been demonstrated in quite a number of cases (see above) a common function of this promoter element in both types of control is controversial. From studies with different promoter mutants it was concluded that stringent regulation and growth rate control have non-identical promoter sequence requirements (Josaitis *et al.*, 1995).

Interestingly, for genes under positive stringent control, like the *his* operon, the sequence corresponding to the stringent discriminator is AT-rich. *In vitro* and *in vivo* analyses revealed that ppGpp is responsible for the activating effect at the *his* promoter. Moreover, mutations within the promoter region support the conclusion that the −10 recognition element and the downstream AT-rich discriminator sequence are both important for ppGpp-dependent regulation (Riggs *et al.*, 1986; Shand *et al.*, 1989). On the other hand, positive stringent control has often been explained as a consequence of a passive increase in RNA polymerase concentration as consequence of the repression of stable RNA promoters. In the *his* operon case, however, studies performed with a coupled transcription translation system did not support passive regulation. Under mixed-template conditions activation was shown to be independent from simultaneous repression (Choy, 2000). Passive regulation as a result of changing RNA polymerase activity may still be important. It requires, however, that RNA polymerase must be limiting, a condition that has not been rigorously proven yet. In a very recent study no stimulation of amino acid promoters could be shown *in vitro*. Instead, amino acid promoters appear to require higher concentrations of RNA polymerase for function *in vitro* and *in vivo* than control promoters. The authors propose that liberated RNA polymerase from stable RNA promoters stimulate transcription from the amino acid promoters. Hence their data supports the passive model (Barker *et al.*, 2001a, b). In conclusion, there is good evidence that also for positive stringent control the promoter structure is crucial for the regulation mechanism. Passive effects may contribute significantly to positive stringent control, however.

Which Steps During the Initiation Cycle are Affected by ppGpp?

Transcription initiation is a multistep process that can formally be divided in four major steps (see Figure 1). In the first step RNA polymerase binds to the promoter DNA forming a binary closed complex (RPc). This complex isomerizes to an open complex (RPo) with about 12 nucleotides of the core promoter sequence in single stranded conformation. When the initial substrate NTPs are bound to this complex a ternary initiating complex is formed (RPinit). This ternary complex can either be converted into an elongating complex (EC) by successive addition of 9 to 12 NTPs. At this point the σ subunit leaves the complex. The ternary initiation complex can alternatively undergo repeated rounds of abortive cycling releasing abortive transcription products between 3 and 12 nucleotides. Depending on the respective promoter, each step during this cycle can be rate limiting and in principle each step may be affected during stringent control.

Promoter melting (open complex formation) is considered to depend on the stability and thus on the GC-content of the promoter and flanking downstream discriminator. The presence of GC- or AT-rich discriminators in positive and negative regulated promoters, respectively have led to the conclusion that open complex formation is affected by ppGpp. In fact, for several promoters under stringent control open complex formation has been shown to be the step which is primarily affected by ppGpp (Ohlson and Gralla, 1992;

Raghavan *et al.*, 1998). We have shown for the *rrnB* P1 promoter that not the amount or the stability of open complexes formed but the rate of formation is affected by high ppGpp concentrations (Jöres and Wagner, unpublished results). For other promoters steps of the initiation cycle later than open complex formation can be rate limiting, however. Hence, open complex formation seems to be at least one but probably not the only step which is influenced by ppGpp (see below).

A number of additional observations at rRNA promoters are consistent with the conclusion that open complex formation is one of the crucial steps affected by ppGpp during transcription initiation. For instance, ribosomal RNA promoters have a characteristically suboptimal 16 bp spacing between the –35 and –10 recognition elements (underwound). The promoters are therefore twist-sensitive and open complex formation or isomerization from the closed to the strand-opened structure is facilitated by negative supercoiling. Interestingly, the *his* promoter under positive stringent control has a suboptimal larger spacer of 18 bp (overwound). It has been proposed that for many promoters the supercoil sensitivity parallels the responsiveness to stringent control (Figueroa-Bossi *et al.*, 1998). Interestingly, supercoiled templates have been found to resist ppGpp-dependent inhibition *in vitro* consistent with the view that open complex formation is now more efficient at high superhelical density and no longer rate limiting, regardless if ppGpp is present or not.

There are examples indicating that steps different from open complex formation may be essential for ppGpp-dependent regulation. This was concluded from a detailed kinetic analysis with the *rrnB* P2 promoter and a P2 promoter variant with high ppGpp sensitivity. The study supported the view that a combination of effects at initiation, promoter clearance and elongation are mediated by ppGpp. During initiation an alternative pathway is triggered by ppGpp involving stabilized initial closed complexes and impeded open complexes. According to this study ppGpp-modified RNA polymerases are trapped in the closed complex which is formed at higher rates in presence of ppGpp. The open complex, on the other hand, is destabilised and the reaction back to the closed complex is facilitated. Hence, discrimination between stringently and non-stringently controlled promoters occurs at the early steps of initiation. This finding has led to the proposal of a *trapping mechanism* (Heinemann and Wagner, 1997). Efficient inhibition is caused at later steps, very likely during promoter clearance where the apparent K_M values for substrate NTP binding are increased by ppGpp (see below).

In continuation of those studies we have performed experiments to answer the question at which step(s) during the initiation pathway at *rrn* P1 promoters the inhibitory effect of ppGpp is exerted (Jöres and Wagner, unpublished). We can show that the formation of the open ternary transcription complex is the crucial step of inhibition. The ternary transcription complex consists of the RNA polymerase at the strand-opened promoter to which the first substrate NTP(s) are bound. Because promoters under negative stringent control share an apparently higher K_M for the first NTP and require the presence of the starting NTPs to form stable initiation complexes this finding supports a model whereupon ppGpp-dependent inhibition could be explained by substrate competition for the active site at steps which are characterized by strikingly high K_M values (see below).

Recent data presented from the Gourse laboratory (Barker *et al.*, 2001a, b), where both positive and negative transcription regulation of the rRNA P1 promoter and a series of inversely growth rate-dependent amino acid biosynthesis/transport promoters had been compared, revealed that the major difference between the two systems is related to the stability of the open complexes formed. This is consistent with the finding that certain RNA polymerase mutations in *rpoB* which destabilize initiation complexes at stringently controlled promoters in the absence of ppGpp behave like normal RNA polymerases in the presence of ppGpp (Zhou and Jin, 1998). Open complexes at inversely growth rate-dependent regulated amino acid promoters are formed more slowly and have longer life times than *rrnB* P1 open complexes. While ppGpp decreased the half-lives of open complexes for all promoters a direct effect of the mediator was only measurable for the negatively regulated *rrnB* P1 promoter which has a very short half-life. No direct effects of ppGpp on the rates of association or escape from amino acid promoters was detected *in vitro*.

While the above study offers a mechanistic explanation for many observations with respect to stringent control there are many questions still unanswered. Comparison of different results obtained *in vitro* have shown that the system depends critically on the particular experimental conditions, indicating that other steps during the initiation cycle may also contribute if the efficiency of the *in vitro* transcription system is different. Therefore, to analyse ppGpp-dependent inhibition in a reliable manner it is essential to adjust the *in vitro* transcription conditions in a defined way. This has been exemplified by varying the concentrations of the NTPs necessary to form the initial open complex. If the concentration of these substrates is too high the regulatory effect of ppGpp can easily be masked by the efficiency of the system. The concentration of the initial NTP substrates is thus essential for the efficient *in vitro* determination of the ppGpp-dependent repression. This observation differs from the proposal that growth rate-dependent regulation in the absence of ppGpp is mediated by limiting substrate NTP concentrations (see below). It only means that ppGpp-dependent inhibition cannot be measured at conditions where *in vitro* transcription is too efficient. The finding is in accordance with the effect described above that supercoiled templates do not support significant inhibition when analyzed *in vitro* (see above). We propose that inhibition by ppGpp is only apparent *in vitro* when the step(s) affected are limiting and not too effective. Suboptimal transcription may be the consequence of a structural alteration of the RNA polymerase in the transcription complex or result from other destabilising or limiting factors which may facilitate a backward reaction instead of yielding productive transcription complexes, for instance.

The Role of ppGpp as Growth Rate Regulator

Already from early studies of bacterial physiology it was concluded that (p)ppGpp represents the link between the physiological state of the cell and the activity of promoters that respond to growth rate changes. The evidence is indirect, however, and largely based on the observation that a linear inverse correlation has been noted between the growth rate and the cellular ppGpp concentration. This observation holds true under almost all conditions (Baracchini and Bremer, 1988; Zacharias *et al.*, 1989). The low ppGpp concentration range between fast and slow growth rates (micromolar) is maintained by a cellular activity different from RelA. The *spoT* gene product was found to be responsible for synthesis and maintenance of the RelA-independent (p)ppGpp synthesis (Hernandez and Bremer, 1991). Hence, SpoT represents a second ppGpp synthetase (PSII) (Figure 1). The involvement of (p)ppGpp in growth rate regulation is further supported by experiments with isogenic strains which have different mutations in the *spoT* gene. These mutations cause different basal cellular ppGpp concentrations. Interestingly, the different SpoT mutations cause different growth rates of otherwise isogenic strains when cultured in the same medium. The growth rates of the different mutants correlate perfectly with the cellular ppGpp concentration present in each mutant (Sarubbi *et al.*, 1988). Moreover, the results outlined above that promoter mutations changing the sensitivity for stringent regulation affect growth rate control in a very similar manner has strengthened the view that (p)ppGpp is the common mediator for both types of control. It is quite obvious, therefore, that ppGpp must play a role in growth rate regulation. On the other hand, several reports have shown that growth rate regulation of rRNA transcription occurs in strains deficient in both the *relA* and *spoT* genes (ppGpp^0 phenotype) (Gaal and Gourse, 1990). This finding means that there must be at least a second (p)ppGpp-independent mechanism for growth rate regulation. A feedback control mechanism had been proposed to explain rRNA shut down. According to the original suggestion repression should be independent of ppGpp and mediated by ribosomes. However, the activity of ribosomes or any other components of the translation machinery as repressors could not be verified. In controversial studies evidence for an increase in ppGpp under the feedback conditions was presented reinforcing the view that ppGpp is also responsible for feedback control. The discussion is not yet settled and attempts to identify specific feedback repressors still continue. In principle, growth rate regulation, in addition to the ppGpp pathway, could be maintained by cellular transcription factors like FIS and H-NS which are known as activator and repressor for rRNA transcription, respectively (Afflerbach *et al.*, 1998; Schröder and Wagner, 2000). It is interesting in this regard that transcription of one of the factors (FIS) shows a strictly growth phase-dependent expression and is itself under ppGpp control (Ninnemann *et al.*, 1992).

Based on the finding that growth rate-dependent regulation can be seen for rRNA promoters in the absence of ppGpp and the fact that rRNA promoters form notoriously unstable transcription initiation complexes an alternative model for growth rate-dependent rRNA synthesis was proposed. Since open complex formation at rRNA promoters *in vitro* requires the presence of the starting substrate NTPs it was assumed that the cellular concentration of the initiating nucleotides for rRNA transcription (ATP and GTP) are responsible for this type of control. This has led to a model according to which the concentration of NTPs is sensed by the instability of the initiation complexes before productive transcription occurs (Gaal *et al.*, 1997). The NTP sensing model rests on the premise that the NTP concentrations are limiting during slow growth and vary notably when the growth rate changes. This could not be verified in different studies, however (Petersen and Møller, 2000).

Transcription Initiation and Elongation are Affected by ppGpp

It is clear today that the effect of ppGpp on transcription is not restricted to the initiation cycle. Measurements *in vitro and in vivo* have revealed that transcription elongation rates are reduced at enhanced ppGpp concentrations. This is consistent with the known coupling between transcription and translation, for instance when the substrates for one of the processes become scarce or if premature termination occurs (e.g. during attenuation or transcription polarity). The reduction in transcription elongation rates at elevated ppGpp concentrations is not simply the consequence of a general increase in the step time for the nucleotide addition reaction (Kingston *et al.*, 1981). Reduced elongation rates are rather brought about by discrete RNA polymerase pausing events at defined template positions (Figure 1). Although the exact mechanism(s) leading to enhanced pausing still need some clarifications several conclusions have been reached with the aid of special techniques to quantify RNA polymerase pausing during *in vitro* transcription. Specifically stalled ternary complexes which were elongated under defined conditions have helped to answer a number of questions (Theißen *et al.*, 1990; Krohn *et al.*, 1992). According to such studies RNA polymerase pausing depends on the nature of the promoter and on the transcription of the closely adjacent sequence downstream of the transcription start site. For instance, strong enhancements of transcriptional pauses predominate for RNA polymerases initiated at rRNA promoters, which have transcribed through the early sequence region. This is true even if the downstream region has been replaced by sequences unrelated to the rRNA transcription unit. On the other hand, pauses are only weakly or moderately affected if transcription is initiated from a promoter which is not ppGpp dependent. Moreover, not all RNA polymerase pausing sites are affected in the same way. Pausing can be enhanced, unchanged or in some rare cases, pausing may even be reduced in presence of ppGpp (Krohn and Wagner, 1996). The same study also showed that in order to enhance transcriptional pauses the presence of ppGpp was only required during elongation when σ^{70} has already left the transcription complex.

The conclusion that transcription elongation is affected by ppGpp is supported by a different set of studies performed *in vivo* (Vogel and Jensen, 1994a, b). In contrast to the above findings these studies indicate that the presence of a *nut*-like boxA sequence element downstream from the transcription start site (as it is naturally found in the leader of all rRNA operons) is crucial for a high elongation rate and reduced pausing (Vogel and Jensen, 1995). Only genes devoid of such a boxA signal (like normal mRNAs) show significantly reduced elongation rates in presence of enhanced ppGpp levels. Since in the studies described above transcription was always started from the same inducible hybrid promoter the results may not be directly comparable with our own *in vitro* data.

In conclusion, altered transcription elongation rates as a consequence of ppGpp-dependent RNA polymerase pausing must be considered in explaining transcriptional control under conditions of stringent or growth rate regulation.

How do the Different rRNA Operons Respond to the Stringent Control and What do We Know from *In Vivo* Studies?

The seven different *E. coli* rRNA operons have very similar promoter sequences. The individual upstream sequences (UAS regions) important for transcription regulation deviate significantly from each other, however. Because the different UAS regions comprise binding sites for the growth rate-dependent transcription factors FIS and H-NS a differential transcription of some of the rRNA operons is feasible. As a consequence, individual rRNA operons not only deviate in their basal transcription efficiencies but also in their response to regulation (Hillebrand and Wagner, unpublished results). In previous studies with the *rrnB* P1 promoter the sequences upstream of the core promoters were not considered to be involved in stringent regulation. The differences in the individual UAS regions have led us to reinvestigate the differential regulation of individual rRNA operons in more detail. This question is currently being addressed by systematic *in vitro* and *in vivo* analyses employing isolated rRNA transcription units. The study is an extension of a previous *in vivo* investigation of rRNA promoter activities. In this study the effects of altered gene dosage (by plasmid copy number), different growth conditions, and strains with a defective *fis* gene were analyzed. In accordance with other investigators (Gafny *et al.*, 1994; Josaitis *et al.*, 1995) we found that not only rRNA P1 promoters but also P2 promoters were downregulated by ppGpp (although to a smaller extent). In the absence of FIS or at high ppGpp concentration almost all rRNA is transcribed from the P2 promoters. Furthermore, clear evidence for differential regulation between the *rrnB* and *rrnD* operons was apparent underlining that individual rRNA operons are controlled differentially (Liebig and Wagner, 1995). With our present study we are able to exactly compare expression of all individual rRNA operons. From this analysis a much more detailed picture of the operon-specific regulatory events under different growth conditions emerges (Hillebrand and Wagner; Pohl and Wagner, unpublished results).

A New Mechanistic Proposal for ppGpp-Dependent Transcriptional Control

According to the recently published three-dimensional structure bacterial core RNA polymerase can be described as a crab claw-shaped molecule with a 27 Å wide internal channel formed by the arms of the claw (Zhang *et al.*, 1999). This channel harbours the upstream and downstream duplex DNA, oriented at a bend angle of about 90° which results from a kink formed between the downstream duplex and the RNA/DNA hybrid. One arm of the claw is primarily the β subunit, the other primarily β′. At the inner end of the claw the two subunits make extensive interactions with each other. Within this base of the channel lies the active center consisting of a highly conserved structure and a chelated Mg^{2+}. A wall-like structure composed of conserved β′ domains F and G bifurcates the main channel into two separate cavities. The smaller secondary channel (10 to 12 Å in diameter, 45 Å in length) is too small to accommodate duplex nucleic acid structures although it might be broad enough to take up backtracked RNA (Korzheva *et al.*, 2000). It is believed that the secondary channel functions in delivering nucleotide substrates to the active center (Zhang *et al.*, 1999). It appears that the crowded nucleic acid elements in the major channel prevent direct access of substrate NTPs to the catalytic site. In contrast, the secondary channel is large enough and it provides a clear path for incoming NTPs. From its dimensions there is room enough for only one NTP at a time to diffuse through the NTP channel, however. Based on this structural considerations I like to propose a model explaining the inhibitory effects of ppGpp during different stages of transcription by preventing free access of substrate NTPs to the secondary channel. The model outlined below is illustrated in Figure 2. Due to its similarity with the NTP substrates (p)ppGpp should be able to enter the secondary channel. With the 3′ diphosphates it is

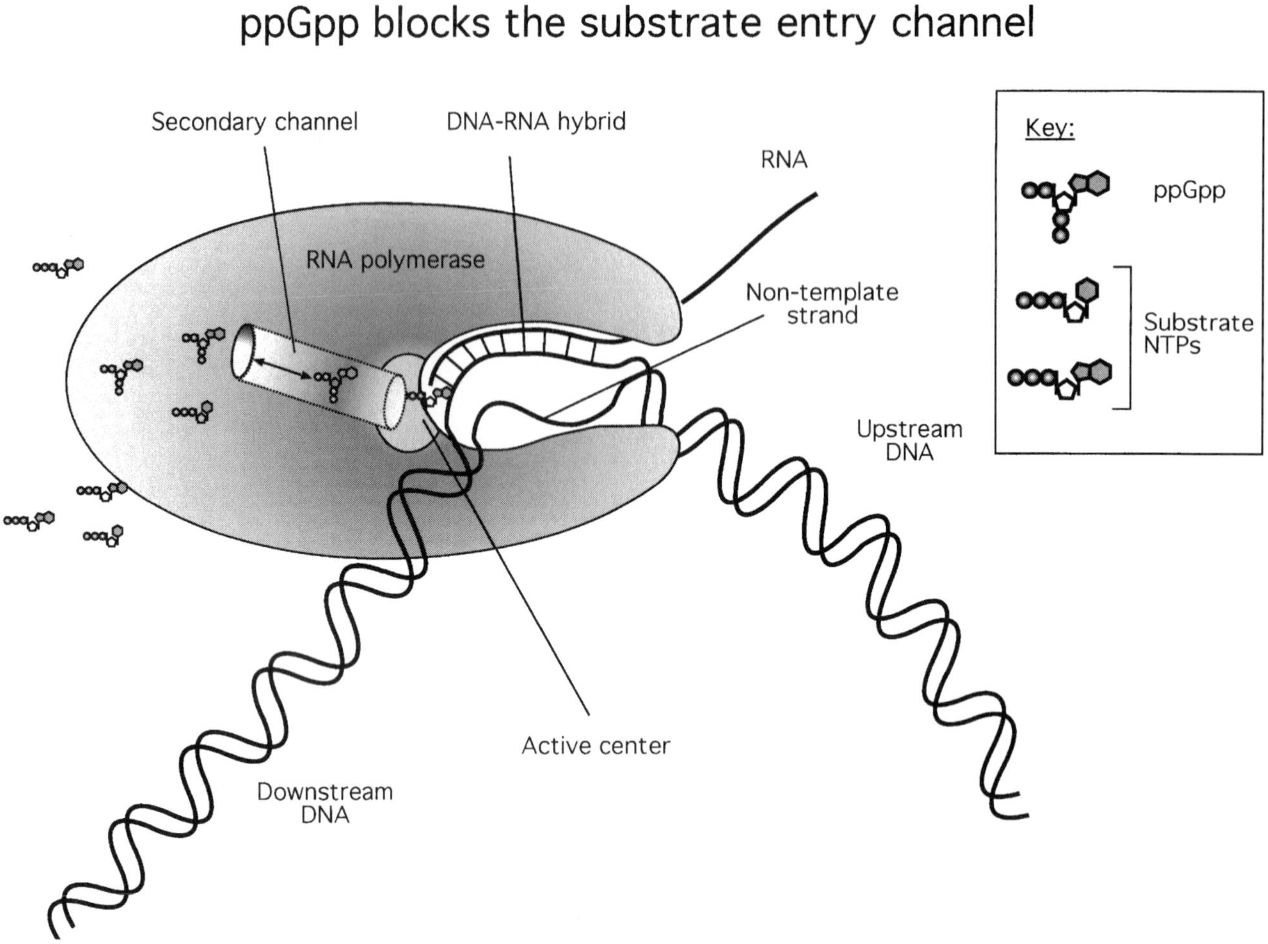

Figure 2. ppGpp-dependent inhibition of RNA polymerase by blockage of the NTP substrate channel. RNA polymerase, the secondary channel as entrance for NTPs are presented as cartoon and indicated. The path of the upstream and downstream DNA duplex, the position of the non-template strand, the active center as well as the DNA-RNA hybrid are schematically illustrated. Substrate NTPs and ppGpp molecules compete for passage through the secondary channel.

unable, however, to gain access into the catalytic center where the nucleotide addition reaction occurs. The specific arrangement of phosphate groups, the negative surface charge and the purine base may provide the specificity for a transient interaction with exposed amino acid side chains of the β' and β subunits. Bound ppGpp could thus temporarily block the channel and reduce the rate of NTP access considerably. This will not stop but slow down the progress of NTP incorporation. The effect becomes dramatic at template positions characterized by a high apparent K_M value for NTP addition. This, for instance, is the case for the first nucleotide position at all stable RNA promoters. The conformation of the secondary channel, and thus its affinity for ppGpp, may be triggered by the promoter structure and therefore be different for stringently or non-stringently controlled genes. The activity of phage polymerases, which do not have a NTP entry channel are not regulated by ppGpp. The model also provides an explanation for the effect of ppGpp during growth rate regulation where the concentration of ppGpp determines the rate of access of NTPs to the substrate channel. Moreover, the model explains ppGpp-dependent inhibition during elongation where extensive RNA polymerase pausing is observed at defined template positions. Note that pausing sites are almost always characterized by a high apparent K_M for the complementary nucleotide at which RNA polymerase is stalled. The model thus explains ppGpp-dependent inhibition during initiation and elongation and equally for genes under stringent control or during growth rate regulation. It even explains the weak general reduction of the transcription efficiency observed for genes not under stringent control. There are additional arguments which support the above model. The property of the secondary channel as entry site for substrate NTPs limiting free access to the catalytic site is supported by the well known fact that the rate of transcription catalyzed by bacterial RNA polymerase is considerably slower as for phage polymerases (5 to 10 times) or even DNA polymerases

(10 to 20 times). These enzymes all exceed NTP incorporation rates at comparable NTP concentrations. The published proximity of the rifampicin resistance site to the ppGpp binding site and the results from cross-linking studies are all consistent with the model. The finding that stable RNA promoters require high concentrations of starting NTPs and the fact that the presence of ppGpp increases the apparent K_M for sensitive genes fit nicely into the model. The chemical and structural similarity of ppGpp with substrate NTPs represents an important requirement to compete for access into the space limiting substrate channel. The 3′ pyrophosphate of ppGpp makes it distinct enough not to block the catalytic center irreversibly. Finally, the concentration range for ppGpp during stringent and growth rate regulation and substrate NTPs are in line with the range of inhibition observed. The model, which basically rests on substrate competition, has been verified by a series of first tests in which substrate NTPs were substituted with nucleotide analogues that are known to enter the active site of RNA polymerase but are unable to serve as donors for a phosphordiester linkage. Analyzing the individual steps of the initiation pathway we employed as substrate NTP competitors the dinucleotide substrates ApA, ApC and ApU, which are accepted as primers for transcription but cannot extend the growing RNA chains in 3′ direction. In full support of the model outlined above it turned out that for all promoters tested (*rrnB* P1, *rrnD* P1, *PtacI*) the substrate analogues caused the same inhibition characteristics at comparable concentrations that had also been shown for the effector ppGpp (Jöres and Wagner, unpublished). Further experiments are open to test the validity of the model!

Acknowledgements

I like to thank all the past and present members of my lab for their countless contributions which have enabled me to write this mini review. I apologise though, that many contributions were not mentioned explicitly and not all the important literature could be cited – the reference list is already much too long. I am grateful to the Deutsche Forschungsgemeinschaft, which has funded the work over the past years, and I greatly acknowledge the Fonds der Chemischen Industrie for financial support.

References

Afflerbach, H., Schröder, O., and Wagner, R. 1998. Effects of the *Escherichia coli* DNA-binding protein H-NS on rRNA synthesis *in vivo*. Mol. Microbiol. 28: 641–654.

Baracchini, E., and Bremer, H. 1988. Stringent and growth control of rRNA synthesis in *Escherichia coli* are both mediated by ppGpp. J. Biol. Chem. 263: 2597–2602.

Barker, M.M., Gaal, T., Josaitis, C.A., and Gourse, R.L. 2001a. Mechanism of regulation of transcription initiation by ppGpp. I. Effects of ppGpp on transcription initiation *in vivo* and *in vitro*. J. Mol. Biol. 305: 673–688.

Barker, M.M., Gaal, T., and Gourse, R.L. 2001b. Mechanism of regulation of transcription initiation by ppGpp. II. Models for positive control based on properties of RNAP mutants and competition for RNAP. J. Mol. Biol. 305: 689–702.

Cashel, M., Gentry, D.R., Hernandez, V.J., and Vinella, D. 1996. The stringent response. *Escherichia coli* and *Salmonella* Cellular and Molecular Biology. Washington, D.C., ASM Press.

Choy, H.E. 2000. The study of guanosine 5′-diphosphate 3′-diphosphate-mediated transcription regulation *in vitro* using a coupled transcription-translation system. J. Biol. Chem. 275: 6783–6789.

Condon, C., Squires, C., and Squires, C.L. 1995. Control of rRNA transcription in *Escherichia coli*. Microbiol. Reviews 59: 623–645.

Davies, I.J., and Drabble, W.T. 1996. Stringent and growth-rate-dependent control of the *gua* operon of *Escherichia coli* K-12. Microbiology 142: 2429–2437.

Figueroa-Bossi, N., Guérin, M., Rahmouni, R., Leng, M., and Bossi, L. 1998. The supercoilling sensitivity of a bacterial tRNA promoter parallels its responsiveness to stringent control. EMBO J. 17: 2359–2367.

Gaal, T., Bartlett, M.S., Ross, W., Turnbough, C., and Gourse, R.L. 1997. Transcription regulation by initiating NTP concentration: rRNA synthesis in bacteria. Science 278: 2092–2097.

Gaal, T., and Gourse, R.L. 1990. Guanosine 3′-diphosphate 5′-diphosphate is not required for growth rate-dependent regulation of rRNA transcription in *Escherichia coli*. Proc. Natl. Acad. Sci. U. S. A. 87: 5533–5537.

Gafny, R., Cohen, S., Nachaliel, N., and Glaser, G. 1994. Isolated P2 rRNA promoters of *Escherichia coli* are strong promoters that are subject to stringent control. J. Mol. Biol. 243: 152–156.

Gallant, J.A. 1979. Stringent control in *E. coli*. Ann. Rev. Genet. 13: 393–415.

Gourse, R.L., Gaal, T., Bartlett, M.S., Appleman, J.A., and Ross, W. 1996. rRNA transcription and growth rate-dependent regulation of ribosome synthesis in *Escherichia coli*. Ann. Rev. Microbiol. 50: 645–677.

Heinemann, M., and Wagner, R. 1997. Guanosine 3′,5′-bis(diphosphate) (ppGpp)-dependent inhibition of transcription from stringently controlled *Escherichia coli* promoters can be explained by an altered initiation pathway that traps RNA polymerase. Eur. J. Biochem. 247: 990–999.

Hernandez, V.J., and Bremer, H. 1991. *Escherichia coli* ppGpp synthetase II activity requires *spoT*. J. Biol. Chem. 266: 5991–5999.

Hernandez, V.J., and Cashel, M. 1995. Changes in conserved region 3 of *Escherichia coli* σ^{70} mediate ppGpp-dependent functions *in vivo*. J. Mol. Biol. 252: 536–549.

Josaitis, C.A., Gaal, T., and Gourse, R.L. 1995. Stringent control and growth-rate-dependent control have nonidentical promoter sequence requirements. Proc. Natl. Acad. Sci. U.S.A. 92: 1117–1121.

Kingston, R.E., Nierman, W.C., and Chamberlin, M.J. 1981. A direct effect of guanosine tetraphosphate on pausing of *Escherichia coli* RNA polymerase during RNA chain elongation. J. Biol. Chem. 256: 2787–2797.

Korzheva, N., Mustaev, A., Kozlov, M., Malhotra, A., Nikiforov, V., Goldfarb, A., and Darst, S.A. 2000. A structural model of transcription elongation. Science 289: 619–625.

Krohn, M., Pardon, B., and Wagner, R. 1992. Effects of the template topology on RNA polymerase pausing during *in vitro* transcription of the *E. coli rrnB* leader region. Mol. Microbiol. 6: 581–589.

Krohn, M., and Wagner, R. 1995. A procedure for the rapid preparation of guanosine tetraphosphate (ppGpp) from *Escherichia coli* ribosomes. Anal. Biochem. 225: 188–190.

Krohn, M., and Wagner, R. 1996. Transcriptional pausing of RNA polymerase in the presence of guanosine tetraphosphate depends on the promoter and gene sequence. J. Biol. Chem. 271: 23884–23894.

Liebig, B., and Wagner, R. 1995. Effects of different growth conditions on the *in vivo* activity of the tandem *Escherichia coli* ribosomal RNA promoters P1 and P2. Mol. Gen. Genet. 249: 328–335.

Little, R., Ryals, J., and Bremer, H. 1983. *rpoB* mutation in *Escherichia coli* alters control of ribosome synthesis by guanosine tetraphosphate. J. Bacteriol. 154: 787–792.

Mizushima-Sugano, J., and Kaziro, Y. 1985. Regulation of the expression of the *tufB* operon: DNA sequences directly involved in the stringent control. EMBO J. 4: 1053–1058.

Ninnemann, O., Koch, C., and Kahmann, R. 1992. The *E. coli fis* promoter is subject to stringent control and autoregulation. EMBO J. 11: 1075–1083.

Ohlsen, K.L., and Gralla, J.D. 1992. Melting during steady-state transcription of the *rrnB* P1 promoter *in vivo* and *in vitro*. J. Bacteriol. 174: 6071–6075.

Petersen, C., and Møller, L.B. 2000. Invariance of the nucleoside triphosphate pools of *Escherichia coli* with growth rate. J. Biol. Chem. 275: 3931–3935.

Raghavan, A., Kameshwari, D.B., and Chatterji, D. 1998. The differential effects of guanosine tetraphosphate on open complex formation at the *Escherichia coli* ribosomal protein promoters *rplJ* and *rpsA* P1. Biophys. Chem. 75: 7–19.

Reddy, P.S., Raghavan, A., and Chatterji, D. 1995. Evidence for a ppGpp-binding site on *Escherichia coli* RNA polymerase: proximity relationship with the rifampicin-binding domain. Mol. Microbiol. 15: 255–265.

Riggs, D.L., Müller, R.D., Kwan, H.-S., and Artz, S.W. 1986. Promoter domain mediates guanosine tetraphosphate activation of the histidine operon. Proc. Natl. Acad. Sci. U.S.A. 83: 9333–9337.

Sarubbi, E., Rudd, K.E., and Cashel, M. 1988. Basal ppGpp level adjustment shown by new *spoT* mutants affect steady state growth rates and *rrnA* ribosomal promoter regulation in *Escherichia coli*. Mol. Gen. Genet. 213: 214–222.

Schröder, O., and Wagner, R. 2000. The bacterial DNA-binding protein H-NS represses ribosomal RNA transcription by trapping RNA polymerase in the initiation complex. J. Mol. Biol. 298: 737–748.

Shand, R.F., Blum, P.H., Müller, R.D., Riggs, D.L., and Artz, S.W. 1989. Correlation between histidine operon expression and guanosine 5′-diphosphate-3′-diphosphate levels during amino acid downshift in stringent and relaxed strains of *Salmonella typhimurium*. J. Bacteriol. 171: 737–743.

Tedin, K., and Bremer, H. 1992. Toxic effects of high levels of ppGpp in *Escherichia coli* are relieved by *rpoB* mutations. J. Biol. Chem. 267: 2337–2344.

Theißen, G., Pardon, B., and Wagner, R. 1990. A quantitative assessment for transcriptional pausing of DNA-dependent RNA polymerase *in vitro*. Anal. Biochem. 189: 254–261.

Toulokhonov, I.I., Shulgina, I., and Hernandez, V.J. 2000. Binding of the transcription effector ppGpp to *E. coli* RNA polymerase is allosteric, modular, and occurs near the N-terminus of the β′-subunit. J. Biol. Chem. 276: 1220–1225.

Travers, A.A. 1984. Conserved features of coordinately regulated *E. coli* promoters. Nucleic Acids Res. 12: 2605–2618.

Vogel, U., and Jensen, K.F. 1994a. Effects of guanosine 3′,5′-bisdiphosphate (ppGpp) on rate of transcription elongation in isoleucine-starved *Escherichia coli*. J. Biol. Chem. 269: 16236–16241.

Vogel, U., and Jensen, K.F. 1994b. The RNA chain elongation rate in *Escherichia coli* depends on the growth rate. J. Bacteriol. 176: 2807–2813.

Vogel, U., and Jensen, K.F. 1995. Effects of the antiterminator boxA on transcription elongation kinetics and ppGpp inhibition of transcription elongation in *Escherichia coli*. J. Biol. Chem. 270: 18335–18340.

Wagner, R. 1994. The regulation of ribosomal RNA synthesis and bacterial cell growth. Archives Microbiol. 160: 100–109.

Wagner, R. 2000. Transcription Regulation in Prokaryotes. Oxford, Oxford University Press.

Zacharias, M., Göringer, H.U., and Wagner, R. 1989. Influence of the GCGC discriminator motif introduced into the ribosomal RNA P2- and tac promoter on growth rate control and stringent sensitivity. EMBO J. 11: 3357–3363.

Zacharias, M., Göringer, H.U., and Wagner, R. 1990. The signal for growth rate control and stringent sensitivity in *E. coli* is not restricted to a particular sequence motif within the promoter region. Nucleic Acids Res. 18: 6271–6275.

Zhang, G., Campbell, E.A., Minakhin, L., Richter, C., Severinov, K., and Darst, S.A. 1999. Crystal structure of *Thermus aquaticus* core RNA polymerase at 3.3 Å resolution. Cell 98: 811–824.

Zhou, Y.N., and Jin, D.J. 1998. The *rpoB* mutants destabilizing initiation complexes at stringently controlled promoters behave like "stringent" RNA polymerases in *Escherichia coli*. Proc. Natl. Acad. Sci. U.S.A. 95: 2908–2913.

Chapter 26

The General Stress Response Regulatory Network in *Escherichia coli*

Regine Hengge-Aronis

Abstract

Many bacterial species exhibit a general stress response that can be induced by numerous very different stress conditions and, phenotypically, renders the cells broadly stress resistant. In *Escherichia coli*, this response is dependent on the σ^S (RpoS) subunit of RNA polymerase. σ^S is a close relative of the vegetative sigma factor σ^{70} (RpoD) and recognizes very similar promoter sequences. In recent years, significant progress has been made with respect to elucidating (i) the molecular mechanisms that control the cellular σ^S level, which include translational regulation as well as intricate control of σ^S proteolysis, and (ii) the molecular function of σ^S as a transcription initiation factor, where a number of σ^S-dependent promoters have been studied in great detail, and the mechanisms that generate σ^S selectivity are now becoming apparent.

Introduction

Bacteria rarely find optimal growth conditions in their natural environments, i.e. being exposed to stress is their normal life-style. Many bacteria have numerous different single stress-induced responses at their disposal, that allow them to cope with specific stress situations by eliminating the stress agent and repairing actual damage. By contrast, the general stress response can be triggered by many different stress conditions and renders bacteria broadly stress resistant, even against stresses that they have not yet experienced. This means that damage is prevented rather than repaired (Storz and Hengge-Aronis, 2000).

The general stress response is usually controlled by a single master regulator. In *E. coli* and related bacteria (i.e. the gamma-branch of proteobacteria), this master regulator is σ^S (RpoS). σ^S is a sigma subunit of RNA polymerase (RNAP) that is a very close relative of the vegetative σ^{70} (RpoD). Rapidly growing cells contain very little, if any, σ^S, but a variety of quite different stresses result in rapid induction of σ^S. This is usually (but not necessarily) accompanied by a reduction of the growth rate or even entry into stationary phase. As a consequence, expression of numerous σ^S-dependent genes is stimulated, many of which have stress-protective functions. More than 70 genes have been described to be under the control of σ^S (Hengge-Aronis, 2000). Recent microarray analyses have indicated that the expression of up to 10 % of the *E. coli* genes is directly or indirectly affected by σ^S (H. Weber and R. H.-A., unpublished results).

In recent years, regulation of the cellular σ^S level, as well as the function of σ^S as a transcription initiation factor, have been studied intensively. This minireview focusses on these recent developments, and the author apologizes in advance to all colleagues whose work in this very active field cannot be appropriately mentioned because of space constraints. For more details, a recently published compilation of reviews may be consulted (Storz and Hengge-Aronis, 2000).

The General Stress Response of *Escherichia coli* as a Regulatory Network

Regulatory networks in living cells are characterized by their extreme connectivity and hierarchical structure. The latter implies the existence of a limited number of top-level master regulators that act as foci of information integration. Thus, a large number of signal transduction pathways often converge to determine the cellular level or activity of the master regulator. When a certain threshold is reached, a complex physiological response based on the coordinated expression of numerous genes is set in motion. Prominent examples of such responses are the Spo0A-dependent initiation of sporulation of *Bacillus subtilis*, the ComK-dependent competence development in the same species, and the σ^S-and σ^B-mediated general stress responses in *E. coli* and *B.subtilis*, respectively (Dubnau and Turgay, 2000; Hengge-Aronis, 2000; Price, 2000; Sonenshein, 2000).

In addition, complex signal integration also occurs at the lower levels of regulatory networks, in particular in the transcriptional control of the many target genes. These structural genes can have complex promoter regions with multiple binding sites, not only for the master regulator, but also for additional regulatory factors that can be subject to intricate control themselves. This complex transcriptional control often provides connections to other regulatory circuits and establishes differential gene expression within large regulons (Hengge-Aronis, 1999a).

Signal Integration in the Control of the Cellular σ^S Level

During rapid growth in rich medium, σ^S is not detectable in *E. coli*. This changes when the cells grow in minimal medium, where low σ^S levels can be observed. In response to acute stresses, such as sudden starvation or shift to hyperosmolarity, acidic pH or non-optimally low or high temperature, σ^S levels increase further between three- and 20fold. Some of these stresses stimulate σ^S synthesis, whereas others interfere with σ^S degradation. Some stresses (e.g. hyperosmolarity) affect both processes, and therefore σ^S levels increase particularly rapidly and drastically (Figure 1). The details of all known regulatory mechanisms involved have recently been summarized (Hengge-Aronis, R., 2002).

Control of σ^S Synthesis

All known acute stresses that affect σ^S levels act at the post-transcriptional level of σ^S control. Therefore, transcriptional regulation of the *rpoS* gene has not been studied in great detail, even though it is clear that the major *rpoS* promoter becomes activated during entry into stationary phase, when cells are grown in rich medium. *rpoS* transcription is subject to complex control by cAMP-CRP (Lange and Hengge-Aronis, 1994). Positive control by a sensor kinase, BarA, has been reported (Mukhopadhyay *et al.*, 2000), although mutating an apparent cognate response regulator, YecA(UvrY), did not affect *rpoS* transcription (G. Kampmann and R.H.-A., unpublished results).

Translation of existing *rpoS* mRNA is stimulated in response to hyperosmotic shift or pH downshift, beyond a certain cell density or during growth at low temperature (Figure 1). It is widely accepted that, under non-stress conditions, translation is inhibited by the formation of secondary structures in which the translational initiation region (TIR) is probably involved in base-pairing and, therefore, not accessible to ribosomes (although the actual secondary structure has not been demonstrated experimentally). An obvious, though still speculative, mechanism of translational activation would be a structural rearrangement somehow triggered by stress conditions. This can easily be envisioned for low temperature induction, where the small regulatory DsrA RNA is induced, which is partially complementary to an *rpoS* mRNA segment that is probably involved in intramolecular base-pairing with the TIR (Sledjeski *et al.*, 1996; Lease *et al.*, 1998; Majdalani *et al.*, 1998). A similar mechanism has been proposed for another small RNA, RprA, which upon overproduction also stimulates *rpoS* translation (although the physiological relevance of this regulation has not been clarified) (Majdalani *et al.*, 2001 and 2002). How a shift to high osmolarity triggers an at least fivefold higher rate of *rpoS* translation within a few minutes (Muffler *et al.*, 1996c) is still unclear. It is clear, however, that the RNA-binding protein Hfq is crucial for *rpoS* translation (Brown and Elliott, 1996; Muffler *et al.*, 1996b) and actually binds with high affinity to *rpoS* mRNA *in vitro* (D. Traulsen and R. H.-A., unpublished results). Also, the histone-like protein HU is important for *rpoS* translation and may act directly since it can also bind to an *rpoS* mRNA fragment *in vitro* (Balandina *et al.*, 2001). By contrast, the histone-like protein H-NS represses *rpoS* translation (Barth *et al.*, 1995; Yamashino *et al.*, 1995), and so does another small regulatory RNA, OxyS, which is induced by oxidative stress and forms a complex with *rpoS* mRNA and Hfq protein (Zhang *et al.*, 1998). In conclusion, it appears that *rpoS* translational control is extremely complex and involves numerous trans-acting factors, but a clear mechanistic model that would integrate all these factors has yet to emerge.

Control of σ^S Proteolysis

During growth in minimal medium, σ^S half-life is between one and several minutes. σ^S is completely stabilized within a few minutes after osmotic upshift (Muffler *et al.*, 1996c). Also C-starvation and acid shift strongly interfere with σ^S degradation (Lange and Hengge-Aronis, 1994; Bearson *et al.*, 1996). Heat shock results in a more moderate increase in σ^S half-life, that takes approximately 20 minutes to develop (Muffler *et al.*, 1997).

The protease responsible for this rapid turnover is ClpXP (Schweder *et al.*, 1996), a multi-subunit ATP-dependent processive protease. In addition, the response regulator RssB is absolutely essential for σ^S degradation (Bearson *et al.*, 1996; Muffler *et al.*, 1996a; Pratt and Silhavy, 1996). RssB acts as a direct

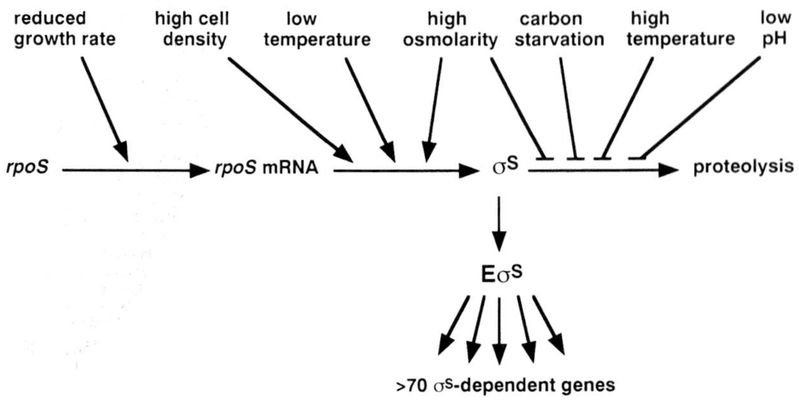

Figure 1. Various stress conditions differentially affect *rpoS* transcription and translation or σ^S proteolysis.

recognition factor for σ^S, whose affinity for σ^S and, therefore, activity in σ^S degradation is dependent on the phosphorylation of its receiver domain (Bouché *et al.*, 1998; Becker *et al.*, 1999; Klauck *et al.*, 2001). A cis-acting region in σ^S, the turnover element, was shown to act as a binding site for RssB, with K173 in σ^S being absolutely essential for RssB interaction *in vitro* and σ^S proteolysis *in vivo* (Becker *et al.*, 1999). Interestingly, K173 is located just downstream of region 2.4 (known to be involved in recognition of the −10 promoter element), and has now been shown to interact directly with a cytosine present at the −13 position in more than 80% of known σ^S-dependent promoters (Becker and Hengge-Aronis, 2001) (see also below). All this indicates that K173 is a crucial amino acid in σ^S, especially for those functions that distinguish σ^S from the otherwise closely related house-keeping sigma factor σ^{70}: σ^{70} is not degraded by the RssB/ClpXP system, and, with a glutamate (E458) at the position corresponding to K173 in σ^S, σ^{70} can contact a guanine at the −13 promoter position (Becker and Hengge-Aronis, 2001).

Once bound to RssB, σ^S is transferred to the ClpXP protease, where, like other Clp protease substrates, it is unfolded and completely degraded by an ATP hydrolysis-dependent mechanism (Figure 2). RssB itself is not a substrate for ClpXP protease *in vivo*, and also its extremely low cellular level (the σ^S:RssB ratio is approximately 20:1) is consistent with a catalytic role in σ^S delivery to ClpXP (Becker *et al.*, 2000; Klauck *et al.*, 2001; Zhou *et al.*, 2001).

Defining the entry points for stress signal pathways into the σ^S recognition and degradation pathway (Figure 2) remains a challenge for future studies. In particular, sensor kinase(s) and/or phosphatase(s) for RssB have not yet been identified. In principle, stress signal input may result in active dephosphorylation of RssB, in inhibition of rephosphorylation of RssB (if dephosphorylation is an obligatory part of the RssB catalytic cycle), and/or in downregulation of the cellular RssB content. It is also possible that not all stress conditions affect RssB activity or cellular level. For carbon starved cells, there is indirect evidence that it is not σ^S-RssB interaction, but a process further downstream that is inhibited (Becker *et al.*, 2000). Moreover, since association with core RNA polymerase protects σ^S against RssB binding and degradation (Zhou *et al.*, 2001), any factor that stimulates core binding of σ^S would also contribute to σ^S stabilization.

In addition, titration of RssB can result in σ^S stabilization (Pruteanu and Hengge-Aronis, 2002). Evidence has been provided that RssB is the overall rate-limiting factor for σ^S proteolysis. As a consequence, the cell has to continuously adjust the expression of RssB to σ^S in order to maintain σ^S proteolysis in growing cells despite variations in the rate of σ^S synthesis. This adjustment operates by a homeostatic feedback mechanism that is provided by *rssB* transcription being controlled by the strictly σ^S-dependent *rssAB* operon promotor (Figure 3). However, strong and rapid increases in σ^S synthesis in response to acute stress (such as osmotic upshift or pH downshift) exceed the compensatory potential of this feedback loop, because the *rssAB* promoter shows a maximal inducibility of approximately three to fourfold only. As a result, σ^S is stabilized because of RssB

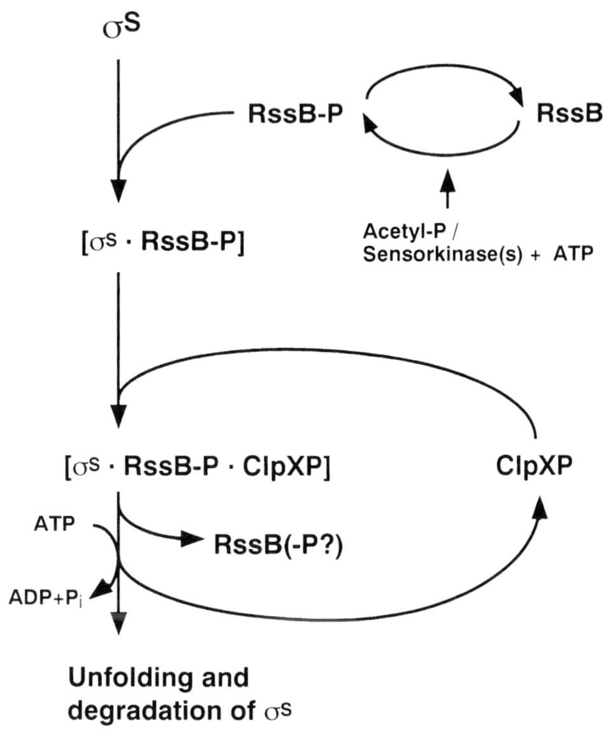

Figure 2. Pathway of σ^S recognition, unfolding and degradation by the response regulator RssB and ClpXP protease.

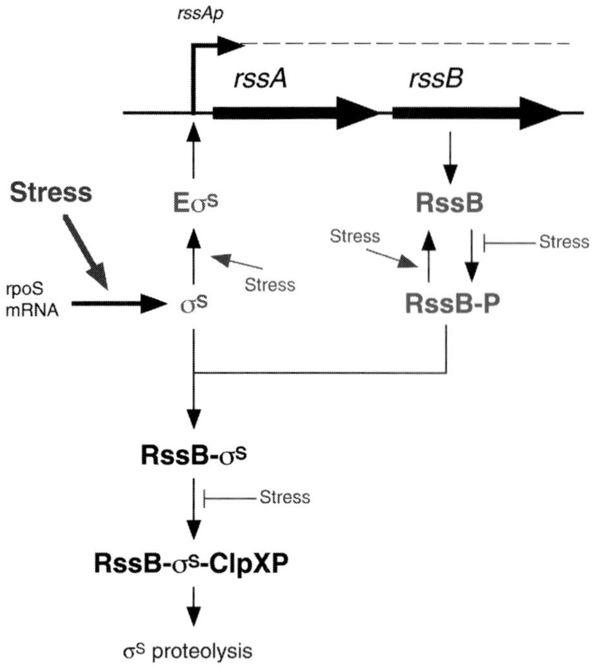

Figure 3. Cellular RssB and σ^S levels are coupled in a homeostatic feedback loop.

titration. Thus, RssB seems to control σ^S proteolysis by a genetic switch, in which (i) the "off" state (low σ^S levels due to proteolysis) is stabilized by a homeostatic negative feedback, and (ii) the threshold for switching to the "on" state (high levels of stable σ^S) is dependent on the cellular level of active, i.e. phosphorylated RssB. This control is highly flexible and information-integrative since the cell obviously can combine RssB titration with alterations in the phosphorylation state of RssB, effects on σ^S/σ^{70} competition for core RNA polymerase and/or more general effects on ClpXP protease (Figure 3).

What is a σ^S-dependent Promoter?

Since the first *in-vitro* transcription assays were performed with σ^S-containing RNAP holoenzyme ($E\sigma^S$), it has puzzled researchers that, *in vitro*, many genes can be transcribed by $E\sigma^S$ as well as by $E\sigma^{70}$, whereas *in vivo*, the same genes are often strongly dependent on σ^S for expression. Therefore, the basic consensus sequence for σ^{70}- and σ^S-controlled promoters must be very similar. But, if there is little promoter-inherent $E\sigma^S/E\sigma^{70}$ selectivity *in vitro*, where does it come from *in vivo*? The currently emerging picture indicates that $E\sigma^S$ selectivity can be generated in different ways at different σ^S-controlled promoters (For a recent summary, see Hengge-Aronis, 2002). A subset of these promoters indeed contain sequence-based features that more or less strongly favor transcription initiation by $E\sigma^S$ (even *in vitro*), whereas at other promoters, additional regulatory factors contribute to or even generate $E\sigma^S$ selectivity (see below). Promotor sequence features that enhance $E\sigma^S$ selectivity are the following:

(i) A degenerate −35 region: most σ^S-controlled promoters have −35 regions with little similarity to the −35(σ^{70}) consensus sequence (TTGACA); Recently it has been shown experimentally that transcription initiation by $E\sigma^S$ is less affected by systematic degeneration of the −35 sequence than initiation by $E\sigma^{70}$ (Gaal *et al.*, 2001).

(ii) An extended −10 region that includes positions (−14) to (−7) with the consensus "KCTATACT" (K stands for T or G): more than 85 % of *in-vivo* $E\sigma^S$-selective promoters contain C(−13); replacing C(−13) or T/G(−14) by other nucleotides reduces expression; and allele-specific suppression data indicate a direct σ^S-specific interaction of C(−13) with K173 in region 2.5 in σ^S (Espinosa-Urgel *et al.*, 1996; Bordes *et al.*, 2000; Becker and Hengge-Aronis, 2001).

(iii) A subset of σ^S-controlled promoter regions feature a distal UP-element half-site located around position −50 (Ballesteros *et al.*, 1998; Germer *et al.*, 2001). A recent study with the *csiD* promoter indicates that a distal UP-element half-site favors transcription initiation by $E\sigma^S$, whereas a full UP-element can be equally used by $E\sigma^S$ and $E\sigma^{70}$ (Germer *et al.*, 2001).

(iv) Nearly all *in-vivo* $E\sigma^S$-selective promoters are AT-rich between the −10 region and the transcriptional start site: this feature was associated with inverse growth rate-correlated expression even before σ^S was discovered (Travers, 1984). For the stringently controlled *tyrT* promoter is was shown that GC-richness downstream of the −10 region results in instability of the open promoter complex (Pemberton *et al.*, 2000). Preliminary data suggest that sequences downstream of the −10 region matter for $E\sigma^S$-selectivity (Ojangu *et al.*, 2000). An interesting exception to this rule is the σ^S-dependent *rssAB* promoter, which is GC-rich downstream of the −10 region. Replacing this region by an AT-rich sequence strongly stimulates σ^S-dependent stationary phase induction of this promoter (Pruteanu and Hengge-Aronis, 2002).

Signal Integration by Multiple Regulators at the Promoters of σ^S-Dependent Genes

Several σ^S-controlled promoters have now been studied in great molecular detail. In all these cases, it has been found that additional regulatory proteins contribute to regulation, which is reflected in quite complicated promoter architectures that involve multiple factor binding sites.

Fine-Tuning Regulation in Response to Additional Signals

A requirement for an additional activator can efficiently limit the expression of a σ^S-dependent gene to a very specific situation. Because it has to be activated by cAMP-CRP, the *csiD* gene is specifically induced by carbon starvation, but not by other conditions that also induce σ^S (Marschall *et al.*, 1998). Simultaneous dependency on σ^S and FIS limits *proP*p2 activity to a small time window during late exponential phase, where both factors are present (Xu and Johnson, 1997b).

In other cases, the spectrum of inducibility can be enlarged by using $E\sigma^S$ as well as $E\sigma^{70}$ for expression. The *osmE* promoter can be recognized by both holoenzymes, with $E\sigma^{70}$ being responsible for hyperosmotic induction, whereas $E\sigma^S$ mediates stationary phase induction (Conter *et al.*, 1997; Bordes *et al.*, 2000). The *osmC* gene features two separate but overlapping promoters p1 and p2, that are dependent on $E\sigma^{70}$ and $E\sigma^S$, respectively. Additional regulators have a modulatory role, with Lrp repressing at p1 and activating at p2, and H-NS repressing both promoters (Bouvier *et al.*, 1998). Also *proP* is under dual control of a σ^{70}-controlled promoter p1 and the above mentioned σ^S-dependent promoter p2 (87 bp apart), that respond to hyperosmolarity and stationary phase, respectively. cAMP-CRP bound to a site that overlaps with the −35 region of p1 acts as a low osmolarity repressor at p1, as well as an upstream co-activator (together with FIS) at p2 (Xu and Johnson, 1997a; McLeod *et al.*, 2000).

Generation of σ^S Selectivity

In vivo σ^S-dependent promoters, that do not have a strong sequence-inherent preference for $E\sigma^S$ (and therefore show expression by $E\sigma^S$ and $E\sigma^{70}$ in standard *in-vitro* transcription assays), must use other means to establish $E\sigma^S$-selectivity *in vivo*. High concentration of salt (especially potassium glutamate, which *in vivo* transiently accumulates upon osmotic upshift) and reduced negative supercoiling of the template have been shown to stimulate (or less inhibit) $E\sigma^S$-mediated *in-vitro* transcription (Ding *et al.*, 1995; Kusano *et al.*, 1996). Promoter-specific enhancement of $E\sigma^S$ selectivity, however, can be due to the effect of additional regulatory proteins, in particular certain abundant nucleoid-associated or histone-like proteins (Hengge-Aronis, 1999b).

This has first been recognized for H-NS, which has a repressing effect on many σ^S-dependent genes (partly, this is due to negative control of σ^S itself). A number of σ^S-controlled genes exhibit higher expression in *hns rpoS* double mutants than in a strain that is defective in *rpoS* alone, i.e. $E\sigma^{70}$ can take over expression only in the absence of H-NS (Olsén *et al.*, 1993; Barth *et al.*, 1995; Yamashino *et al.*, 1995).

The *osmY* promoter is negatively modulated by cAMP-CRP, Lrp and IHF (Lange *et al.*, 1993). For all three factors it could be demonstrated *in vitro* that they interfere more strongly with $E\sigma^{70}$-mediated than with $E\sigma^S$-mediated transcriptional initiation. By adding these factors to the *in-vitro* assays for open complex formation and run-off transcription, $E\sigma^S$ selectivity could be generated (Colland *et al.*, 2000). At the *osmC*p1 promoter, Lrp selectively interferes with $E\sigma^S$-initiated transcription, thus making this promoter σ^{70}-dependent *in vivo* (Bouvier *et al.*, 1998).

The *csiD* promoter exhibits sequence-inherent $E\sigma^S$ preference, which is apparent when basal transcription in the absence of the activator cAMP-CRP is tested *in vitro* (Marschall *et al.*, 1998; Germer *et al.*, 2001; Becker *et al.*, 2001). However, also cAMP-CRP contributes to $E\sigma^S$ selectivity, since it is bound at a position (−68.5), which in the context of the *csiD* promoter region, is unproductive for $E\sigma^{70}$-mediated expression, but which can activate $E\sigma^S$-initiated transcription. Moreover, H-NS activates this promoter by a novel mechanism (Germer *et al.*, 2001).

In all these cases, the RNAP holoenzymes have to deal with complex nucleoprotein structures formed in the promoter regions. Recent reports suggest subtle differences in the modes of DNA binding and open complex formation by $E\sigma^S$ and $E\sigma^{70}$ (Bown *et al.*, 1999; Colland *et al.*, 1999). These properties may differentially affect the ability of the two holoenzymes to interact productively with a specific promoter-protein complex and thereby strongly contribute to or even generate sigma factor selectivity (Hengge-Aronis, 1999b).

Perspectives

Taken together, significant progress has been made in recent years with respect to the regulation and the mechanistic and physiological functions of the σ^S subunit of RNAP. Many important questions, however, remain to be solved. How do the signal transduction pathways function that relay so many stress conditions to alterations in the rates of *rpoS* transcription or translation or σ^S proteolysis? What is the molecular role of the abundant nucleoid-associated protein H-NS in σ^S translation and proteolysis as well as at many σ^S-dependent promoters? What are the exact structural and functional similarities and differences between σ^S and σ^{70} alone and in the complex with RNAP core? What is the physiological function of all σ^S-controlled genes (many of which are just being identified now)? How is the σ^S-mediated stress response interconnected with other cellular networks, e.g. other stress responses or cell division control? With σ^S and the general stress response now being under intense investigation, we can be confident to get the answers before long.

Acknowledgments

This review is also a summary of the work for which the author received generous funding (project He 1556/5) from the Deutsche Forschungsgemeinschaft within the "Regulatory Networks in Bacteria" priority program.

References

Balandina, A., Claret, L., Hengge-Aronis, R., and Rouvière-Yaniv, J. 2001. The *Escherichia coli* histone-like protein HU regulates *rpoS* translation. Mol. Microbiol. 39: 1069–1079.

Ballesteros, M., Kusano, S., Ishihama, A., and Vicente, M. 1998. The *ftsQ*1p gearbox promoter of *Escherichia coli* is a major σ^S-dependent promoter in the *ddlB-ftsA* region. Mol. Microbiol. 30: 419–430.

Barth, M., Marschall, C., Muffler, A., Fischer, D., and Hengge-Aronis, R. 1995. A role for the histone-like protein H-NS in growth phase-dependent and osmotic regulation of σ^S and many σ^S-dependent genes in *Escherichia coli*. J. Bacteriol. 177: 3455–3464.

Bearson, S.M.D., Benjamin W.H., Jr., Swords, W.E., and Foster, J.W. 1996. Acid shock induction of RpoS is mediated by the mouse virulence gene *mviA* of *Salmonella typhimurium.* J. Bacteriol. 178: 2572–2579.

Becker, G., and Hengge-Aronis, R. 2001. What makes an *Escherichia coli* promoter σ^S-dependent? Role of the −13/−14 nucleotide promoter positions and region 2.5 of σ^S. Mol. Microbiol. 39: 1153–1165.

Becker, G., Klauck, E., and Hengge-Aronis, R. 1999. Regulation of RpoS proteolysis in *Escherichia coli*: The response regulator RssB is a recognition factor that interacts with the turnover element in RpoS. Proc. Natl. Acad. Sci. USA 96: 6439–6444.

Becker, G., Klauck, E., and Hengge-Aronis, R. 2000. The response regulator RssB, a recognition factor for σ^S proteolysis in *Escherichia coli*, can act like an anti-σ^S factor. Mol. Microbiol. 35: 657–666.

Bordes, P., Repoila, R., Kolb, A., and Gutierrez, C. 2000. Involvement of differential efficiency of transcription by $E\sigma^S$ and $E\sigma^{70}$ RNA polymerase holenzymes in growth phase regulation of the *Escherichia coli osmE* promoter. Mol. Microbiol. 35: 845–853.

Bouché, S., Klauck, E., Fischer, D., Lucassen, M., Jung, K., and Hengge-Aronis, R. 1998. Regulation of RssB–dependent proteolysis in *Escherichia coli*: a role for acetyl phosphate in a response regulator–controlled process. Mol. Microbiol. 27: 787–795.

Bouvier, J., Gordia, S., Kampmann, G., Lange, R., Hengge-Aronis, R., and Gutierrez, C. 1998. Interplay between global regulators of *Escherichia coli*: effect of RpoS, H-NS and Lrp on transcription of the gene *osmC*. Mol. Microbiol. 28: 971–980.

Bown, J.A., Owens, J.T., Meares, C.F., Fujita, N., Ishihama, A., Busby, S.J.W., and Minchin, S.D. 1999. Organisation of open complexes at *Escherichia coli* promoters: location of promoter DNA sites close to region 2.5 of the σ^{70} subunit of RNA polymerase. J. Biol. Chem. 274: 2263–2270.

Brown, L., and Elliott, T. 1996. Efficient translation of the RpoS sigma factor in *Salmonella typhimurium* requires host factor I, an RNA-binding protein encoded by the hfq gene. J. Bacteriol. 178: 3763–3770.

Colland, F., Barth, M., Hengge-Aronis, R., and Kolb, A. 2000. Sigma factor selectivity of *Escherichia coli* RNA polymerase: a role for CRP, IHF and Lrp transcription factors. EMBO J. 19: 3028–3037.

Colland, F., Fujita, N., Kotlarz, D., Ishihama, A., and Kolb, A. 1999. Positioning of σ^S, the stationary phase σ factor, in *Escherichia coli* RNA polymerase-promoter open complexes. EMBO J. 18: 4049–4059.

Conter, A., Menchon, C., and Gutierrez, C. 1997. Role of DNA supercoiling and RpoS sigma factor in the osmotic and growth phase–dependent induction of the gene *osmE* of *Escherichia coli* K12. J. Mol. Biol. 273: 75–83.

Ding, Q., Kusano, S., Villarejo, M., and Ishihama, A. 1995. Promoter selectivity control of *Escherichia coli* RNA polymerase by ionic strength: differential recognition of osmoregulated promoters by $E\sigma^D$ and $E\sigma^S$ holoenzymes. Mol. Microbiol. 16: 649–656.

Dubnau, D., and Turgay, K. 2000. Regulation of competence in *Bacillus subtilis* and its relation to stress. In: Bacterial Stress Responses. Storz, G. and Hengge-Aronis, R., ed. ASM Press, Washington, D.C. p. 249–260.

Eisenstark, A., Calcutt, M.J., Becker-Hapak, M., and Ivanova, A. 1999. Role of *Escherichia coli rpoS* and associated genes in defense against oxidative damage. Free Rad. Biol. Med. 21: 975–993.

Espinosa-Urgel, M., Chamizo, C., and Tormo, A. 1996. A consensus structure for σ^S-dependent promoters. Mol. Microbiol. 21: 657–659.

Gaal, T., Ross, W., Estrem, S.T., Nguyen, L.H., Burgess, R.R., and Gourse, R.L. 2001. Promoter recognition and discrimination by $E\sigma^S$ RNA polymerase. Mol. Microbiol. 42: 939–954.

Germer, J., Becker, G., and Hengge-Aronis, R. 2001. Role of activator site position and a distal UP–element half–site for sigma factor selectivity and a novel role for H-NS at a σ^S-dependent promoter in *Escherichia coli.* Mol. Microbiol. 41: 705–716.

Hengge-Aronis, R. 1999a. Integration of control devices: A global regulatory network in *Escherichia coli.* In: Prokaryotic gene expression. Baumberg, S., ed. Oxford University Press, Oxford. p. 169–193.

Hengge-Aronis, R. 1999b. Interplay of global regulators in the general stress response of *Escherichia coli.* Curr. Op. Microbiol. 2: 148–152.

Hengge-Aronis, R. 2000. The general stress response in *Escherichia coli.* In: Bacterial Stress Responses. Storz, G. and Hengge-Aronis, R., ed. ASM Press, Washington, D.C. p. 161–178.

Hengge-Aronis, R. 2002a. Signal transduction and regulatory mechanisms involved in control of the σ^S subunit of RNA polymerase in *Escherichia coli.* Microbiol. Molec. Biol. Rev. 66: 373–395.

Hengge-Aronis, R. 2002b. Stationary phase gene regulation: what makes an *Escherichia coli* promoter σ^S-dependent? Curr. Op. Microbiol. 5: 591–595.

Klauck, E., Lingnau, M., and Hengge-Aronis, R. 2001. Role of the response regulator RssB in σ^S recognition and initiation of σ^S proteolysis in *Escherichia coli.* Mol. Microbiol. 40: 1381–1390.

Kusano, S., Ding, Q.Q., Fujita, N., and Ishihama, A. 1996. Promoter selectivity of *Escherichia coli* RNA polymerase $E\sigma^{70}$ and $E\sigma^{38}$ holoenzymes – Effect of DNA supercoiling. J. Biol. Chem. 271: 1998–2004.

Lange, R., Barth, M., and Hengge-Aronis, R. 1993. Complex transcriptional control of the σ^S-dependent stationary phase-induced and osmotically regulated *osmY* (*csi-5*) gene suggests novel roles for Lrp, cyclic AMP (cAMP) receptor protein-cAMP complex and integration host factor in the stationary phase response of *Escherichia coli.* J. Bacteriol. 175: 7910–7917.

Lange, R., and Hengge-Aronis, R. 1994. The cellular concentration of the σ^S subunit of RNA-polymerase in *Escherichia coli* is controlled at the levels of transcription, translation and protein stability. Genes Dev. 8: 1600–1612.

Lease, R.A., Cusick, M.E., and Belfort, M. 1998. Riboregulation in *Escherichia coli*: DsrA RNA acts by RNA:RNA interaction at multiple loci. Proc. Natl. Acad. Sci. USA 95: 12456–12461.

Majdalani, N., Cunning, C., Sledjeski, D., Elliott, T., and Gottesman, S. 1998. DsrA RNA regulates translation of RpoS message by an anti-antisense mechanims, independent of its action as an antisilencer of transcription. Proc. Natl. Acad. Sci. USA 95: 12462–12467.

Majdalani, N., Chen, S., Murrow, J., St. John, K., and Gottesman, S. 2001. Regulation of RpoS by a novel small RNA: the characterization of RprA. Mol. Microbiol. 39: 1382–1394.

Majdalani, N., Hernandez, D., and Gottesman, S. 2002. Regulation and mode of action of the second small RNA activator of RpoS translation, RprA. Mol. Microbiol. 46: 813–826.

Marschall, C., Labrousse, V., Kreimer, M., Weichart, D., Kolb, A., and Hengge-Aronis, R. 1998. Molecular analysis of the regulation of *csiD*, a carbon starvation-inducible gene in *Escherichia coli* that is exclusively dependent on σ^S and requires activation by cAMP-CRP. J. Mol. Biol. 276: 339–353.

McLeod, S.M., Xu, J., and Johnson, R.C. 2000. Coactivation of the RpoS-dependent *proP* P2 promoter by Fis and cyclic AMP receptor protein. J. Bacteriol. 182: 4180–4187.

Muffler, A., Barth, M., Marschall, C., and Hengge-Aronis, R. 1997. Heat shock regulation of σ^S turnover: a role for DnaK and relationship between stress responses mediated by σ^S and σ^{32} in *Escherichia coli.* J. Bacteriol. 179: 445–452.

Muffler, A., Fischer, D., Altuvia, S., Storz, G., and Hengge-Aronis, R. 1996a. The response regulator RssB controls stability of the σ^S subunit of RNA polymerase in *Escherichia coli.* EMBO J. 15: 1333–1339.

Muffler, A., Fischer, D., and Hengge-Aronis, R. 1996b. The RNA-binding protein HF-I, known as a host factor for phage Qβ RNA replication, is essential for the translational regulation of *rpoS* in *Escherichia coli.* Genes Dev. 10: 1143–1151.

Muffler, A., Traulsen, D.D., Lange, R., and Hengge-Aronis, R. 1996c. Posttranscriptional osmotic regulation of the σ^S subunit of RNA polymerase in *Escherichia coli.* J. Bacteriol. 178: 1607–1613.

Mukhopadhyay, S., Audia, J.P., Roy, R.N., and Schellhorn, H.E. 2000. Transcriptional induction of the conserved alternative sigma factor RpoS in *Escherichia coli* is dependent on BarA, a probable two-component regulator. Mol. Microbiol. 37: 371–381.

Ojangu, E.-L., Tover, A., Teras, R., and Kivisaar, M. 2000. Effects of combination of different −10 hexamers and downstream sequences on stationary phase-specific sigma factor σ^S-dependent transcription in *Pseudomonas putida.* J. Bacteriol. 182: 6707–6713.

Olsén, A., Arnqvist, A., Hammar, M., Sukupolvi, S., and Normark, S. 1993. The RpoS sigma factor relieves H-NS-mediated transcriptional repression of *csgA*, the subunit gene of fibronectin binding curli in *Escherichia coli.* Mol. Microbiol. 7: 523–536.

Pemberton, I.K., Muskhelishvili, G., Travers, A.A., and Buckle, M. 2000. The G+C-rich discriminator region of the *tyrT* promoter antagonises the formation of stable preinitiation complexes. J. Mol. Biol. 299: 859–864.

Pratt, L.A., and Silhavy, T.J. 1996. The response regulator, SprE, controls the stability of RpoS. Proc. Natl. Acad. Sci. USA 93: 2488–2492.

Price, C.W. 2000. Protective function and regulation of the general stress response in *Bacillus subtilis* and related gram-positive bacteria. In: Bacterial Stress Responses. Storz, G. and Hengge-Aronis, R., ed. ASM Press, Washington, D.C. p. 179–197.

Pruteanu, M., and Hengge-Aronis, R. 2002. The cellular level of the recognition factor RssB is rate-limiting for σ^S proteolysis: Implications for RssB regulation and signal transduction in σ^S turnover in *Escherichia coli*. Mol. Microbiol. 45: 1701–1714.

Schweder, T., Lee, K.-H., Lomovskaya, O., and Matin, A. 1996. Regulation of *Escherichia coli* starvation sigma factor (σ^S) by ClpXP protease. J. Bacteriol. 178: 470–476.

Sledjeski, D.D., Gupta, A., and Gottesman, S. 1996. The small RNA, DsrA, is essential for the low temperature expression of RpoS during exponential growth in *E. coli*. EMBO J. 15: 3993–4000.

Sonenshein, A.L. 2000. Bacterial sporulation: a response to environmental signals. In: Bacterial Stress Responses. Storz, G. and Hengge-Aronis, R., ed. ASM Press, Washington, D.C. p. 199–215.

Storz, G., and Hengge-Aronis, R. 2000. Bacterial Stress Responses. ASM Press, Washington, D.C.

Tanaka, K., Kusano, S., Fujita, N., Ishihama, A., and Takahashi, H. 1995. Promoter determinants for *Escherichia coli* RNA polymerase holoenzyme containing σ^{38} (the *rpoS* gene product). Nucl. Acids Res. 23: 827–834.

Travers, A.A. 1984. Conserved features of coordinately regulated *E.coli* promoters. Nucl. Acids Res. 12: 2605–2618.

Xu, J., and Johnson, R.C. 1997a. Cyclic AMP receptor protein functions as a repressor of the osmotically inducible promoter *proP* P1 in *Escherichia coli*. J. Bacteriol. 179: 2410–2417.

Xu, J.M., and Johnson, C. 1997b. Activation of RpoS-dependent *proP* P2 transcription by the Fis protein *in vitro*. J. Mol. Biol. 270: 346–359.

Yamashino, T., Ueguchi, C., and Mizuno, T. 1995. Quantitative control of the stationary phase-specific sigma factor, σ^S, in *Escherichia coli*: involvement of the nucleoid protein H-NS. EMBO J. 14: 594–602.

Zhang, A., Altuvia, S., Tiwari, A., Argaman, L., Hengge-Aronis, R., and Storz, G. 1998. The OxyS regulatory RNA represses *rpoS* translation and binds the Hfq (HF-I) protein. EMBO J. 17: 6061–6068.

Zhou, Y., Gottesman, S., Hoskins, J.R., Maurizi, M.R., and Wickner, S. 2001. The RssB response regulator directly targets σ^S for degradation by ClpXP. Genes Dev. 15: 627–637.

Chapter 27

Positive Regulation of Gene Expression by the Catabolite Control Protein CcpA in *Bacillus subtilis*

Holger Ludwig, Hans-Matti Blencke, Matthias Schmalisch, Christian Detsch, Matthias Merzbacher and Jörg Stülke

Abstract

In *Bacillus subtilis* and other Gram-positive bacteria, carbon catabolite control is mediated by the pleiotropic regulatory protein CcpA. In addition to loss of catabolite repression, *ccpA* mutants exhibit a severe growth defect. This growth defect may result from loss of expression of several genes that are activated by CcpA. Gene activation by CcpA has been studied at different levels such as proteome and transcriptome analysis and by investigation of the regulation of individual genes in wild type and *ccpA* mutant strains. Important cellular functions such as glycolysis, overflow metabolism to excrete excess carbon from the cell, and ammonium assimilation depend on a functional CcpA. While CcpA can act directly as a transcriptional activator to allow expression of *ackA* and *pta* genes, its role is indirect for genes of glycolysis. In this case, the accumulation of an intracellular inducer cannot occur in *ccpA* mutants due to a defect in sugar transport by the phosphoenolpyruvate:sugar phosphotransferase system. Several mutations were isolated that exhibit loss of catabolite repression due to the *ccpA* mutation but that do not cause a growth defect. These mutations were identified within the *ccpA* gene or are extragenic suppressors.

Introduction

Carbon metabolism in bacterial cells is controlled in a way that allows the utilization of a specific carbon source only if no preferred carbon source is present in the growth medium. This control can be achieved both at the level of enzymatic activities and enzyme synthesis. In bacteria, the concerted action of several regulatory mechanisms results in the sequential expression of genes and operons encoding catabolic enzymes. This phenomenon, called carbon catabolite repression, has been the subject of intensive investigation since the classical observation of a glucose-lactose diauxie in *Escherichia coli* (Monod, 1942; for reviews see Saier, 1998; Stülke and Hillen, 1999; Brückner and Titgemeyer, 2002; Titgemeyer and Hillen, 2002).

In *Bacillus subtilis* several mutations resulting in loss of glucose repression have been isolated. They affected the *ccpA*, *ccpB*, and *glcR* repressor genes or components of the phosphotransferase system (PTS) (Henkin *et al.*, 1991; Chauvaux *et al.*, 1998; Stülke *et al.*, 2001; Frisby and Zuber, 1994; Stülke *et al.*, 1997). The Catabolite control protein A (CcpA) is central to carbon catabolite repression in *B. subtilis* and other Gram-positive bacteria (Stülke and Hillen, 1999; Titgemeyer and Hillen, 2002). Signal transduction to CcpA is mediated by components of the PTS: in the presence of excess glucose the HPr protein of the PTS can be phosphorylated at a regulatory seryl residue by the metabolite controlled HPr kinase (Reizer *et al.*, 1998; Galinier *et al.*, 1998; Stülke and Hillen, 1998). Seryl-phosphorylated HPr (or its regulatory paralogue Crh) can form a complex with CcpA and this complex binds to catabolite responsive elements (*cre*) in the promoter regions of catabolic genes and operons to prevent their transcription in the presence of repressing carbon sources (Deutscher *et al.*, 1995; Fujita *et al.*, 1995; Galinier *et al.*, 1997; Galinier *et al.*, 1999; Martin-Verstraete *et al.*, 1999).

B. subtilis ccpA mutants exhibit a severe growth defect in minimal media (Martin *et al.*, 1989; Krüger *et al.*, 1993; Lindner *et al.*, 1994). By transformation, this growth defect was shown to be a direct consequence of the *ccpA* mutation (Wray *et al.*, 1994). Using a mutant strain carrying a point mutation in the *ccpA* ribosomal binding site, Miwa *et al.* (1994) isolated revertants that were able to grow at the same rate as the wild type. Several independent revertants had the wild type sequence of the ribosomal binding site. These sets of experiments demonstrate that a *ccpA* mutation or a deficiency in CcpA synthesis both result in slow growth on minimal media. The growth defects exhibited by *ccpA* mutants strains are not observed on complex media, however, they are clearly visible on minimal media. Moreover, *ccpA* mutants are not able to grow on minimal media with glucose and ammonia as the only sources of carbon and nitrogen, respectively (Wray *et al.*, 1994; Faires *et al.*, 1999).

The growth defect of *ccpA* mutants was subsequently observed for several bacteria such as *B. megaterium*, *Staphylococcus xylosus*, *Lactococcus lactis*, *Lactobacillus casei*, *Listeria monocytogenes*, *Enterococcus faecalis*, *Streptococcus mutans* and *S. thermophilus* (Hueck *et al.*, 1995; Egeter and Brückner, 1996; Luesink *et al.*, 1998; Monedero *et al.*, 1997; Behari and Youngman, 1998; Leboeuf *et al.*, 2000; Simpson and Russell, 1998; van den Bogaard *et al.*, 2000). An exception was *Lactobacillus pentosus*. The absence of a observable growth defect of a *ccpA* mutant correlates, however, with the requirement to grow *L. pentosus* on complex media (Lokman *et al.*, 1997).

The observation that *ccpA* mutants show a growth defect that is not directly related to carbon catabolite repression led to the idea that CcpA might fulfill functions distinct from catabolite repression (Henkin, 1996). In this review, we discuss the findings on the role of CcpA as a positive regulator of gene expression and correlate the implication of CcpA in different metabolic pathways to its function in allowing balanced growth of bacteria. The growth defect of *ccpA* mutants on minimal media might have different reasons: (i) the unbalanced expression of catabolic genes and operons even if their products are not necessary may result in longer generation times or (ii) some genes that are important for rapid growth may require a functional CcpA for their expression. Finally, a combination of both variants seems possible. The different approaches that were used to address this problem are reviewed below.

Genetic Dissection of *ccpA* Mutant Phenotypes

B. subtilis ccpA mutant strains have two basic phenotypes: they are deficient in carbon catabolite repression, and they exhibit a severe growth defect. Two genetic approaches were used to separate the two phenotypes from each other: Second-site mutations that restore growth of *ccpA* mutants on minimal media were isolated and studied by Faires *et al.* (1999), whereas *ccpA* mutant alleles causing normal catabolite repression but resulting in a growth defect (and *vice versa*) were analyzed by Küster *et al.* (1999).

Several *trans*-acting mutations suppressing the growth defect (*sgd* mutations) were studied. While all mutations allowed the *ccpA* mutant strain to grow on minimal media supplemented with glucose and ammonium as single sources of carbon and nitrogen, they did not restore carbon catabolite repression of β-xylosidase synthesis (Faires *et al.*, 1999). Thus, the two phenotypes of the *ccpA* mutant are not directly related to each other.

This idea is supported by the isolation of *B. megaterium ccpA* alleles that are proficient in carbon catabolite repression but which result in a growth defect. Moreover, other *ccpA* alleles were found not to result in impaired growth but to cause a loss of catabolite repression of the *xylAB* operon. The sequence analysis of these mutants revealed that all *ccpA* gene sequences resulting in separation of the CcpA functions carry mutations affecting the N-terminal DNA-binding domain of CcpA (Küster *et al.*, 1999). This finding may have two implications: First, the different functions may result from binding to distinct, specialized, *cre* sequences, or DNA binding might be necessary for catabolite repression but not for growth control. It was therefore proposed that growth control might be exerted by protein-protein interaction (Stülke and Hillen, 2000). To test this hypothesis, a *B. subtilis* strain expressing a CcpA protein devoid of its N-terminal domain was constructed (Ludwig and Stülke, 2001). The C-terminal domain of CcpA expressed by this strain is sufficient for the interaction with the cofactor, HPr(Ser-P) (Jones *et al.*, 1997). The analysis of this strain revealed, however, that it exhibited both phenotypes also found in *ccpA* disruption mutants (Ludwig and Stülke, 2001). It can thus be concluded that all known functions of CcpA require the presence of the DNA-binding domain and, probably, DNA-binding activity of CcpA.

Moreover, the two approaches used to separate the functions of CcpA provide compelling evidence that the growth defect of *ccpA* mutants may result from the dependence of some important genes on a functional CcpA rather than from unbalanced expression of catabolic genes and operons. It was thus a challenge to identify these important CcpA-dependent genes.

Identification of the Growth Requirements of *ccpA* Mutant Strains

Two observations suggested that CcpA might be required for the biosynthesis of amino acids. First, the mutant strains are capable of growing in minimal media if provided with glutamate or other organic nitrogen sources (Faires *et al.*, 1999). Second, the absence of a growth defect in media containing mixtures of amino acids such as caseine hydrolysate also indicated a deficiency in amino acid metabolism. A rigorous analysis of the growth requirements of *ccpA* mutant strains revealed that they are auxotrophic for glutamate and therefore are indistinguishable from *gltAB* mutants which are unable to synthesize glutamate. Since CcpA is a transcriptional regulator, the expression of the *gltAB* operon was studied in wild type and *ccpA* mutant cells. In contrast to the wild type, no *gltAB* expression occurred in the *ccpA* mutant in the presence of glucose. Thus, CcpA is a positive regulator involved in the expression of the *gltAB* operon (Faires *et al.*, 1999). The requirement for CcpA for the expression of the *gltAB* operon is circumvented by the above-mentioned *sgd* suppressor mutations which simultaneously allow growth on minimal media in the absence of glutamate and expression of the *gltAB* operon even in a *ccpA* mutant background (Faires *et al.*, 1999). Unfortunately, the identity of the *sgd* mutations has not yet been uncovered.

However, even the addition of glutamate to minimal medium does not allow the *ccpA* mutant to grow as fast as the wild type strain. We suspected therefore that the synthesis of other amino acids might depend on a functional CcpA protein. Indeed, the addition of methionine and the branched-chain amino acids isoleucine, valine and leucine is sufficient to completely suppress the growth defect in minimal media containing glucose and glutamate as carbon and nitrogen source, respectively (Ludwig *et al.*, 2002a). Insufficient expression of the *ilv-leu* operon encoding enzymes of branched-chain amino acid synthesis in the *ccpA* mutant accounts for the growth defect in the absence of these amino acids. As demonstrated for *gltAB*, the expression of the *ilv-leu* operon is induced by glucose and involves a functional CcpA. However, while *gltAB* is not expressed in *ccpA* mutants, the effect of the *ccpA* mutation on the expression of the *ilv-leu* operon on CcpA is rather weak. This explains why *ccpA* mutants are auxotrophic for glutamate but not for the branched-chain amino acids (Ludwig *et al.*, 2002a).

Global Analysis of Protein Synthesis and Gene Expression in *ccpA* Mutant Strains

The role of CcpA in the control of functions different from repression of transcription of catabolic genes and operons was addressed by studying gene expression in wild type and *ccpA* mutant strains of *B. subtilis* after growth in the presence or absence of glucose as a trigger of CcpA activity.

Proteome analyses (Tobisch *et al.*, 1999) revealed that the enzymes of one of the central pathways of sugar catabolism, glycolysis, are induced by glucose and that this induction depends on a functional CcpA. Moreover, enzymes of overflow metabolism that are involved in the excretion of excess carbon are induced in the presence of glucose in a *ccpA*-dependent manner. This finding is substantiated by the observation that the consumption of glucose results in acidification of the medium. However, no acidification was observed in the *ccpA* mutant. In addition to the enzymes of central carbon catabolism, several proteins are expressed in the wild type but not in the *ccpA* mutant suggesting that their synthesis is directly or indirectly subject to positive control by CcpA. Unfortunately, the identities and functions of these proteins are so far unknown (Tobisch *et al.*, 1999).

With the completion of bacterial genome sequences the analysis of the transcription profile of bacterial cells became possible. Among the first regulatory phenomema to be addressed by transcriptome analysis was regulation of gene expression by glucose and CcpA in *B. subtilis* (Yoshida *et al.*, 2001; Moreno *et al.*, 2001). While Yoshida *et al.* studied genes repressed by glucose, Moreno *et al.* analyzed both glucose repression and activation of gene expression. They found that among about 3,900 genes about 306 are activated more than two-fold and 100 even more than three-fold by glucose. Transcriptional activation of the majority of these genes by glucose was lost in the *ccpA* mutant suggesting that CcpA is involved in activation of their expression. Among the genes activated by glucose are genes coding for enzymes of overflow metabolism such as the *alsSD* and *lctEP* operons encoding enzymes for acetoin and lactate synthesis, respectively. Moreover, *narH* and *ilvB* encoding the β-subunit of nitrate reductase and the large subunit of acetolactate synthase are activated by glucose. Activation of all the above-mentioned genes requires a functional CcpA. However, *cre* sites that may serve as target for CcpA binding are not associated with these genes suggesting that the role of CcpA may be indirect (Moreno *et al.*, 2001).

Despite all advantages, the proteome and transcriptome analyses described above have three major drawbacks: first, they were performed with cells grown in minimal media which result in very different growth rates of wild type and *ccpA* mutant strains (Tobisch *et al.*, 1999) or in complex media (Yoshida *et al.*, 2001; Moreno *et al.*, 2001). Thus, it was not possible to cultivate the wild type and *ccpA* mutant strains in chemically defined media with the same generation time. Second, and even more important, these techniques allow the identification of many genes that are controlled by CcpA but they do not identify the genes that are relevant for the observed biological effects such as the growth defect. Third, proteome and transcriptome analyses do not reveal by which mechanism(s) CcpA may control the expression of the genes. This makes a careful physiological analysis of *ccpA* mutants and of individual CcpA-dependent genes inevitable even in the post-genomic era.

Having identified the growth requirements of *ccpA* mutants, we were able to design a defined minimal medium which allows growth of the wild type and the *ccpA* mutant strains at very similar rates. This medium contains glutamate as a source of nitrogen, methionine, and the branched-chain amino acids (Blencke *et al.*, 2003). An analysis of the transcriptional profile of the two strains after growth with or without glucose revealed that 18 genes are induced more than threefold by glucose. As observed in previous studies (Tobisch *et al.*, 1999; Moreno *et al.*, 2001), the genes encoding enzymes of glucose transport and glycolysis are significantly induced by glucose, as well as the *pdhABCD* operon encoding pyruvate dehydrogenase. While induction of the *ptsGHI* operon encoding the glucose-specific PTS is independent from CcpA, induction of the glycolytic *gapA* operon and of the *pdh* operon requires a functional CcpA. As suggested by the studies on the growth defect of *ccpA* mutants (Faires *et al.*, 1999), the expression of the *gltAB* operon encoding enzymes for glutamate biosynthesis is strongly induced by glucose in a CcpA-dependent manner. Similarly, expression of the *ilv-leu* operon was subject to a weak CcpA-dependent induction by glucose. In addition to the genes and operons already known to be under positive control by CcpA, the transcriptome analysis revealed CcpA-dependent induction of *lysC* encoding aspartokinase II and of genes involved in purine biosynthesis. Moreover, several genes encoding unknown proteins are subject to induction by CcpA (Blencke *et al.*, 2003).

Positive Control of Gene Expression by Direct Binding of CcpA to *cre* Sites

As observed for genes repressed by CcpA in the presence of glucose, several of the genes under positive control by CcpA contain *cre* sites in their promoter regions that serve as target for CcpA binding and transcription activation. Among these genes are *pta* and *ackA* encoding phosphotransacetylase and acetate kinase, respectively (Grundy *et al.*, 1993; Shin *et al.*, 1999; Presecan-Siedel *et al.*, 1999). These enzymes allow the utilization of acetyl-CoA to excrete acetate and to perform an additional substrate level phosphorylation to obtain two extra molecules of ATP per molecule of glucose catabolized.

In the case of genes repressed by CcpA the *cre* sites are located in the promoter regions or downstream of the transcriptional start point. Binding of CcpA regions is thought to prevent transcription initiation or to interfere with transcription elongation (for review see Stülke and Hillen, 2000). Only in the case of the levanase operon, CcpA binds to a site upstream of the promoter, and it was suggested that CcpA binding prevents the interaction between the transcription activator LevR and the RNA polymerase containing the alternative σ factor, σ^{54} (Martin-Verstraete *et al.*, 1995; Martin-Verstraete *et al.*, 1999). Similarly, the *cre* sites of the *pta* and *ackA* genes are located about 55 bp upstream of the transcriptional start site (Grundy *et al.*, 1993; Shin *et al.*, 1999; Presecan-Siedel *et al.*, 1999). Originally two *cre* sequences were proposed to be involved in *ackA* activation by CcpA. Site-directed mutagenesis revealed, however, that only the site closer to the promoter

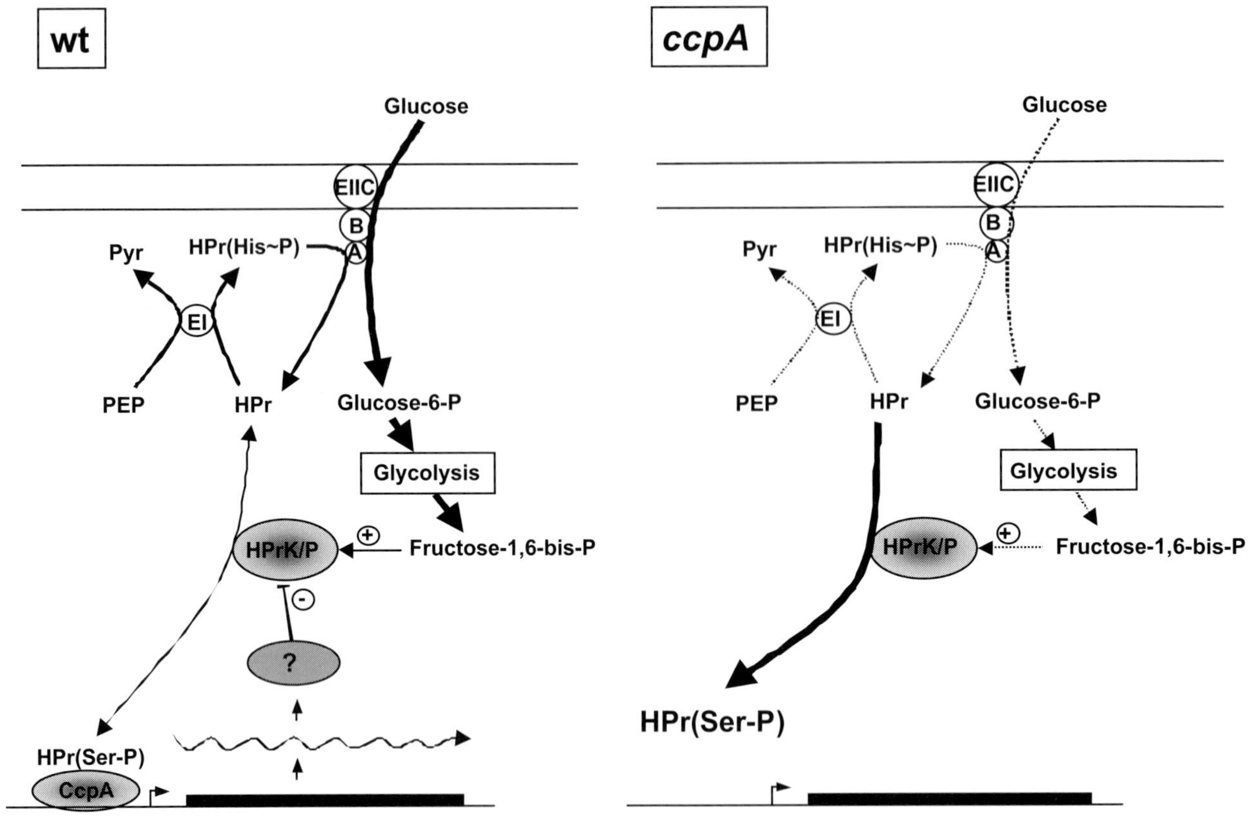

Figure 1. Negative feedback regulation of HPr kinase in *B. subtilis*. Upon sugar transport and glycolytic catabolism, HPr kinase (HPrK/P) activity is stimulated by the accumulation of glycolytic intermediates such as fructose-1,6-bisphosphate. In the wild type strain (wt), an as yet unidentified factor switches off HPr kinase activity once a sufficient amount of HPr(Ser-P) is formed. The expression of this factor is thought to be under positive control by CcpA. The negative feedback of HPr kinase activity prevents complete phosphorylation of HPr on the regulatory site, Ser-46 and the resulting depletion of the free HPr pool necessary for sugar transport. In the *ccpA* mutant strain, the negative effector of HPr kinase activity cannot be expressed. Therefore, HPr kinase is active until all HPr is converted to HPr(Ser-P). This form of HPr is not active in PTS phosphorylation resulting in a significant reduction of sugar transport.

is required (Turinsky *et al.*, 1998). Similarly, the existence of a *cre* sequence located upstream of the promoter was proposed in the case of the glycolytic *gapA* operon (Miwa *et al.*, 2000; Moreno *et al.*, 2001). However, a mutational analysis revealed that this proposed sequence is not involved in positive regulation of the *gapA* operon by CcpA (Ludwig *et al.*, 2001). Thus, the only documented examples of *cre* sites mediating direct transcription activation are those centered at position -55. This may be significant as we can expect that a productive interaction of CcpA with RNA polymerase may require a proper spacing and orientation of the two protein complexes (remember that CcpA binds as a dimer together with one of its cofactors, HPr(Ser-P) or Crh(Ser-P)). Indeed, a detailed study of the sequence and position requirements for the *cre* site of the *B. subtilis ackA* gene revealed that the position of CcpA on the DNA helix is crucial since insertions of 10 bp between the *cre* and the promoter did not interfere with induction whereas a 5 bp insertion prevented transcription activation by CcpA. Moreover, mutations that convert the biologically inactive pseudo-*cre* site in front of the *ackA* gene to the consensus did not increase the induction by CcpA suggesting that an additional *cre* at an unfavourable position does not affect transcription activation (Turinsky *et al.*, 1998).

A Novel Mechanism of CcpA-Dependent Positive Control of Gene Expression: Control of Inducer Accumulation

Control of gene expression by CcpA due to the binding of the regulator to its *cre* target site is well established (see above, Stülke and Hillen, 2000; Brückner and Titgemeyer, 2002). However, several genes controlled by CcpA do not possess obvious *cre* sites or suggested *cre* sequences are biologically inactive (Moreno *et al.*, 2001; Faires *et al.*, 1999; Ludwig *et al.*, 2001; Turinsky *et al.*, 2000).

The mechanisms by which CcpA controls the induction of gene expression in the absence of a *cre* site was studied to some detail for the *B. subtilis* glycolytic *gapA* operon. This operon encodes enzymes required for the conversion of glyceraldehyde-3-phosphate to phosphoenolpyruvate, among them the key step of glycolysis, the oxidation of glyceraldehyde-3-phosphate. While the main transcripts of this glycolytic gene cluster start with the *gapA* gene, there is no promoter present in front of this gene. The promoter of the operon is located upstream of the *cggR* gene which is located upstream of *gapA*. The *cggR* gene encodes the repressor of the *gapA* operon (Fillinger *et al.*, 2000; Ludwig *et al.*, 2001). Transcription of the *gapA* operon is not only induced by glucose but also by several other carbohydrates that give rise to triose phosphates that are catabolized by glycolysis. An analysis of the role of CcpA in induction showed that CcpA was not required for induction of the operon by glucose in *cggR* mutants suggesting an indirect role for CcpA (Ludwig *et al.*, 2001). In addition, only a subset of sugars induces the expression of the *gapA* operon in a CcpA-dependent manner (Fillinger *et al.*, 2000; Ludwig *et al.*, 2002b). These sugars have a common characteristic - they are all transported by the PTS. In contrast, non-PTS sugars induce the *gapA* operon irrespective of the functional state of CcpA. This finding suggests that CcpA might control the PTS transport capacity in *B. subtilis*. Indeed, the uptake of PTS sugars is severely impaired in *ccpA* mutant strains. This might result from insufficient expression of genes encoding PTS components or from reduced activity of the PTS proteins. A detailed analysis of the HPr phosphorylation pattern of cells grown in the presence of glucose provided compelling evidence for the latter idea. The reduced PTS activity in *ccpA* mutants results from a shift of the phosphorylation of the phosphotransfer protein HPr of the PTS. In the wild type, this protein exists in its non-phosphorylated form, and the two phosphorylated forms (HPr(His~P) and HPr(Ser-P)) are present. In the *ccpA* mutant, however, nearly all HPr is phosphorylated on Ser-46. This form of HPr is a very poor substrate of Enzyme I of the PTS and therefore not competent for PTS-dependent sugar transport (Deutscher *et al.*, 1984). An altered phosphorylation pattern of HPr was also observed in a *ccpA* mutant of *E. faecalis* (Leboeuf *et al.*, 2000). The defects of a *ccpA* mutant with respect to HPr phosphorylation by Enzyme I, glucose transport and expression of the *gapA* operon can all be suppressed by a mutation (*ptsH1*) that prevents phosphorylation of HPr by the HPr kinase (Ludwig *et al.*, 2002b). We can therefore conclude that induction of the *gapA* operon is not possible in the *ccpA* mutant due to the insufficient accumulation of an inducer for CggR. However, the question immediately arising from this finding is by which mechanism CcpA switches off HPr kinase activity once a certain amount of HPr(Ser-P) and Crh(Ser-P) is formed. Among the possibilities is a direct interaction between CcpA and HPr kinase. However, as concluded from experiments with a variant of CcpA devoid of its N-terminal DNA-binding domain there is no evidence for any activity of CcpA that would not involve DNA binding (Ludwig and Stülke, 2001). Moreover, biochemical studies demonstrate that such a proposed interaction does not occur. In addition, the concentration of the metabolites that affect HPr kinase activity is not altered in a way to account for the increased kinase activity (Ludwig *et al.*, 2002b). We may therefore propose that a novel, as yet unknown protein which is expressed in a CcpA-dependent manner, is a negative effector of HPr kinase activity (Figure 1).

Insufficient production of the intracellular inducer, acetate, was also proposed to be the reason for the lack of induction of the *alsSD* operon encoding enzymes for acetoin synthesis by glucose in *ccpA* mutants (Renna *et al.*, 1993). However, point mutations in CcpA were isolated that prevent the induction of *alsS* but not acetate production. The molecular mechnism of induction of acetoin synthesis remains therefore unknown (Turinsky *et al.*, 2000). It is interesting to note that *B. subtilis* can perform a kind of fermentation in the presence of glucose that results in the formation of acetate and extra ATP. If a certain amount of acetate is reached, the cells switch to the formation of acetoin. This resembles strongly the switch from butyrate to solvent formation in *Clostridium acetobutylicum* (Dürre *et al.*, 2002).

Perspectives

While the research conducted over the last few years has provided significant new insights in both the extent and the mechanisms of positive regulation by CcpA there remain several unsolved questions to be answered to fully understand the molecular mechanisms by which CcpA acts as a positive regulator of gene expression. First, it will be interesting to find out how CcpA interacts with RNA polymerase to activate transcription at the *ackA* and *pta* promoters. Second, the mechanism by which CcpA controls the activity of HPr kinase is of high interest. Third, the identification of the *sgd* mutations allowing growth of *ccpA* mutants on minimal media will improve our understanding of the physiological balance of the cell. Finally, the unravelling of the mechanism(s) by which CcpA controls glutamate synthesis and thus ammonium assimilation is of high priority.

Acknowledgements

We acknowledge the contributions of Christoph Meinken, Nicole Popp, Ingrid Wacker, and Anastasija Matin to this work. The work in our lab was supported by grants from the Deutsche Forschungsgemeinschaft (Stu 214/2-1; Stu 214/2-2; SFB473) and the Fonds der Chemischen Industrie.

References

Behari, J., and Youngman, P. 1998. A homolog of CcpA mediates catabolite control in *Listeria monocytogenes* but not carbon source regulation of virulence genes. J. Bacteriol. 180: 6316-6324.

Blencke, H.-M., Homuth, G., Ludwig, H., Hecker, M., and Stülke, J. Transcriptional profiling of gene expression in response to glucose in *Bacillus subtilis*: regulation of the central metabolic pathways. subm.

Brückner, R., and Titgemeyer, F. 2002. Carbon catabolite repression in bacteria: choice of the carbon source and autoregulatory limitation of sugar utilization. FEMS Microbiol. Lett. 209: 141-148.

Chauvaux, S., Paulsen, I. T., and Saier Jr., M. H. 1998. CcpB, a novel transcription factor implicated in catabolite repression in *Bacillus subtilis*. J. Bacteriol. 180: 491-497.

Deutscher, J., Kessler, U., Alpert, C. A., and Hengstenberg, W. 1984. Bacterial phosphoenolpyruvate-dependent phosphotransferase system: P-ser-HPr and its possible regulatory function. Biochemistry 23: 4455-4460.

Deutscher, J., Küster, E., Bergstedt, U., Charrier, V., and Hillen, W. 1995. Protein kinase-dependent HPr/CcpA interaction links glycolytic activity to carbon catabolite repression in Gram-positive bacteria. Mol. Microbiol. 15: 1049-1053.

Dürre, P., Böhringer, M., Nakotte, S., Schaffer, S., Thormann, K., and Zickner, B. 2002. Transcriptional regulation of solventogenesis in *Clostridium acetobutylicum*. J. Mol. Microbiol. Biotechnol. 4: 295-300.

Egeter, O., and Brückner, R. 1996. Catabolite repression mediated by the catabolite control protein CcpA in *Staphylococcus xylosus*. Mol. Microbiol. 21: 739-749.

Faires, N., Tobisch, S., Bachem, S., Martin-Verstraete, I., Hecker, M., and Stülke, J. 1999. The catabolite control protein CcpA controls ammonium assimilation in *Bacillus subtilis*. J. Mol. Microbiol. Biotechnol. 1: 141-148.

Fillinger, S., Boschi-Muller, S., Azza, S., Dervyn, E., Branlant, G., and Aymerich, S. 2000. Two glyceraldehyde-3-phosphate dehydrogenases with opposite physiological roles in a nonphotosynthetic bacterium. J. Biol. Chem. 275: 14031-14037.

Frisby, D., and Zuber, P. 1994. Mutations in *pts* cause catabolite-resistant sporulation and altered regulation of *spo0H* in *Bacillus subtilis*. J. Bacteriol.

176: 2587-2595.

Fujita, Y., Miwa, Y., Galinier, A. and Deutscher, J. 1995. Specific recognition of the *Bacillus subtilis gnt cis*-acting catabolite-responsive element by a protein complex formed between CcpA and seryl-phosphorylated HPr. Mol. Microbiol. 17: 953-960.

Galinier, A., Haiech, J., Kilhoffer, M.C., Jaquinod, M., Stülke, J., Deutscher, J., and Martin-Verstraete, I. 1997. The *Bacillus subtilis crh* gene encodes a HPr-like protein involved in carbon catabolite repression. Proc. Natl. Acad. Sci. USA 94: 8439-8444.

Galinier, A., Kravanja, M., Engelmann, R., Hengstenberg, W., Kilhoffer, M.-C., Deutscher, J., and Haiech, J. 1998. New protein kinase and protein phosphatase families mediate signal transduction in bacterial catabolite repression. Proc. Natl. Acad. Sci. USA 95: 1823-1828.

Galinier, A., Deutscher, J., and Martin-Verstraete, I. 1999. Phosphorylation of either Crh or HPr mediates binding of CcpA to the *Bacillus subtilis xyn cre* and catabolite repression of the *xyn* operon. J. Mol. Biol. 286: 307-314.

Grundy, F.J., Waters, D.A., Allen, S.H., and Henkin, T.M. 1993. Regulation of the *Bacillus subtilis* acetate kinase gene by CcpA. J. Bacteriol. 175: 7348-7355.

Henkin, T. M., Grundy, F. J., Nicholson, W. L., and Chambliss, G. H. 1991. Catabolite repression of α-amylase gene expression in *Bacillus subtilis* involves a *trans*-acting gene product homologous to the *Escherichia coli lacI* and *galR* repressors. Mol. Microbiol. 5: 575-584.

Henkin, T.M. 1996. The role of the CcpA transcriptional regulator in carbon metabolism in *Bacillus subtilis*. FEMS Microbiol. Lett. 135: 9-15.

Hueck, C.J., Kraus, A., Schmiedel, D. and Hillen, W. 1995. Cloning, expression and functional analyses of the catabolite control protein CcpA from *Bacillus megaterium*. Mol. Microbiol. 16: 855-864.

Jones, B.E., Dossonnet, V., Küster, E., Hillen, W., Deutscher, J. and Klevit, R.E. 1997. Binding of the catabolite repressor protein CcpA to its DNA target is regulated by phosphorylation of its corepressor HPr. J. Biol. Chem. 272: 26530-26535.

Krüger, S., Stülke, J., and Hecker, M. 1993. Carbon catabolite repression of ß-glucanase synthesis in *Bacillus subtilis*. J. Gen. Microbiol. 139: 2047-2054.

Küster, E., Hilbich, T., Dahl, M., and Hillen, W. 1999. Mutations in catabolite control protein CcpA separating growth effects from catabolite repression. J. Bacteriol. 181: 4125-4128.

Leboeuf, C., Leblanc, L., Auffray, Y., and Hartke, A. 2000. Characterization of the *ccpA* gene of *Enterococcus faecalis*: identification of starvation-inducible proteins regulated by CcpA. J. Bacteriol. 182: 5799-5806.

Lindner, C, Stülke, J, and Hecker, M. 1994. Regulation of xylanolytic enzymes in *Bacillus subtilis*. Microbiology 140: 753-757.

Lokman, B. C., Heerikhuisen, M., Leer, R. J., van den Broek, A., Borsboom, Y., Chaillou, S., Postma, P. W., and Pouwels, P. H. 1997. Regulation of expression of the *Lactobacillus pentosus xylAB* operon. J. Bacteriol. 179: 5391-5397.

Ludwig, H., and Stülke, J. 2001. The *Bacillus subtilis* catabolite control protein CcpA exerts all its regulatory functions by DNA-binding. FEMS Microbiol. Lett. 203: 125-129.

Ludwig, H., Homuth, G., Schmalisch, M., Dyka, F. M., Hecker, M., and Stülke, J. 2001. Transcription of glycolytic genes and operons in *Bacillus subtilis*: evidence for the presence of multiple levels of control of the *gapA* operon. Mol. Microbiol. 41: 409-422.

Ludwig, H., Meinken, C., Matin, A., and Stülke, J. 2002a. Insufficient expression of the *ilv-leu* operon encoding enzymes of branched-chain amino acid biosynthesis limits growth of a *Bacillus subtilis ccpA* mutant. J. Bacteriol. 184: 5174-5178.

Ludwig, H., Rebhan, N., Blencke, H.-M., Merzbacher, M., and Stülke, J. 2002b. Control of the glycolytic *gapA* operon by the catabolite control protein A in *Bacillus subtilis*: a novel mechanism of CcpA-mediated regulation. Mol. Microbiol. 45: 543-553

Luesink, E.J., van Herpen, R.E., Grossiord, B.P., Kuipers, O.P., and de Vos, W.M. 1998. Transcriptional activation of the glycolytic *las* operon and catabolite repression of the *gal* operon in *Lactococcus lactis* are mediated by the catabolite control protein CcpA. Mol. Microbiol. 30: 789-798.

Martin, I., Débarbouillé, M., Klier, A., and Rapoport, G. 1989. Induction and metabolite regulation of levanase synthesis in *Bacillus subtilis*. J. Bacteriol. 171: 1885-1892.

Martin-Verstraete, I., Deutscher, J., and Galinier, A. 1999. Phosphorylation of HPr and Crh by HprK, early steps in the catabolite repression signalling pathway for the *Bacillus subtilis* levanase operon. J. Bacteriol. 181: 2966-2969.

Martin-Verstraete, I., Stülke, J., Klier, A., and Rapoport, G. 1995. Two different mechanisms mediate catabolite repression of the *Bacillus subtilis* levanase operon. J. Bacteriol. 177: 6919-6927.

Miwa, Y., Nakata, A., Ogiwara, A., Yamamota, M., and Fujita, Y. 2000. Evaluation and characterization of catabolite-responsive elements (*cre*) of *Bacillus subtilis*. Nucl. Acids Res. 28: 1206-1210.

Miwa, Y., Saikawa, M., and Fujita, Y. 1994. Possible function and some properties of the CcpA protein of *Bacillus subtilis*. Microbiology 140: 2567-2575.

Monedero, V., Gosalbes, M. J., and Perez-Martinez, G. 1997. Catabolite repression in *Lactobacillus casei* ATCC 393 is mediated by CcpA. J. Bacteriol. 179: 6657-6664.

Monod, J., 1942. Recherches sur la croissance des cultures bacteriennes. Paris, Hermann et Cie.

Moreno, M. S., Schneider, B. L., Maile, R. R., Weyler, W., and Saier, M. H., Jr. 2001. Catabolite repression mediated by the CcpA protein in *Bacillus subtilis*: novel modes of regulation revealed by whole-genome analyses. Mol. Microbiol. 39: 1366-1381.

Presecan-Siedel, E., Galinier, A., Longin, R., Deutscher, J., Danchin, A., Glaser, P., and Martin-Verstraete, I. 1999. The catabolite regulation of the *pta* gene as part of carbon flow pathways in *Bacillus subtilis*. J. Bacteriol. 181: 6889-6897.

Reizer, J., Hoischen, C., Titgemeyer, F., Rivolta, C., Rabus, R., Stülke, J., Karamata, D., Saier Jr., M. H., and Hillen, W. 1998. A novel protein kinase that controls carbon catabolite repression in bacteria. Mol. Microbiol. 27: 1157-1169.

Renna, M. C., Najimudin, N., Winik, L. R., and Zahler, S. A. 1993. Regulation of the *Bacillus subtilis alsS*, *alsD*, and *alsR* genes involved in post-exponential production of acetoin. J. Bacteriol. 175: 3863-3875.

Saier, M. H., Jr. 1998. Multiple mechanisms controlling carbon metabolism in bacteria. Biotechnol. Bioeng. 58: 170-174.

Shin, B.- S., Choi, S.-K., and Park, S.-H. 1999. Regulation of the *Bacillus subtilis* phosphotransacetylase gene. J. Biochem. 126: 333-339.

Simpson, C. L., and Russell, R. R. B. 1998. Identification of a homolog of CcpA catabolite repressor protein in *Streptococcus mutans*. Infect. Immun. 66: 2085-202.

Stülke, J., Martin-Verstraete, I., Zagorec, M., Rose, M., Klier, A., and Rapoport, G. 1997. Induction of the *Bacillus subtilis ptsGHI* operon by glucose is controlled by a novel antiterminator, GlcT. Mol. Microbiol. 25: 65-78.

Stülke, J., and Hillen, W. 1998. Coupling physiology and gene regulation in bacteria: the phosphotransferase sugar uptake system delivers the signals. Naturwissenschaften 85: 583-592.

Stülke, J., and Hillen, W. 1999. Carbon catabolite repression in bacteria. Curr. Opin. Microbiol. 2: 195-201.

Stülke, J., and Hillen, W. 2000. Regulation of carbon catabolism in *Bacillus* species. Annu. Rev. Microbiol. 54: 849-880.

Stülke, J., Martin-Verstraete, I., Glaser, P., and Rapoport, G. 2001. Characterization of glucose repression resistant mutants of *Bacillus subtilis*: identification of the *glcR* gene. Arch. Microbiol. 175: 441-449.

Titgemeyer, F., and Hillen, W. 2002. Global control of sugar metabolism: a Gram-positive solution. Antonie van Leeuwenhoek 82: 59-71.

Tobisch, S., Zühlke, D., Bernhardt, J., Stülke, J., and Hecker, M. 1999. Role of CcpA in regulation of the central pathways of carbon catabolism in *Bacillus subtilis*. J. Bacteriol. 181: 6996-7004.

Turinsky, A. J., Grundy, F. J., Kim, J. H., Chambliss, G. H., and Henkin, T. M. 1998. Transcriptional activation of the *Bacillus subtilis ackA* gene requires sequences upstream of the promoter. J. Bacteriol. 180: 5961-5967.

Turinsky, A. J., Moir-Blais, T. R., Grundy, F. J., and Henkin, T. M. 2000. *Bacillus subtilis ccpA* gene mutants specifically defective in activation of acetoin synthesis. J. Bacteriol. 182:5611-5614.

van den Bogaard, P. T. C., Kleerebezem, M., Kuipers, O. P., and de Vos, W. M. 2000. Control of lactose transport, β-galactosidase activity, and glycolysis by CcpA in *Streptococcus thermophilus*: Evidence for carbon catabolite repression by a non-phosphoenolpyruvate-dependent phosphotransferase system sugar. J. Bacteriol. 182: 5982-5989.

Wray, L. W., Jr., Pettengill, F. K., and Fisher, S. H. 1994. Catabolite repression of the *Bacillus subtilis hut* operon requires a *cis*-acting site located downstream of the transcription initiation site. J. Bacteriol. 176: 1894-1902.

Yoshida, K.-I., Kobayashi, K., Miwa, Y., Kang, C.-M., Matsunaga, M., Yamaguchi, H., Tojo, S., Yamamoto, M., Nishi, R., Ogasawara, N., Nakayama, T., and Fujita, Y. 2001. Combined transcriptome and proteome analysis as a powerful approach to study genes under glucose repression in *Bacillus subtilis*. Nucl. Acids Res. 29: 683-692.

Index